Non-marine organic geochemistry

CAMBRIDGE EARTH SCIENCE SERIES

Editors: W. B. HARLAND (*General Editor*), S. O. AGRELL, D. DAVIES, N. F. HUGHES

Non-marine organic geochemistry

FREDERICK M. SWAIN

Professor of Geology
University of Minnesota
University of Delaware

CAMBRIDGE

at the University Press, 1970

CAMBRIDGE UNIVERSITY PRESS
Cambridge, New York, Melbourne, Madrid, Cape Town, Singapore,
São Paulo, Delhi, Dubai, Tokyo, Mexico City

Cambridge University Press
The Edinburgh Building, Cambridge CB2 8RU, UK

Published in the United States of America by Cambridge University Press, New York

www.cambridge.org
Information on this title: www.cambridge.org/9780521155106

First published 1970
First paperback edition 2010

A catalogue record for this publication is available from the British Library

Library of Congress Catalogue Card Number: 78-103798

ISBN 978-0-521-07757-6 Hardback
ISBN 978-0-521-15510-6 Paperback

Contents

Contents

Contents

Contents

Preface

The plan for this book developed over a period of years as a result of work by the writer, his students and associates on lacustrine sediments and on organic residues in sedimentary rocks and fossils. It seemed that there was a need to summarize knowledge of non-marine organic geochemistry in a form more readily accessible to students and workers in related disciplines than is presently available in books in organic geochemistry, sedimentology or limnology. It will be immediately evident to the reader that knowledge of non-marine organic geochemistry, except for soils, is extremely fragmentary.

In chapter 1 are given brief introductions to organic compounds of geochemical interest and some organic reactions, that occur or might be expected to occur in the geochemical environment. Chapter 2 describes some field and laboratory procedures in organic geochemistry, not in an exhaustive way but as a guide to those who may engage in such studies.

Chapter 3 presents some of the sedimentary characteristics of non-marine sediments, particularly lake sediments. Enough descriptive material is given to acquaint the reader with the wide variety of sedimentary-organic features that one may encounter in the biogeochemistry of such deposits.

Chapters 4–8 deal with the organic geochemistry of bitumens, amino acids, carbohydrates, pigments and organic acids in non-marine deposits. These are not treated exhaustively but aim to provide a summary of present knowledge and the close relationship of source materials and trophication stages to the organic residues. The theme of the book basically centers around this close relationship and the relatively less important effects of post-depositional changes.

In chapters 9 and 10, an attempt is made to summarize knowledge of the organic residues of soil humus and the geology and geochemistry of coal. Current literature, including several compilations, has been drawn upon for the bulk of these chapters as noted in the discussions. The information cited was selected to provide as much background as possible for the organic geochemist who may wish to pursue problems in these fields.

The possible developmental history of non-marine organisms as it is reflected in organic residues in sediments and rocks is briefly outlined in chapter 11.

The tools for further research in non-marine organic geochemistry are available and there are many problems of geological interest that await investigation. It is one of the main purposes of this book to encourage students to undertake work on these problems.

Preface

Acknowledgements. The writer is particularly indebted to the many former students and laboratory and field assistants who have contributed so greatly to his research in organic geochemical studies: Ausma Blumentals, Ruta Millers, Gunta Pakalns, Inara Porietis, Dace Bakuzis, Judy M. Bratt, Shirley Kraemer, N. P. Prokopovich, K. A. Dickinson, J. G. Palacas, M. A. Rogers, F. T. Manheim, Iwan Tkachenko, F. T. Ting, G. A. Paulsen, Robert Meader, James Cornell, Paul Engel, John C. Kraft, D. B. Roberts, Harvey Meyer, G. B. Morey, Richard N. Benson, Peter Fleischer, Dennis Deischl, Ronald Evans, Robert W. Wolfe, Gregory Gohn, David Rowland, Christian Engen, Charles Rhoades, Vernon DeRuyter, James Niehaus and Frederick Boethling. Fred Smith, Samuel Kirkwood, H. E. Wright, R. O. Megard, Joseph Shapiro and G. C. Speers have contributed freely of their time and advice. Chapters 1 and 4 were read by W. E. Noland and James Janke, chapters 2 and 8 by Joseph Shapiro, chapter 3 by E. J. Cushing, chapters 5 and 9 by E. L. Schmidt, chapter 6 by Samuel Kirkwood, chapter 7 by R. O. Megard and Jon Sanger, chapter 10 by F. T. Ting and chapter 11 by E. J. Cushing. The first draft was read by G. C. Speers who contributed many helpful comments. V. R. Murthy provided advice on the radioactive nuclides discussed in chapter 11. The writer also acknowledges with sincere thanks the assistance over the past 15 years provided by grants from the Graduate School, University of Minnesota, the National Science Foundation, the U.S. Bureau of Reclamation, the Petroleum Research Fund of the American Chemical Society, the National Aeronautics and Space Administration, and the Minnesota Geological Survey.

F. M. SWAIN

November 1968
Minneapolis, Minnesota

1 Introduction to study of organic sediments

1.1 INTRODUCTION

The study of Holocene organic sediments as well as organic sedimentary rocks has increased greatly in the past two decades owing to activities in oil exploration, oceanographic and other hydrologic studies, government-sponsored research programs and greatly improved field and laboratory facilities and instrumentation. Great strides forward have been made, backed by industrial and governmental research, in such things as sampling and coring devices, pH–Eh meters, oxygen meters, dissolved salt-meters and other devices for measurement of sediment-properties and freeze-dryers for preservation of samples. Laboratory instrumentation for study of sedimentary organic matter has developed at a phenomenal rate. Among such instruments or processes the following can be mentioned: improved methods of extraction of organic materials, such as ultrasonic extraction, counter-current extraction, and new solvents or combinations of solvents; adsorption chromatography including paper, column, and gas–liquid chromatography; infrared, ultraviolet, and visible spectrophotometers; mass spectrometers; electron spin resonance and nuclear magnetic resonance apparatus; automatic amino acid analyzers; and enzymatic methods.

The purposes of the organic sediment studies have been almost as varied as the methods. Petroleum chemists (Smith, 1952, 1954; Stevens *et al.* 1956) have examined the hydrocarbon contents of Holocene sediments to understand more about the potential of the sediments as petroleum source beds. Other organic geochemists with oil companies have investigated the composition and isotopic properties of crude petroleum in relationship to their geologic age, sources, and diagenetic history (Bray & Evans, 1961; Dunning *et al.* 1954; Erdman, 1961; Forsman & Hunt, 1958; Meinschein & Kenny, 1957; Silverman & Epstein, 1958).

Marine chemists and sedimentologists have analyzed offshore sediments for hydrocarbons, pigment contents, amino acids, sugars, organic acids, and other components in an effort to learn more about their relationships to environments of deposition and post-depositional changes (Emery & Rittenberg, 1952; Fox, 1937; Jeffrey *et al.* 1964; Plunkett, 1957; Prashnowsky *et al.* 1961; Degens *et al.* 1961).

Freshwater sedimentologists and limnologists have found the organic materials in lake and bog deposits to be highly varied and to reflect reasonably well the principal source organisms and sequence of events in the accumulation of sediments (Vallentyne, 1956; Swain & Prokopovich, 1954; Vallentyne & Craston, 1957; Anderson & Gundersen, 1955). Perhaps the most diverse organic geochemical studies have been in fresh water deposits, and the writer has decided to restrict himself here to these kinds of sediments and sedimentary rocks.

The organic geochemistry of sedimentary rocks has been studied mainly by petroleum chemists, and in fact the investigations have centered around petroleum source-bed evaluation and the organic chemistry of the Eocene Green River Oil Shales (Abelson *et al.* 1964; Blumer, 1950; Hunt & Jamieson, 1956; Eglinton *et al.* 1966*a, b, c*). More recently several geochemists have analyzed some of the carbonaceous Precambrian rocks and have reported the presence of biogenic hydrocarbons, carbohydrates, organic acids, and perhaps amino acids (Meinschein, 1965; Eglinton *et al.* 1966*a, b, c*; Swain, Bratt & Kirkwood, 1968; Schopf *et al.* 1968).

The purpose of this book which is written from the viewpoint of a geologist rather than a chemist is to discuss the occurrence of organic residues in freshwater sediments and sedimentary rocks and to evaluate the importance of these residues in paleoenvironmental and related problems, and their role as biochemical fossils. Other books in the field of organic chemistry and geochemistry such as those listed below, and in References, should be consulted for more purely chemical treament:

Abelson, *Researches in Geochemistry*, I (1959).
Abelson, *Researches in Geochemistry*, II (1968).
Breger, *Organic Geochemistry* (1963).
Colombo & Hobson, *Advances in Organic Geochemistry* (1964).
Degens, *Geochemistry of Sediments* (1965).
Degering, *Organic Chemistry*, 6th ed. (1962).
Eglinton & Murphy, *Organic Geochemistry—Methods and Results* (1969).
Geissman, *Principles of Organic Chemistry*, 3rd ed. (1968).
Given, *Coal Science* (1965).
Hobson & Louis, *Advances in Organic Geochemistry 1964* (1966).
Hobson & Speers, *Advances in Organic Geochemistry 1966* (1969).
Morrison & Boyd, *Organic Chemistry*, 2nd ed. (1966).
Noller, *Chemistry of Organic Compounds* (1951).

1.2 THE GENERAL NATURE OF ORGANIC COMPOUNDS

1.2.1 *Introduction*

A brief discussion of the electronic theory with special reference to carbon and carbon-like elements is given as an introduction to naturally occurring organic compounds.

The number of electrons in the outer or valence electron shell governs many properties of the elements. Thus alkali metals (Li, Na, K, Rb, Ce) have one valence electron; alkaline earth metals (Ca, Mg) have two; oxygen family members (O, S, Se, Te) have six; halogens have seven; rare gases except helium (N, A, Kr, Xe, Rd) have eight. The kernel is that part of the atom inside the valence shell. The pair (for an *s* valence shell) and the octet (for an *s* and *p* valence shell) are unusually stable electron arrangements.

The following fundamental properties possessed by many atoms are important: there is a tendency for electrons to pair; metallic, alkali and alkaline earth atoms tend to yield one or more electrons and form positive ions; non-metallic atoms tend to gain one or more electrons and become negative ions; reactions tend to occur that will result in a decrease in kernel repulsions.

Ionic bonds are formed when one or more electrons transfer from the valence shell of one atom to that of another atom. Examples are (dots indicate valence electrons):

$$\text{Na} + :\ddot{\text{Cl}}: \longrightarrow \text{Na}^+ + :\ddot{\text{Cl}}:^- ; \tag{1}$$

$$\text{Ba}: + \ddot{\text{O}}: \longrightarrow \text{Ba}^{++} + :\ddot{\text{O}}:^- \tag{2}$$

Ionic bonding results in polar compounds that are salts and have such properties as: high melting point, high boiling point, easy solubility in water, are good electrolytes, and as solutions react rapidly with solutions of other salts.

Covalent bonds are the result of a sharing of electrons between atoms. The shared-electrons or Lewis bond forms between atoms that are too alike in affinity for one atom to completely acquire an electron from the other

$$\text{H}:\text{H} \ , \ :\ddot{\text{Cl}}:\ddot{\text{Cl}}: \ , \ \text{H}:\ddot{\text{O}}:\text{H} \ , \ :\ddot{\text{Cl}}:\ddot{\text{P}}:\ddot{\text{Cl}}: \ , \ :\ddot{\text{O}}::\text{C}::\ddot{\text{O}}:$$

also in complex ions,

$$:\ddot{\text{O}}::\text{N}^+:\ddot{\text{O}}:^-$$
$$:\ddot{\text{O}}:^-$$

Assuming that each valence shell has a complete octet of electrons, the number of electrons shared between atoms can be determined as follows:

$$n = x - \epsilon, \tag{3}$$

where n is the number of electrons shared, x is the number of electrons necessary to give each atom a separate octet in the valence shell, and ϵ is the sum of valence electrons of all the atoms of the molecule; examples are:

$$Cl_2,\ 2\times8-(2\times7) = 2 \tag{4}$$
$$CO_2,\ 3\times8-(4+2\times6) = 8 \tag{5}$$
$$NO_3^-,\ 4\times8-(1+5+3\times6) = 8. \tag{6}$$

The number of shared electrons in covalent bonds may be 1, 2, 4 or 6. The first is rare (B_2H_6). The two-electron bond ('single bond'), the four electron ('double bond') and six-electron ('triple bond') are common in organic chemistry.

Owing to the repulsion of atomic nuclei and of electron shells for each other, the atomic nuclei are held apart; this is the interatomic distance which is measurable in crystalline solids by means of X-ray analysis. The radius of an atom in an ionic compound typically is different than the radius of the same atom undergoing electron sharing. Examples of single normal covalent radii are, in Ångström units ($= 10^{-8}$ cm): H, 0·29, C, 0·77, N, 1·70, O, 0·66, Si, 1·17, P, 1·10 and S, 1·04.

Double-bonded atoms typically have smaller inter-atomic distances ($\sim 13\%$) than the single-bond distance. In triple-bonded atoms the distance is 22% smaller than the single-bond distance.

For some substances it is possible to show two or more equivalent electronic structures; the more such possible configurations, the more stable the substance. Three equivalent structures for carbonate ion are:

$$\begin{array}{ccc}
:\ddot{O}: & :O: & :\ddot{O}: \\
O::C:O: & :\ddot{O}:C:O: & :\ddot{O}:C::\ddot{O}:
\end{array}$$

The characteristic is known as resonance and the actual structure is a hybrid or average condition of the electronic structures drawn for a particular substance, for example: carbonate ion, nitrate ion (*a*), vinyl chloride (*b*), guanidinium ion (*c*), acetate ion (*d*), and nitro group (*e*)

$$\begin{array}{ccc}
\overset{+}{:O::N\ \ddot{O}:} & H:C::C:\ddot{Cl}: & \overset{+}{\ \ H} \\
:O: & H\ \ \ H & H:N::C:N:H \\
& & H\ :N:N \\
& & H
\end{array}$$

$$\qquad (a) \qquad\qquad\qquad (b) \qquad\qquad\qquad (c)$$

$$\begin{array}{cc}
\ddot{O}::C:\ddot{O}:^- & \overset{+}{\ddot{O}::N:\ddot{O}:^-} \\
H:C:H & H:C:H \\
H & H
\end{array}$$

$$\qquad (d) \qquad\qquad (e)$$

The importance of resonance is seen in the benzene molecule, C_6H_6. This molecule consists of six carbon atoms and attached hydrogen atoms lying in a plane with the carbon atoms placed at the angles of a regular hexagon and with the atoms connected by σ bonds. The remainder of the molecule consists of p orbitals one for each of the remaining electrons of the individual carbon atoms; the p orbitals overlap to form π bonds. The overlapping orbitals result in π 'clouds' above and below the plane of the ring. The overlapping of the p orbitals in both directions, and the resulting participation of each electron of the orbital in several types of bonds is an example of a resonance hybrid of two structures. The so-called Kekulé structure of benzene may be shown in two forms (I, II) and the resulting resonance structure is shown by III:

I   II   III

The straight lines represent the σ bonds joining carbon atoms, and the circle represents the cloud of six 'delocalized' π electrons.

The bond energy that results from the formation of a gram molecular weight of a substance, or that required to dissociate it varies as in the following examples, in kcal per mole at 25 °C:

H—H	102·36	C=O	175·0
C—C	83·87	C—O	81·8
C=C	150·0	C=C	198·8
H—C	97·7	H—O	109·94
C—N	66·4	C—S	73·5

The electron-pair bond between two unlike atoms may be a completely covalent bond, i.e. equal sharing of the electron pair, as C—O, or it may have ionic character, i.e. unequal sharing of the electron pair, as BrCl or HF. The amount of ionic character, however, varies widely in terms of bond energies and is due to the degree of electronegativity of the ionic-bond atoms.

Electronegativity, or the tendency for atoms to attract electrons, affects the relative polarity of covalent bonds; the greater the difference in electronegativity the greater will be the polarity of the bond. Of the substances likely to be found in organic compounds fluorine has the highest electronegativity, followed in turn by oxygen, nitrogen, chlorine, bromine, and carbon. The most electronegative elements are those in the upper right-hand corner of the Periodic Table.

Introduction to study of organic sediments

1.2.2 General classification of organic compounds

(A) Open chain, aliphatic compounds. Homologous examples of this type
of compound are ethane (CH_3CH_3), ethyl alcohol or ethanol (CH_3CH_2OH),
acetaldehyde or ethanal (CH_3CHO), acetic (ethanoic) acid (CH_3COOH).

(B) Closed chain or ring compounds.

(1) Alicyclic compounds such as:

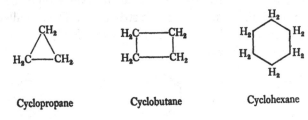

| Cyclopropane | Cyclobutane | Cyclohexane |

by convention cyclohexane may be drawn ⬡ .

(2) Aromatic compounds (carbon atoms occur at corners of structure)

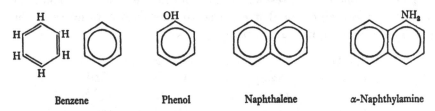

| Benzene | Phenol | Naphthalene | α-Naphthylamine |

(3) Heterocyclic compounds

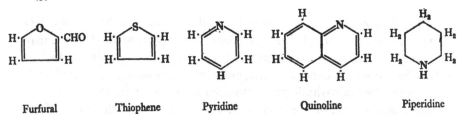

| Furfural | Thiophene | Pyridine | Quinoline | Piperidine |

1.2.3 Mechanisms of organic reactions

The tendency of atoms to attract electrons (electronegativity) depends on
their electronic structure and size; small atoms such as fluorine, which lack
only one electron to complete its outer inert gas shell, are more electro-

negative than larger atoms, such as oxygen which furthermore needs two electrons to complete its stable outer shell.

Within a molecule the strengths of the bonds will be affected by the number, distribution and relative negativity of electronegative groups. The effects of these groups on relative distribution of electrons in a molecule may be shown on an electronegativity map. If hydrogen on such a map is assigned a value of $+1$, sulfur or iodide is o, bromide is -0.75, chlorine -1.25, nitrogen -1.25, single-bonded oxygen atom -2.5 and double-bonded oxygen -5. In an aliphatic hydrocarbon, i.e. methane, electrons are distributed in a condition of maximum stability. Distortion of the methane-type electron distribution, owing to replacement of hydrogens by more or less electronegative groups results in a decrease in stability in proportion to the distortion. The distortion decreases away from points of electronegative substitution.

Amongst molecular forces: ionic, dipolar, van der Waals, and directed bonding forces such as the hydrogen bond, the van der Waals forces are the most important in crystalline organic compounds. These forces are believed to result from interaction of molecular electronic fields that produces polarization and attractive forces.

The rate of a chemical reaction is expressed in the following way:

$$\text{rate} = \text{collision frequency} \times \text{energy factor} \times \text{probability factor.} \quad (7)$$

The collision frequency is dependent upon the size of the particles, their degree of concentration or crowding, and their speed, which in turn depends upon their weight and the temperature. The energy factor depends upon temperature and upon the activation energy E_{act} of the reaction, i.e. a measure of the fraction of collisions that have sufficient energy for the reaction to take place. The greater the value of E_{act} the smaller will be the fraction of collisions that have that energy. The equation for the relationship between energy of activation and the fraction of collisions with that energy is (Morrison & Boyd, 1966):

$$e^{-E_{act}/RT} = \text{fraction of collisions with energy greater than } E_{act} \quad (8)$$

$e = 2.718$ (base of natural logarithms), $R = 1.986$ (gas constant), $T =$ absolute temperature. The probability factor in the rate equation is the fraction of collisions that have proper orientation. This depends on the geometry of the particles and the type of reaction, and for closely related reactions is similar.

In reactions that involve the breaking of covalent bonds, the bond may break in two ways: (1) either one electron goes to each atom joined by the bond, i.e. homolytic cleavage, or (2) the pair of electrons stays with one or another of the two atoms, i.e. heterolytic cleavage.

In the first case, free radicals result. In the second case, if the resulting group contains a carbon atom with a positive charge on it, it is called a carbonium ion; if the resulting group is negative, it is called a carbanion.

(A) Examples of a free radical mechanism are:

(1) Pyrolysis of hydrocarbons: thermal decomposition of propane.

$$CH_3CH_2CH_3 \rightarrow CH_3 \cdot + CH_3CH_2 \cdot \tag{9}$$

(2) Catalytic reduction: metals can supply an unpaired electron to a molecule. When ethylene is absorbed on platinum,

$$H_2C-CH_2 = [H_2C-CH_2] \tag{10}$$

an unpaired electron on each carbon results from the breaking of the carbon–carbon bond. A process known as chemisorption is one by which the unpaired electrons are paired with electrons at the surface of the metal. Ethane is produced when chemisorbed hydrogen atoms and ethylene di-radicals react in the catalytic hydrogenation of ethylene.

(3) Thermal or photochemical halogenation: processes by which saturated hydrocarbons react with chlorine or bromine at elevated temperatures or in presence of short wavelength light at room temperature. The reactions involve first the formation of halogen atoms which react with hydrocarbon molecules to produce free hydrocarbon radicals that in turn collide with halogen molecules to form alkyl halide and more halogen atoms, i.e. a chain reaction.

(4) Polymerizations catalyzed by oxygen molecules, acting as free radicals; the two unpaired electrons of like spin of the oxygen molecule combine with one carbon atom of an unsaturated compound, i.e. ethylene,

$$:\overset{..}{O}:\overset{..}{O}:+CH_2 = CH_2 \rightarrow :\overset{..}{O}:\overset{..}{O}:CH_2-CH_2 \cdot \tag{11}$$

to form an intermediate product than can combine with another molecule of ethylene, resulting in a polymeric chain.

(B) Examples of ionic reaction mechanisms.

(1) Alkyl halides are examples of basic electron-rich reagents called nucleophilic reagents (Gr., nucleus loving, from its reluctance to share its electrons), and a typical reaction of alkyl halides is nucleophilic substitution. One kind of nucleophilic substitution is referred to as an Sn2 (substitution nucleophilic bimolecular) reaction such as that between methyl bromide and hydroxide ion to yield methanol.

$$\left.\begin{array}{l} CH_3Br + OH^- \rightarrow CH_3OH + Br^- \\ \text{rate} = k\,[CH_3Br]\,[OH^-] \end{array}\right\} \tag{12a}$$

This is a second-order reaction as the rate depends on the concentration of both reactants.

(2) Another type of ionic reaction mechanism is referred to as Sn1 (substitution nucleophilic unimolecular), such as the reaction between *tert*-butyl bromide and hydroxide ion to yield *tert*-butyl alcohol.

$$
\left.
\begin{array}{c}
\underset{\underset{\displaystyle \text{Br}}{|}}{\overset{\overset{\displaystyle \text{CH}_3}{|}}{\text{CH}_3\!-\!\text{C}\!-\!\text{CH}_3}} + \text{OH}^- \rightarrow \underset{\underset{\displaystyle \text{OH}}{|}}{\overset{\overset{\displaystyle \text{CH}_3}{|}}{\text{CH}_3\!-\!\text{C}\!-\!\text{CH}_3}} + \text{Br}^- \\[2em]
\text{rate} = k\,[\text{RBr}]
\end{array}
\right\} \qquad (12b)
$$

It is a first-order reaction because the rate depends only on the concentration of one reactant, *tert*-butyl bromide.

(3) There are two methods for the dehydrohalogenation of alkyl halides, referred to as E_2 and E_1 mechanisms, respectively. In the E_2 mechanism (bimolecular elimination) the base pulls a hydrogen ion away from the carbon and a halide ion separates simultaneously. In the E_1 mechanism (unimolecular elimination) alkyl halides, especially tertiary ones, can dissociate into halide ions and carbonium ions; the carbonium ion can then lose a hydrogen ion to a base to form an alkene.

(4) Dehydration of alcohols: the first step is formation of oxonium salt and a water molecule is subsequently lost.

(5) Additions of halogens and of acids to the double bond; halogens are added stepwise to unsaturated hydrocarbons through the formation of positively charged halonium ions which react with a halide ion from the solution with a Walden inversion of one of the carbon atoms.

$$(13)$$

In normal addition of acids to the double bond a two step reaction takes place:

$$
\left.
\begin{array}{c}
:\text{CH}_2\!-\!\text{CH}_2 + \text{H}\,:\!\ddot{\text{Br}}\!: \longrightarrow [\text{CH}_3\!-\!\text{CH}_2] + [:\!\ddot{\text{Br}}\!:^-] \\[0.5em]
\text{Ethonium} \\
\text{ion} \\[1em]
[\text{CH}_3\!-\!\overset{+}{\text{CH}_2}] + \text{Br}^- \longrightarrow \text{CH}_3\!-\!\text{CH}_2\,\text{Br}
\end{array}
\right\} \qquad (14)
$$

Because ethylene is a symmetrical molecule the product is the same no matter which carbon atom combines with bromide ion. If the unsaturated hydrocarbon is asymmetrical two products may form; the predominant one is that in which the proton has combined with the carbon atom that carries the most hydrogens. This phenomenon is referred to as Markovnikov's Rule.

(6) Acid-catalyzed polymerization; a process by which olefins undergo self addition under influence of acid catalysts which results in compounds having 2, 3, 4, or many times the molecular weight of the original compound.

(7) Addition of Grignard reagents to the carbon–oxygen double bond: owing to the polarized nature of the carbon–metal bond in organometallic compounds, with the metal at the positive end of the dipole, when this bond is cleaved the alkyl group will take the pair of electrons with it. In carbonyl compounds, such as CO_2, aldehydes and ketones, this group is polarized with the C atom at the positive end of the dipole. Therefore, when Grignard reagents add to a carbonyl group, alkyl group adds to carbon and metal to oxygen.

1.3 ORGANIC COMPOUNDS OF GEOLOGIC IMPORTANCE

Organic chemical compounds that contribute to the contents of sediments and sedimentary rocks and are useful in interpretations of source materials and diagenetic history are: (1) hydrocarbons, alcohols, ketones, fatty acids, organic acids, steroids, esters; and aliphatic and aromatic sulfur compounds; (2) proteins, peptides, amino acids and their derivatives; (3) carbohydrates, polyhydric alcohols and aldehydes; (4) heteroaromatic compounds, including organic pigments; and (5) humic complexes formed from a variety of the foregoing materials.

1.3.1 *Hydrocarbons, alcohols, organic acids, fatty acids, steroids and their esters and derivatives; sulfur compounds*

The *hydrocarbons* are classified on the basis of their degree of saturation with hydrogen and arrangement of the carbon and hydrogen atoms. The principal groups are: *alkanes*, type formula C_nH_{2n+2}, saturated, aliphatic or chain hydrocarbons (table 1); *alkenes*, unsaturated chain compounds, type formula C_nH_{2n} (table 2), acetylenes or diolefins, unsaturated chain compounds, type formula C_nH_{2n-2}, with double-bonded carbon atoms (diolefines) or with one set of triple-bonded carbon atoms (acetylenes); *terpenes*, unsaturated combination chain and ring hydrocarbons, type for-

TABLE I   *Nomenclature of alkanes (Noller, 1951)*

C_5	Pentanes	C_{16}–C_{19}	Etc.
C_6	Hexanes	C_{20}	Eicosanes
C_7	Heptanes	C_{21}	Heneicosanes
C_8	Octanes	C_{22}	Docosanes
C_9	Nonanes	C_{23}	Tricosanes
C_{10}	Decanes	C_{24}–C_{29}	Etc.
C_{11}	Undecanes	C_{30}	Triacontanes
C_{12}	Dodecanes	C_{31}	Hentriacontanes
C_{13}	Tridecanes	C_{32}–C_{39}	Etc.
C_{14}	Tetradecanes	C_{40}	Tetracontanes
C_{15}	Pentadecanes		Etc.

TABLE 2   *Nomenclature of alkenes (Noller, 1951)*

Formula	Common name	By Geneva system
CH_2=CH_2	Ethylene	Ethene
CH_3CH=CH_2	Propylene	Propene
CH_3CH_2CH=CH_2	α-Butylene	1-Butene
CH_3CH=$CHCH_2$	β-Butylene	2-Butene
CH_3C=CH_2 $\quad\vert$ $\quad CH_3$	Isobutylene	Methylpropene
C_2H_5CH=$CHCH_3$	β-Amylene	2-Pentene
$\overset{6}{C}H_3\overset{5}{C}H_2\overset{4}{C}HCH_3$ $\qquad\vert$ $\quad \overset{3}{C}H_3\overset{2}{C}=\overset{1}{C}HCH_3$	None	3,4-Dimethyl-2-hexene

mula C_nH_{2n-4} (figs. 1–6) (see carotenoid pigments below), *aromatic hydro-carbons*, type formula C_nH_{2n-6}, unsaturated ring compounds having alternate double-bonded and single-bonded carbon atoms including *benzenes*, C_nH_{2n-6} (figs. 7, 8), and *naphthalenes*, C_nH_{2n-12} (figs. 9, 10); alicyclic hydrocarbons having type formula C_nH_{2n} and including cycloalkanes and cycloalkenes C_nH_{2n-2}.

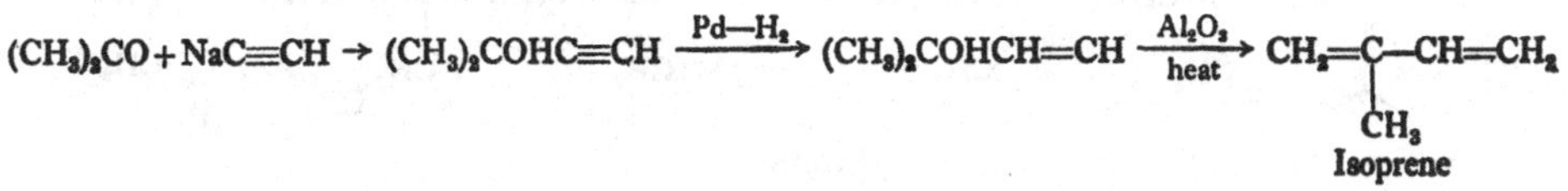

$$(CH_3)_2CO + NaC\equiv CH \rightarrow (CH_3)_2COHC\equiv CH \xrightarrow{Pd-H_2} (CH_3)_2COHCH=CH \xrightarrow[heat]{Al_2O_3} CH_2=C-CH=CH_2$$
$$\qquad\qquad\qquad\qquad\qquad\qquad\qquad\qquad\qquad\qquad\qquad\qquad\qquad\qquad\qquad\qquad\qquad\vert$$
$$\qquad\qquad\qquad\qquad\qquad\qquad\qquad\qquad\qquad\qquad\qquad\qquad\qquad\qquad\qquad\qquad\quad CH_3$$
$$\qquad\qquad\qquad\qquad\qquad\qquad\qquad\qquad\qquad\qquad\qquad\qquad\qquad\qquad\qquad\qquad\text{Isoprene}$$

FIGURE I   Terpenes: isoprene, a hemiterpene synthesized from acetone.

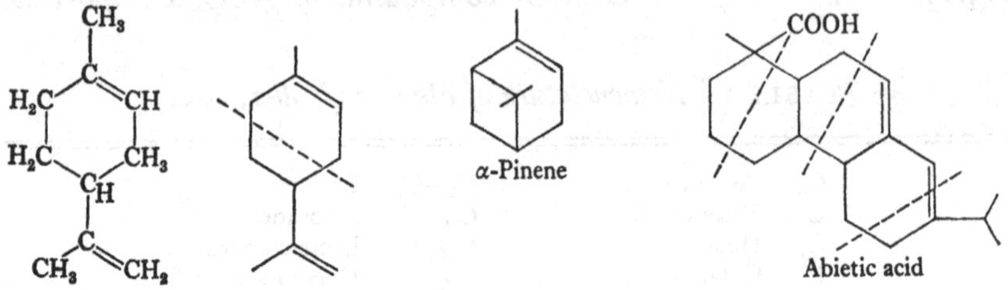

Limonene
α-Pinene
Abietic acid

FIGURE 2. Terpenes: limonene and α-pinene are common constituents of plants; abietic acid is the chief constituent of rosin.

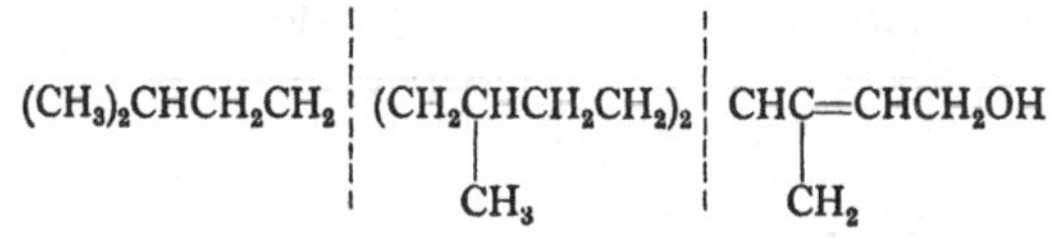

$(CH_3)_2CHCH_2CH_2$ | $(CH_2CHCH_2CH_2)_2$ | $CHC=CHCH_2OH$
CH_3 | CH_2

Phytol

FIGURE 3 Terpenes: phytol makes up about one-third of the chlorophyll molecule; it yields the isoprenoid hydrocarbon phytane.

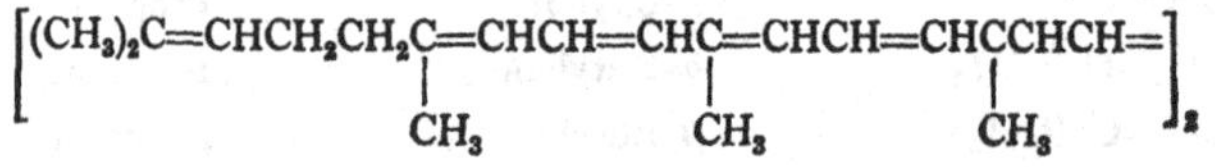

Lycopene

FIGURE 4 Terpenes: lycopene is a red pigment of fruit; it yields perhydroly-copane $C_{40}H_{82}$, a saturated hydrocarbon, on catalytic hydrogenation.

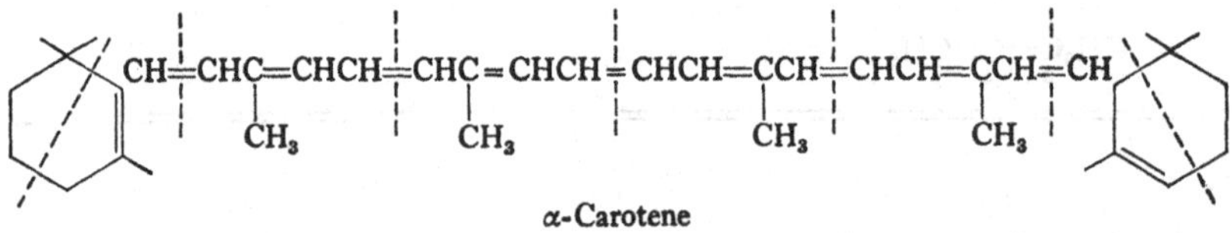

α-Carotene

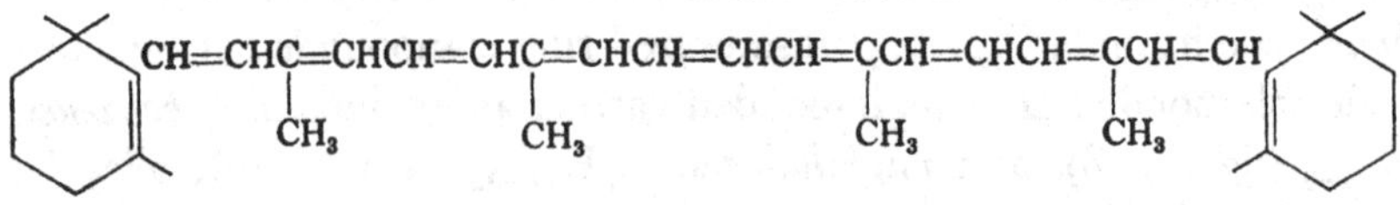

β-Carotene

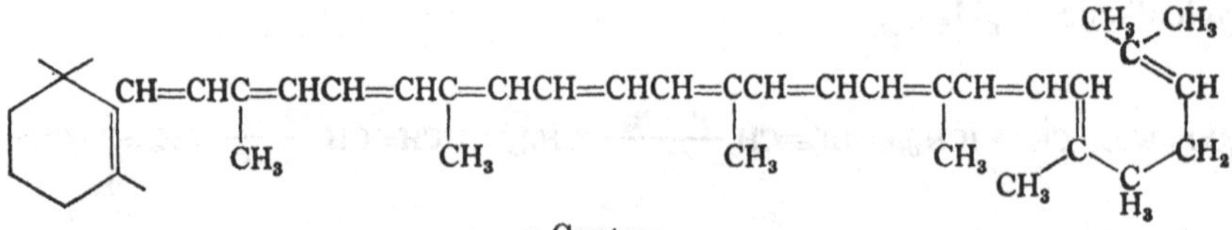

γ-Carotene

FIGURE 5 Terpenes: carotenoid pigments.

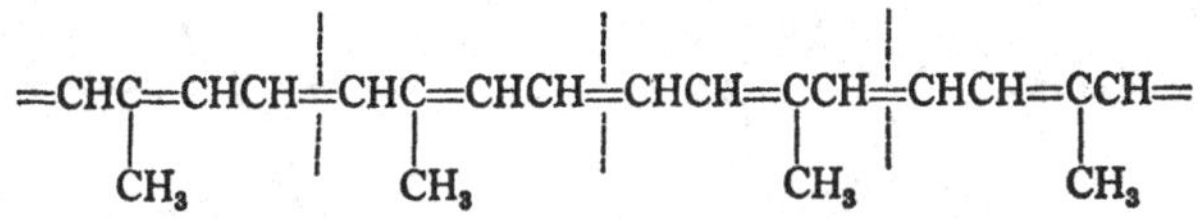

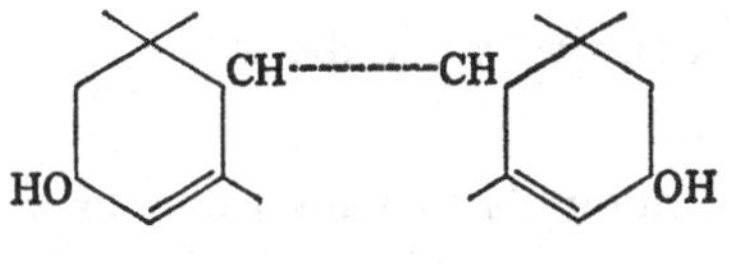

$$=CHC=CHCH=CHC=CHCH=CHCH=CCH=CHCH=CCH=$$

with CH₃ groups below:

=CHC̶=CHCH̶=CHC̶=CHCH̶=CHCH=CCH̶=CHCH=CCH=
$\quad$ CH₃ $\qquad$ CH₃ $\qquad$ CH₃ $\qquad$ CH₃

Lutein

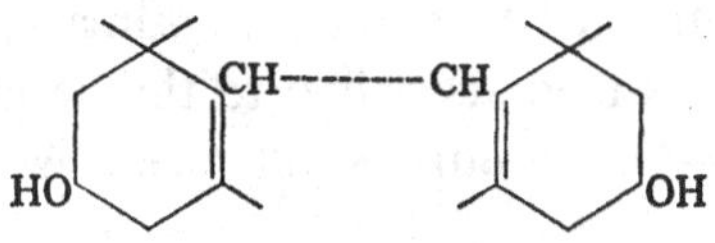

Zeaxanthin

FIGURE 6 Terpenes: xanthophyll pigments in leaf (lutein) and corn (zeaxanthin).

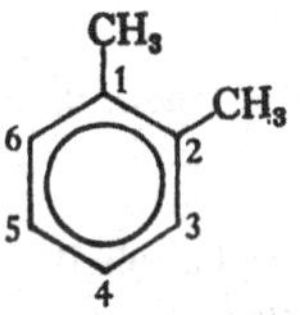

o-Xylene 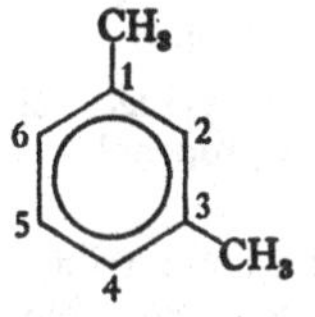 $\qquad$ m-Xylene 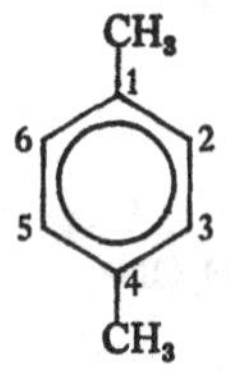 $\qquad$ p-Xylene

FIGURE 7 Aromatic hydrocarbons.

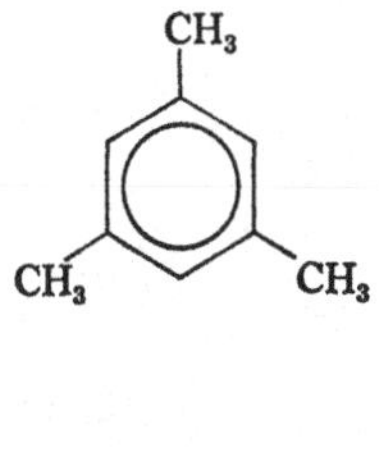

Mesitylene 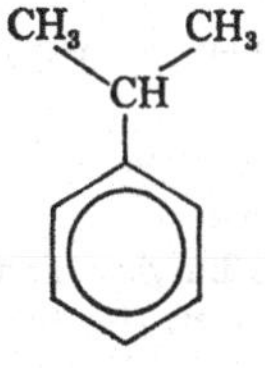 $\qquad$ Cumene 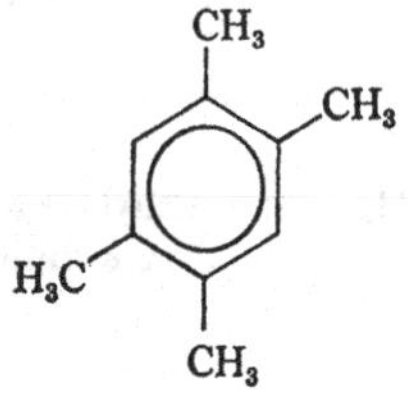 $\qquad$ Durene 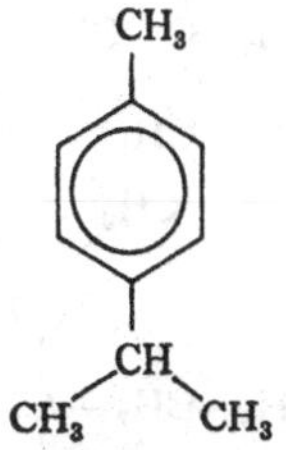 $\qquad$ p-Cymene

FIGURE 8 Aromatic hydrocarbons.

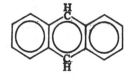

Naphthalene

Anthracene

FIGURE 9   Aromatic hydrocarbons.

The alcohols (table 3) are compounds in which one or more of the hydrogen atoms of the hydrocarbon analogue is replaced by an —OH group. Carbon atoms are classified according to the number of other carbon atoms attached to them, i.e. primary, secondary and tertiary, and alcohols are classified according to the designation of the carbon atoms to which the —OH group is attached. Monohydroxy alcohols are called primary

TABLE 3   *Nomenclature of alcohols (Noller, 1951)*

CH_3OH	Methyl alcohol, carbinol, methanol
C_2H_5OH	Ethyl alcohol, methylcarbinol, ethanol
$CH_3CH_2CH_2OH$	Normal propyl alcohol (*n*-propyl alcohol), ethylcarbinol, 1-propanol

CH_3CHCH_3
   |
   OH
	Isopropyl alcohol (*i*-propyl alcohol), dimethylcarbinol, 2-propanol

$CH_3CH_2CH_2CH_2OH$	Normal butyl alcohol (*n*-butyl alcohol), *n*-propylcarbinol, 1-butanol

$CH_3CH_2CHCH_3$
     |
     OH
	Secondary butyl alcohol (*s*-butyl alcohol), methylethylcarbinol, 2-butanol

CH_3CHCH_2OH
   |
   CH_3
	Isobutyl alcohol (*i*-butyl alcohol), *i*-propylcarbinol, 2-methyl-1-propanol

CH_3
    |
CH_3—C—OH
    |
    CH_3
	Tertiary butyl alcohol (*t*-butyl alcohol), trimethylcarbinol, 2-methyl-2-propanol

CH_3 CH_3
  \\ /
   CH
   |
CH_3—CH_2—CH—OH
      |
      CH
     / \\
CH_3—CH_2 CH_3
	Ethyl-*i*-propyl-*s*-butylcarbinol, 2,4-dimethyl-3-ethyl-3-hexanol

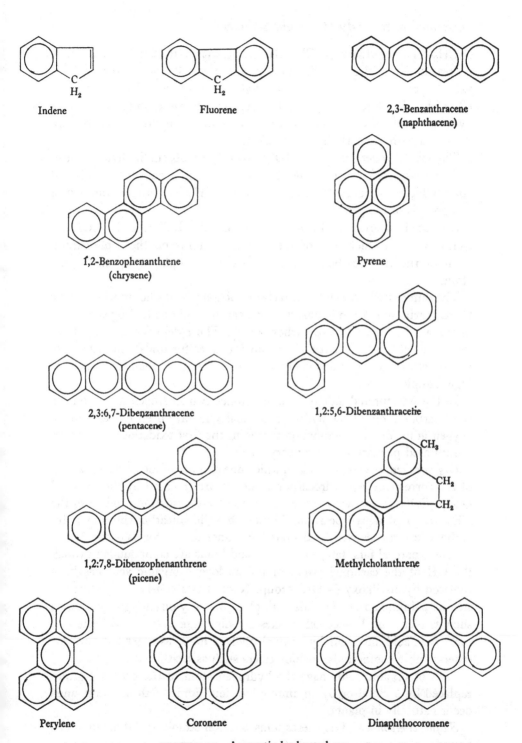

Indene

Fluorene

2,3-Benzanthracene
(naphthacene)

1,2-Benzophenanthrene
(chrysene)

Pyrene

2,3:6,7-Dibenzanthracene
(pentacene)

1,2:5,6-Dibenzanthracene

1,2:7,8-Dibenzophenanthrene
(picene)

Methylcholanthrene

Perylene

Coronene

Dinaphthocoronene

FIGURE 10   Aromatic hydrocarbons.

(RCH$_2$OH), secondary (R$_2$CHOH), and tertiary (R$_3$COH), in which R represents an alkyl group. An alkyl group is any aliphatic saturated univalent hydrocarbon radical, e.g. methyl, CH$_3$·, ethyl CH$_3$CH$_2$·, propyl, CH$_3$CH$_2$CH$_2$·, butyl, CH$_3$CH$_2$CH$_2$CH$_2$·, etc. Some examples of monohydroxy alcohols are given in Table 3. The polyhydroxy alcohols are discussed with the carbohydrates below.

The *ethers*, general formula ROR, are rare or absent in living matter and in most geologic materials, but occur in some petroleums. Methyl ether (CH$_3$OCH$_3$) and ethyl ether (CH$_3$CH$_2$OCH$_2$CH$_3$) are the commonest compounds of the group.

Saturated *aldehydes* and *ketones*, type formulas RCOH and RCOR are compounds in which two of the hydrogen atoms on the same carbon atom of the hydrocarbon analogue have been replaced by an oxygen atom.

The difference between the two classes of compounds lies in whether the C=O (carbonyl) group is linked to one carbon and one hydrogen atom as in the aldehydes or to two carbon atoms. The aldehydes are important geochemically as derivatives of carbohydrates (as furfurals), but otherwise neither class of compounds is known to occur in more than traces in geologic samples.

In the unsaturated aldehydes and ketones two hydrogen atoms on the same carbon atom of the hydrocarbon analogue have been replaced by an oxygen atom or the compounds represent the first oxidation products of unsaturated primary and secondary alcohols.

Organic acids are compounds in which one or more of the hydrogen atoms of the corresponding hydrocarbon analogue are replaced by the carboxyl group (COOH), or they may represent the third oxidation product of the CH$_3$ group of a hydrocarbon. Monocarboxylic, dicarboxylic and polycarboxylic forms as well as saturated and unsaturated forms occur.

Esters, natural fats, fatty oils, waxes and lipids are compounds in which the OH of the carboxyl group of the analogous organic acid has been replaced by an alkoxy (—OR) group. Natural fats consist of mixtures of stearic, palmitic and oleic acids with glycerol: fatty oils are glyceryl esters such as olein and linseed oil. Waxes are high molecular weight esters of carboxylic acid and monohydroxy alcohols. Lipids, as a general term for ether-soluble compounds, include esters such as lecithin.

Salts of organic acids have the hydrogen atom of the carboxyl group replaced by a metal or by an ammonium ion. Some of these compounds occur naturally in plants.

Sulfur compounds. The mercaptans or alkanethiols are aliphatic compounds which form as metabolic products of yeasts and other micro-organisms and are found in coal and petroleum. They are analogous to alcohols in which the oxygen atom has been replaced by a sulfur atom.

Correspondingly, aromatic sulfur compounds called sulfonic acids, are analogous to aromatic hydrocarbons in which one or more hydrogen atoms have been replaced by a sulfonic acid group. They occur as metabolic waste and degradation products of protein.

1.3.2 *Ammonia, amides, imides, amines, amino acids, peptides and proteins*
Ammonia is an analogue of water in which the amino (NH_2) group replaces the hydroxyl group. Similarly, in *amides* the OH of metallic hydroxides is replaced by NH_2 and in *imides* the oxygen of metallic oxides is replaced by NH groups. *Primary amines* are analogues of primary alcohols (ROH) or phenols (ArOH) in which the OH is replaced by NH_2 and *secondary amines* (RNHR′) are analogues of ethers where NH replaces O. In the case of *tertiary amines* (R_3N) there is no oxygen analogue. Amines can be converted to amides by treatment with an acid chloride. Acid amides, such as phthalamic acid, on heating loses a molecule of water, a ring forms, and two acyl groups become attached to the nitrogen to form an imide, i.e. phthalamide.

Amino acids are composed of amino and carboxyl derivatives of alkanes with a general type formula $RCH(NH_2)(CH_2)_xCOOH$ in which x is typically zero. The amino acids serve as building blocks for *proteins* (fig. 11)

$$\begin{matrix} O \\ \| \end{matrix}$$

in which they are joined by —NHC-linkages or *peptide* bonds. Acidic amino acids contain two or more carboxyl groups, the basic amino acids contain two or more amino groups while the neutral amino acids contain equal numbers, generally one each of these groups. Twenty-two amino acids have been obtained as hydrolysis products of proteins. All of the protein amino acids are of the alpha-type in which the amino group lies adjacent to the carboxyl group. Optical activity* is shown by all the natural amino acids except glycine. The L-configuration is represented in all the protein amino acids. The protein amino acids are listed in table 4.

* Optically active organic compounds that rotate the plane of polarized light in equal but opposite directions are mirror image *isomers* or *enantiomers*. Such compounds are dissymmetric that is they nearly all contain carbon atoms to which four different groups are attached (asymmetric carbon atoms). The designations L- (or −) and D- (or +) are used for the direction of rotation of polarized light to the left or to the right, levo- and dextro-rotatory, respectively. Mirror-image isomers or enantiomers are one type of stereoisomers.

The configuration of a stereoisomer is its characteristic arrangement of atoms. The terms L- and D- are arbitrary designations for individual groups of compounds. Glyceraldehyde, $CH_2OHCHOHCHO$, was chosen as a reference and (+)-glyceraldehyde was designated D-, and (−)-glyceraldehyde was designated L-glyceraldehyde.

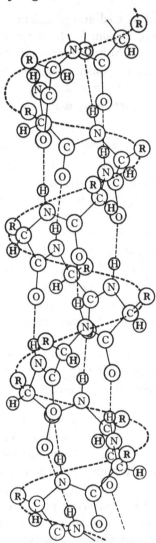

FIGURE II Structure of proteins: the right-handed α-helix with 3·6 amino-acid
residues per turn proposed for α-keratin *b*, Pauling (1951). 6 terminal amino-acid
residues at end of peptide chain having a free alpha amino group; C, terminal
amino-acid residues having a free carboxyl group; bold-face type H and R, side
chains of glycine and other amino acids respectively; light-face type H, hydrogen
bonds between adjacent peptide chains (Morrison & Boyd, 1966).

Proteins or polypeptides (fig. 11) are complex compounds made up of
α-L-amino acids in which the molecular weight ranges from $(10^4$ to $10^6)x$;
carbon, hydrogen, oxygen and nitrogen occur, plus sulfur, and phosphorus.
The proteins are classified as simple, conjugated and derived. Simple

TABLE 4   *Protein amino acids (Fruton & Simonds,* 1958; *Morrison &)*
Boyd 1966

Aliphatic amino acids

  Monoaminomonocarboxylic acids

Glycine: CH_2COO^-
$|$
$^+NH_3$

Alanine: CH_3CHCOO^-
$|$
$^+NH_3$

Valine: $(CH_3)_2 CHCHCOO^-$
$|$
$^+NH_3$

Leucine: $(CH_3)_2CHCH_2CHCOO^-$
$|$
$^+NH_3$

Isoleucine: $CH_3CH_2CH (CH_3) CHCOO^-$
$|$
$^+NH_3$

Serine: $HOCH_2CHCOO^-$
$|$
$^+NH_3$

Threonine: $CH_3CHOHCHCOO^-$
$|$
$^+NH_3$

  Sulfur-containing amino acids

Cysteine: $HSCH_2CHCOO^-$
$|$
$^+NH_3$

Cystine: $-OOCCHCH_2S-SCH_2CHCOO^-$
$|$ $|$
$^+NH_3$ $^+NH_3$

Methionine: $CH_3SCH_2CH_2CHCOO^-$
$|$
$^+NH_3$

Monoaminodicarboxylic acids and their amides

Aspartic acid: $HOOCCH_2CHCOO^-$
$|$
$^+NH_3$

Asparagine: $H_2NCOCH_2CHCOO^-$
$|$
$^+NH_3$

Glutamic acid: $HOOCCH_2CH_2CHCOO^-$
$|$
$^+NH_3$

Glutamine: $H_2NCOCH_2CH_2CHCOO^-$
$|$
$^+NH_3$

TABLE 4 (*cont.*)

Basic amino acids

Lysine: $^+H_3NCH_2CH_2CH_2CH_2CHCOO^-$
$\qquad\qquad\qquad\qquad\quad | $
$\qquad\qquad\qquad\qquad\quad NH_2$

Hydroxylysine: $^+H_3NCH_2CHCH_2CH_2CHCOO^-$
$\qquad\qquad\qquad\qquad\quad |\qquad\qquad\; |$
$\qquad\qquad\qquad\qquad\quad OH\qquad\quad NH_2$

Arginine: $H_2NCN\ HCH_2CH_2CH_2CHCOO^-$
$\qquad\qquad\quad \|\qquad\qquad\qquad\qquad |$
$\qquad\qquad\quad ^+NH_2\qquad\qquad\qquad NH_2$

Histidine:

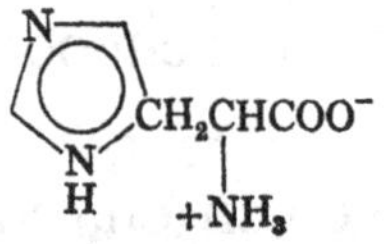

Aromatic amino acids

Phenylalanine:

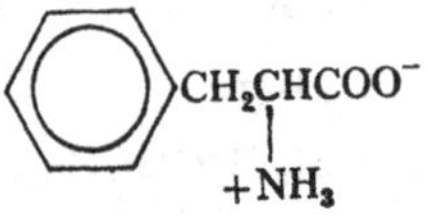

Tyrosine:

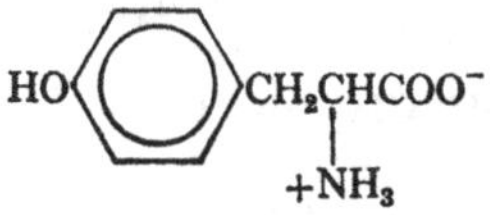

Diiodotyrosine: (in corals and other marine organisms)

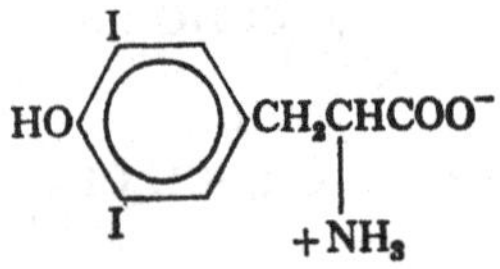

Dibromotyrosine: (in corals)

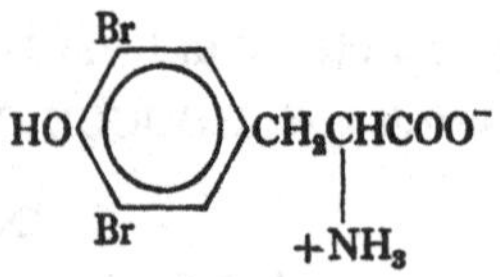

Thyroxine:

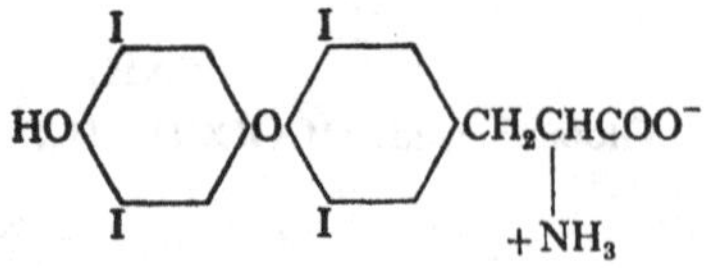

TABLE 4 (*cont.*)

Heterocyclic amino acids

Tryptophan:

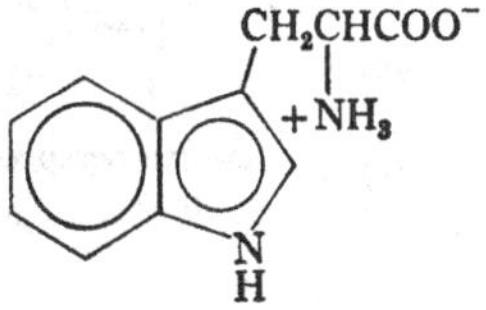

Proline:

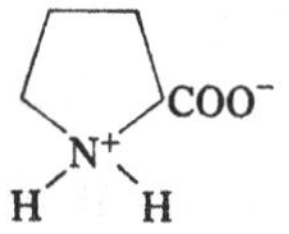

Hydroxyproline:

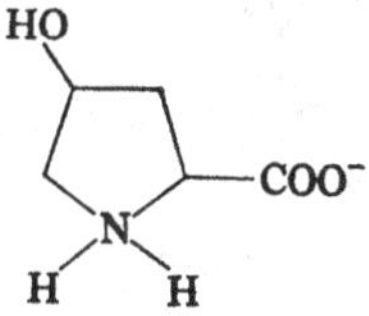

proteins such as albumin yield only α-amino acids on hydrolysis; conjugated proteins, such as casein and haemoglobin give other substances
besides α-amino acids; derived proteins, such as proteoses and peptones
are formed on partial hydrolysis of other proteins.

Other nitrogenous compounds that have some geochemical importance
are *alkyl cyanides* and nitrites, of which hydrogen cyanide is the most
important; their general formula is RCN.

1.3.3 *Carbohydrates and polyhydric alcohols*

The carbohydrates are polyhydroxy aldehydes and polyhydroxy ketones
having a general formula $(C_x[H_2O]_x) - (n-1) H_2)$ in which x is the number
of carbon atoms in a carbohydrate unit and n is the number of carbohydrate
units in a saccharide molecule. The carbohydrates are classified as mono-,
di-, tri-, tetra-, and polysaccharides (figs. 12–14). In the monosaccharides
(fig. 12) a subdivision has been made on the number of oxygen atoms:
diose, triose, tetrose, pentose, hexose, heptose, octose, nonose, and glucosides; of these only pentoses, hexoses and glucosides are important in
natural substances. The commonly occurring carbohydrates are listed in
table 5.

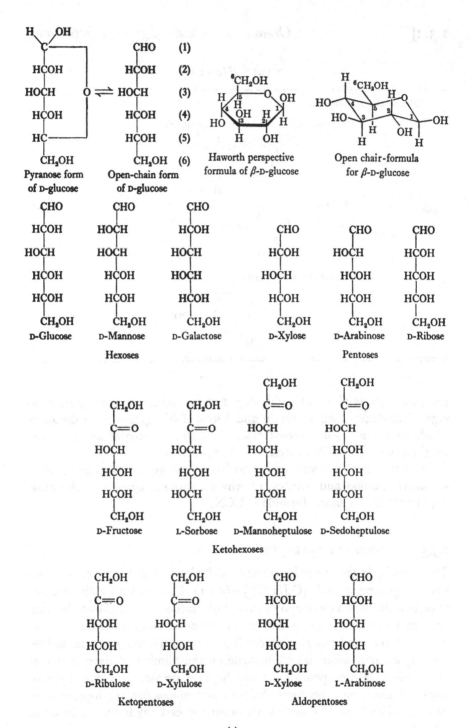

FIGURE 12(a)   Structural formulas of monosaccharides.

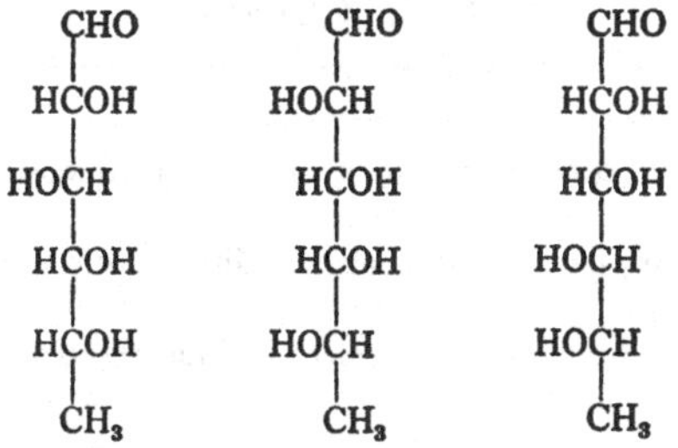

CHO   CHO   CHO
HCOH   HOCH   HCOH
HOCH   HCOH   HCOH
HCOH   HCOH   HOCH
HCOH   HOCH   HOCH
CH_3 CH_3 CH_3

6-Deoxy-D-glucose   L-Fucose   L-Rhamnose

Methylpentoses

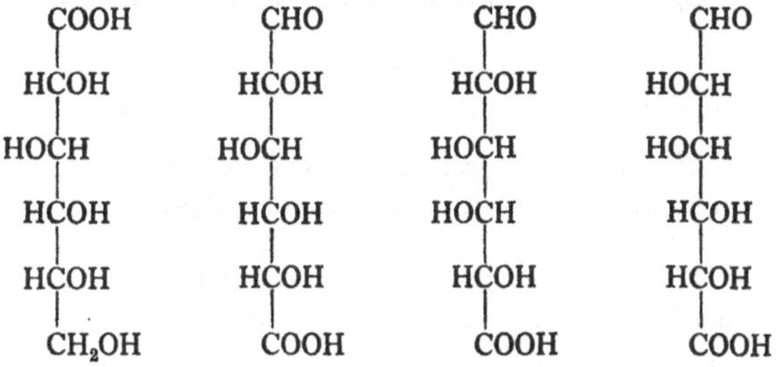

COOH   CHO   CHO   CHO
HCOH   HCOH   HCOH   HOCH
HOCH   HOCH   HOCH   HOCH
HCOH   HCOH   HOCH   HCOH
HCOH   HCOH   HCOH   HCOH
CH_2OH COOH   COOH   COOH

D-Gluconic   D-Glucuronic   D-Galacturonic   D-Mannuronic
acid   acid   acid   acid

Sugar acids

(*b*)

FIGURE 12(*b*)   Structural formulas of monosaccharides.

TABLE 5   *Carbohydrates (Pigman, 1957)*

Monosaccharides

(*a*) Aldohexoses

   D-Glucose, $C_6H_{12}O_6$: in blood of all animals and in sap of plants; the structural unit of the most important polysaccharides

   D-Mannose, $C_6H_{12}O_6$: a constituent of certain polysaccharides and of glycoproteins, differs from D-glucose in its configuration about carbon atom 2

   D-Galactose, $C_6H_{12}O_6$: a component of disaccharide lactose and of several polysaccharides; differs from glucose in its configuration about carbon atom 4

(*b*) Ketohexoses

   D-Fructose, $C_6H_{12}O_6$: in seminal fluid, as a constituent of polysaccharide inulin

   L-Sorbose, $C_6H_{12}O_6$

   D-Mannuloheptose $C_7H_{14}O_7$: in avocado

   D-Sedoheptulose $C_7H_{14}O_7$: in plants of *Sedum* family

TABLE 5 (*cont.*)

(*c*) Aldopentoses

D-Ribose, $C_5H_{10}O_5$: a constituent of ribonucleic acid and of several nucleotides (ATP, DPN)

D-Xylose, $C_5H_{10}O_5$: in plant polysaccharides; differs from D-Ribose in its configuration about carbon 3

L-Arabinose $C_5H_{10}O_5$: in combined form in various plant products

(*d*) Ketopentoses

D-Ribulose, $C_5H_{10}O_5$

D-Xylulose, $C_5H_{10}O_5$

(*e*) Methylpentoses

6-Deoxy-D-glucose $C_6H_{12}O_5$: as a glycoside in bark of species of *Chichona*

L-Fucose, $C_6H_{12}O_5$: in bound form in marine algae and in some mammalian polysaccharides

L-Rhamnose, $C_6H_{12}O_5$: as a glycoside in plants

(*f*) Sugar acids

Gluconic acid $C_6H_{12}O_7$: from oxidation of gluconolactone

Glucuronic acid $C_6H_{10}O_7$: a constituent of certain polysaccharides and an excretion product in animals

Galacturonic acid, $C_6H_{10}O_7$: a constituent of fruit pectins

Mannuronic acid, $C_6H_{10}O_7$: found in polysaccharides of brown marine algae

(*g*) Polyhydroxy alcohols

D-Mannitol, $C_6H_{14}O_6$: an important constituent of brown algae

D-Sorbitol, $C_6H_4O_6$: occurs in fruits

D-Ribotol, $C_5H_{12}O_5$: sugar residue in riboflavin

Meso-Inositol, $C_6H_{12}O_6$: in many plants, animals

Scyllitol, $C_6H_{12}O_6$: in cartilage, several plants, dogfish liver

(*h*) Amino sugars

D-Glucosamine (2-amino-2-deoxy-D-glucose), $C_6H_{13}O_5N$: in several polysaccharides, chitin

D-Galactosamine (2-amino-2-deoxy-D-galactose), $C_6H_{13}O_5N$: in several polysaccharides

Oligosaccharides

(*a*) Disaccharides

Maltose, (4(α-D-glucopyranosyl)-D-glucopyranose) $C_{12}H_{21}O_{11}$: product of partial degradation of starch, not found as such in nature; yields glucose

D-Lactose, (4-(β-D-galactopyranosyl)-D-glucopyranose) $C_{12}H_{21}O_{11}$: in milk; yields D-glucose and D-galactose

Cellobiose, (4(β-D-glucopyranosyl)-D-glucopyranose) $C_{12}H_{21}O_{11}$: forms on degradation of cellulose

Sucrose, (α-D-glucopyranosyl-β-D-fructofuranoside) $C_{12}H_{22}O_{11}$: not a reducing sugar; in all photosynthetic plants

Trehalose, (1-(α-D-glycopyranosyl)-α-D-glucopyranoside) $C_{12}H_{22}O_{11}$: not a reducing sugar; in fungi and yeasts, in blood of some insects

---

TABLE 5 (*cont.*)

(*a*) Disaccharides (*cont.*)

Gentiobiose (6-(β-D-glucopyranosyl)-D-glucopyranose) $C_{12}H_{22}O_{11}$: a constituent of the glycoside amygdalin; ($1 \rightarrow 6$) glycosidic linkage

Melibiose, (6-(α-D-galactopyranosyl)-D-glucopyranose $C_{12}H_{22}O_{11}$: a constituent of the trisaccharide raffinose in certain plant products; ($1 \rightarrow 6$) glycosidic linkage

Primeverose, (6-(β-D-xylopyranosyl)-D-glucopyranose, $C_{11}H_{21}O_{10}$: ($1 \rightarrow 6$) linkage

Vicianose (6-(β-L-arabopyranosyl)-D-glucopyranose), $C_{11}H_{21}O_{10}$: ($1 \rightarrow 6$) linkage

Rutinose (6-(β-L-rhamnosyl)-D-glucopyranose, $C_{11}H_{21}O_{10}$: ($1 \rightarrow 6$) linkage

(*b*) Trisaccharides

Gentianose, $C_{18}H_{32}O_{16}$: contains two molecules of glucose and one of fructose: only in plants

Raffinose, $C_{18}H_{32}O_{16}$: contains glucose, fructose and galactose; only in plants

Polysaccharides

(*a*) Nutrient polysaccharides

Starch: a mixture of amylose and amylopectin both of which yield glucose and glycosidic bonds are principally ($1 \rightarrow 4$); mol. wt. 400–500×10^3; in plants

Inulin: yields D-fructose; β ($2 \rightarrow 1$)-glycosidic linkages; in plants

Glycogen: animal reserve polysaccharide, 11–18 D-glucopyranose units in straight-chain arrays in $\alpha(1 \rightarrow 4)$ glycosidic linkage, cross-linked by means of $\alpha(1 \rightarrow 6)$-glycosidic bonds

(*b*) Structural polysaccharides

Cellulose: a linear array of D-glucopyranose units having $\beta(1 \rightarrow 4)$-glycosidic bonds; particle weight 100–2000×10^3; mol. wt. of native cellulose $\sim 35 \times 10^3$; main structural polysaccharide of plants

Mannan: D-mannose units linked by ($1 \rightarrow 2$)-and ($1 \rightarrow 6$)-glycosidic bonds; in yeasts, cell walls; mannan of ivory nut has $\beta(1 \rightarrow 4)$-bonds

Xylan: D-xylose in chains of 20–40 D-xylopyranose units joined by $\beta(1 \rightarrow 4)$-glycosidic bonds, with ($1 \rightarrow 3$)-glycosidic cross-linkages; in wood tissues; some xylan also contains arabinose

Pectic acids: long chains of D-galacturonic acid units joined in $\alpha(1 \rightarrow 4)$-glycosidic linkages; mol. wt. 25–100×10^3; components of plant pectins which also contain galactan and araban polysaccharides; plant tissues and fruit

Alginic acids: D-mannuronic acid joined by $\beta(1 \rightarrow 4)$-glycosidic linkages to make long chains; also contains L-guluronic acid; in brown algae (also see algal polysaccharides)

Hemicellulose: contains D-glucuronic acid and D-xylose; in woody tissue

Chitin: units of N-acetyl-D-glucosamine joined by $\beta(1 \rightarrow 4)$-glycosidic bonds; in shells of lobsters and crabs and other arthropods

Hyaluronic acid: one of the structural mucopolysaccharides of animals; composed of units of D-glucuronic acid and N-acetyl-d-glucosamine in the form of a linear polymer in which the disaccharide N-acetylhyalobiuronic acid is the principal repeating unit; particle weight 10×10^4 to 4×10^6; in various animal body tissues

Chondroitin sulfate (A, B, C) sulfated mucopolysaccharides: A is formed in cartilage, adult bone and cornea; B in skin, tendons, heart valves; C in cartilage and tendons;

TABLE 5 *(cont.)*

(*b*) Structural polysaccharides *(cont.)*

A and C yield ~ equal amounts of D-glucuronic acid, D-galactosamine, acetic acid and sulfuric acid; the disaccharide chondrosine also occurs; B contains L-iduronic acid; a few non-sulfated chondroitins occur

Heparin: mucopolysaccharides of animal tissues, in liver, lungs and spleen; inhibitors of blood coagulation; contain glucuronic acid, glucosamine, acetic acid and sulfuric acid, the last occurring as a sulfamic acid group (—NHSO$_4$OH); particle weight ~ 17×10^3

(*c*) Plant exudates

Gum arabic: a complex polysaccharide containing D-galactopyranose, L-arabofuranose, L-rhamnopyranose and D-glucuronic acid; a commercial gum from genus *Acacia* of semi-arid regions of Sudan and Senegal

Tragacanth: a complex gum polysaccharide separable into a water-soluble fraction tragacanthin which contains 51 % uronic acid anhydride and 44 % L-arabinose, and a water-insoluble fraction dassorin which contains arabinose, xylose, fucose and galactose

(*d*) Algal polysaccharides

Agar: principally a calcium salt of a sulfuric ester of a linear galactan in which there are about nine $1 \to 3$ links for each $1 \to 4$ link; both L- and D-galactose occur; extracted from *Gelidium amansii* and other red algae

Carageenan: a mixture of polysaccharides of D-galactose having both β-$1 \to 4$ and α-$1 \to 3$ links; extracted with hot water from *Chrondrus crispus*, Irish moss

Alginic acid: principally a linear polymer of β-$1 \to 4$ linked D-mannuronic acid; extracted from *Macrocystis pyrifera* with sodium carbonate solution (also under (b)).

Laminaran: essentially a regular chain of β-$1 \to 3$ linked D-glucopyranose units; large amounts occur in species in the brown alga *Laminaria*

(*e*) Fungal and bacterial polysaccharides

Aspergillus: a linear chain of D-glucopyranose units alternately joined by α-$1 \to 4$ and α-$1 \to 5$ bonds, found in common black mold *Aspergillus niger*

Leuconostoc: branched molecules composed of D-glucopyranose units, principally with α-$1 \to 6$ bonds, but also with $1 \to 4$ and $1 \to 3$ bonds; form dextrans (ropy slimes) produced by *L. mesenteroides* growing in a sucrose solution

(Glycoproteins): see amino-acid compounds

Pneumococcus polysaccharides: Type I of *Pneumococcus* yields glucosamine; Type II contains glucose, glucuronic acid and L-rhamnose, in a highly branched structure in which is found both $(1 \to 4)$- and $(1 \to 6)$-glycosidic bonds; Type III is composed of 3-(glucopyranosyl) glucuronic acid units joined by $(1 \to 4)$-glycosidic bonds; Type IV contains D-glucose, D-galactose and D-glucuronic acid residues; *levans* are polysaccharides of *Bacillus* spp. that are composed of D-fructofuranose units linked mostly by $(2 \to 6)$-glycosidic bonds

(*f*) Crown-gall polysaccharide: a low molecular weight D-glucose polysaccharide predominantly with $1 \to 2$ linkages, produced by the crown-gall organism *Phytomonas tumefaciens*

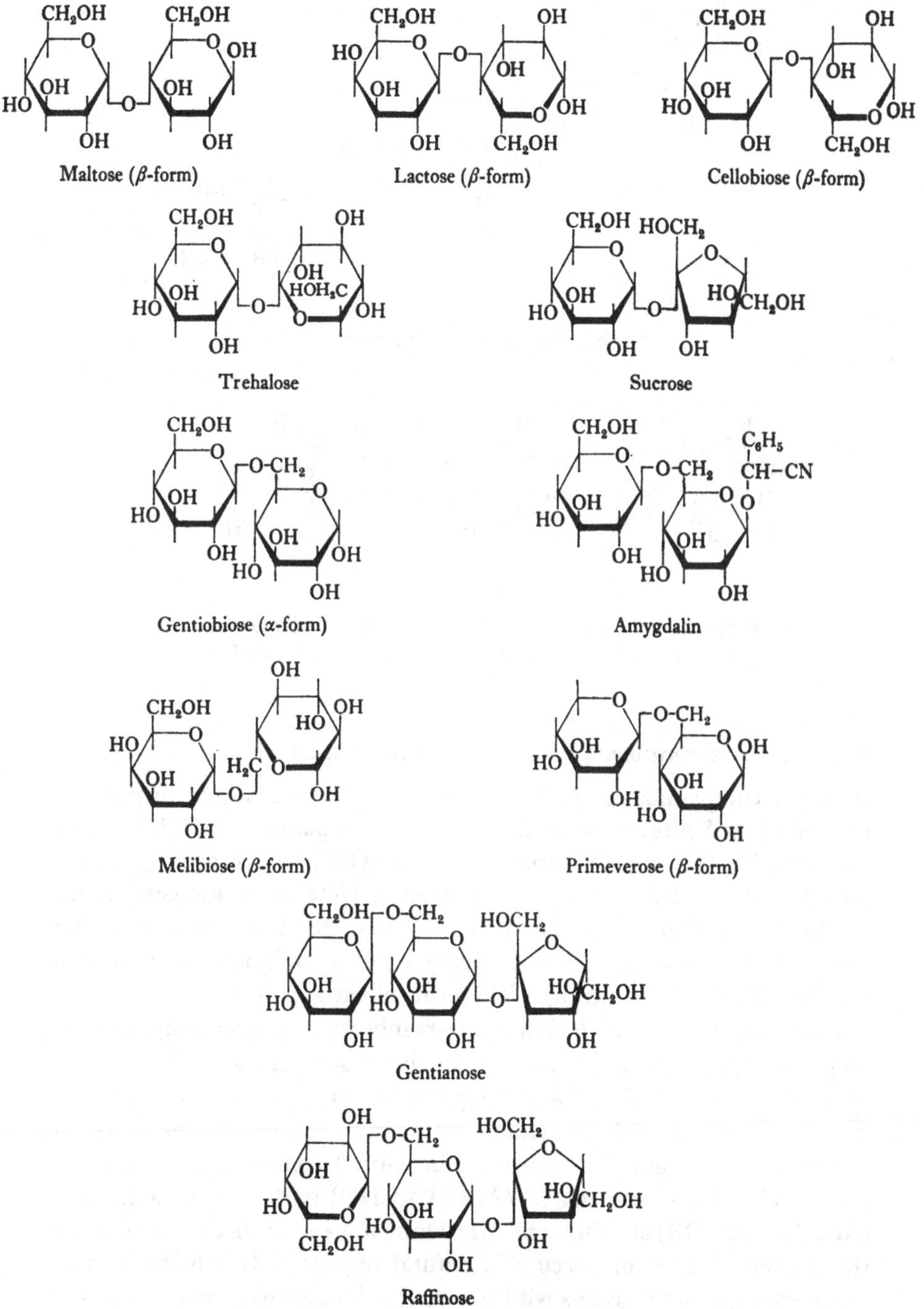

Maltose (β-form)

Lactose (β-form)

Cellobiose (β-form)

Trehalose

Sucrose

Gentiobiose (α-form)

Amygdalin

Melibiose (β-form)

Primeverose (β-form)

Gentianose

Raffinose

FIGURE 13   Structural formulas of oligosaccharides.

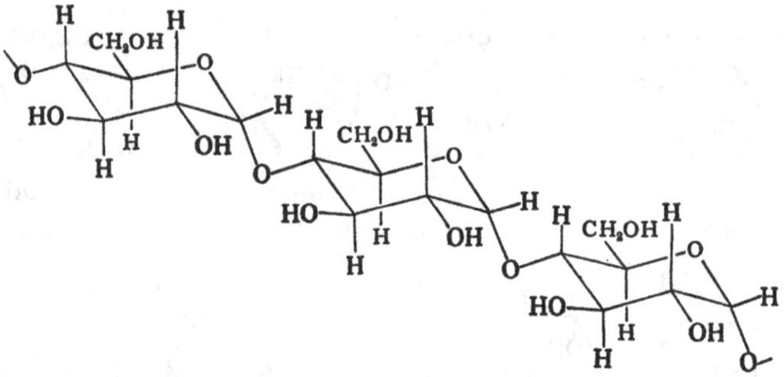

Amylose (chair conformations assumed)

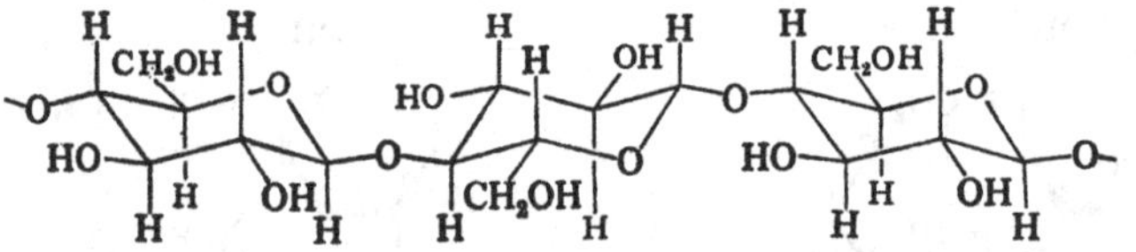

Cellulose

FIGURE 14 Structure of amylose (starch) (α-D-($+$)-glucose units) and cellulose (β-D-($+$)-glucose units), Morrison & Boyd, 1966.

1.3.4 *Non-hydrocarbon aromatic and heteroaromatic compounds*

Aromatic compounds in which one or more hydrogen atoms of an aromatic nucleus of the hydrocarbon analogue have been replaced by OH (hydroxyl), Ar (aryl), R (alkyl), COOH (carboxyl) or SO₂OH (sulfonic acid) group are included in the non-hydrocarbon aromatics. Heteroaromatic compounds are those in which one of the carbon atoms of the aromatic nucleus has been replaced by another atom, usually N, O or S. The aromatic nuclei may be either 5- or 6-membered in terms of carbon atoms.

Phenols and *aromatic alcohols* are 6-membered aromatic compounds in which OH has replaced one or more hydrogen atoms of the nucleus (phenols) or has replaced a hydrogen atom in a side chain (alcohols). These are represented in natural compounds. Aromatic ethers are all synthetic. Aromatic aldehydes occur in natural compounds and include aryl (ArCHO) and aryl-alkanal (Ar(CH₂)ₓCHO) types. Aromatic ketones (ArCOAr, ArCOR) are only synthetic. Quinones are synthetic products but the quinonoid structure occurs in natural pigments. It consists of a 6-membered aromatic nucleus with two double-bonded oxygen atoms linked to the *o*- or *p*-positions. *Aromatic acids* (fig. 15) have the carboxyl group replacing one or more of the hydrogen atoms of an aromatic nucleus. These compounds occur in coal tar and honeystone, in addition to living materials. The *sulfonic acids* have been discussed previously under sulfur-bearing organic compounds. The aromatic nitrogen derivatives (nitrobenzene, etc.) are all synthetic.

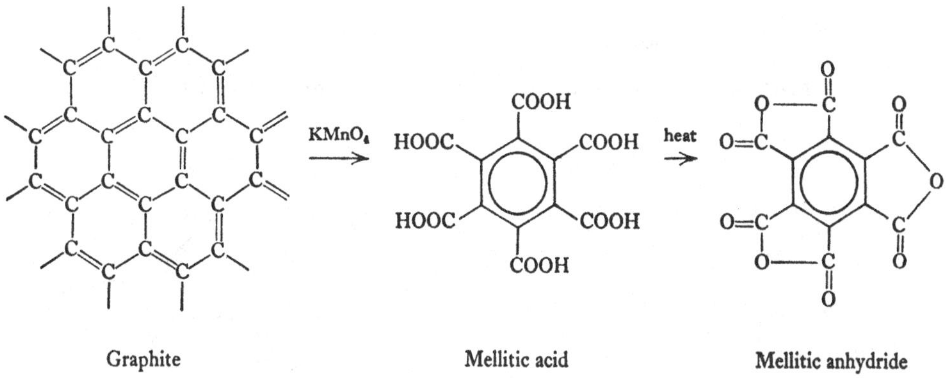

Graphite   Mellitic acid   Mellitic anhydride

FIGURE 15 Aromatic acids: formation of mellitic acid from graphite.

Heterocyclic or *heteroaromatic* compounds (figs. 16–21) include both 5 and 6-membered structures having N-, S- and O- as ring members. Among 5-membered compounds furan (O-substitution), thiophene (S-substitution) and pyrrole (N-substitution) are found in coal tars and petroleum. In addition there are naturally occurring 5-membered hetero-cyclics that are fused to one (coumarone, indole) or two (carbazole) benzene rings.

Heterocyclic six-membered compounds are represented by pyridine (N-substitution) and pyran (O-substitution) among naturally occurring materials, the former an important organic solvent and the latter the basic structure of the hexose carbohydrates. Other 6-membered heterocyclics are those in which the ring is fused to one (benzopyran, quinoline) or to two (acridine) benzene rings.

Organic *dyes* and *pigments* derived from natural materials are principally of heteroaromatic type. Included in this group of dye compounds are: carmine, an acridine; flavones, O-substituted 6-membered heteroaromatic; indigo, an N-substituted 5-membered compound; and berberine, a quino-line. The phthalocyanines, carotenes, xanthophylls, chlorophylls, porphy-rins and flavins are pigments represented in living material. All contain aromatic or heteroaromatic nuclei in association with hydrocarbons and alcohols in some cases. Several of the pigments are important as geochemical residues.

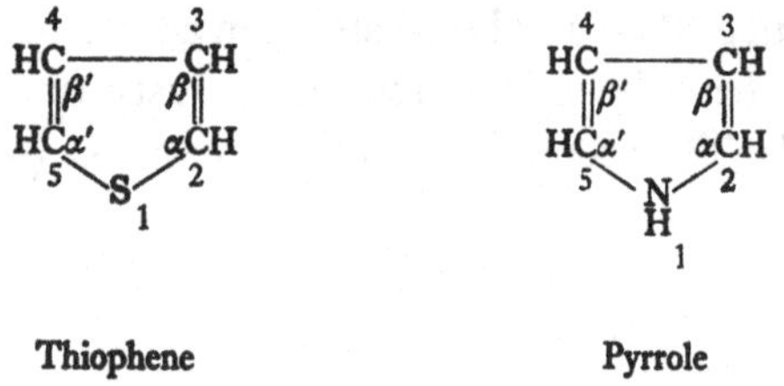

Thiophene Pyrrole

FIGURE 16 Heterocyclic compounds: five-membered structures.

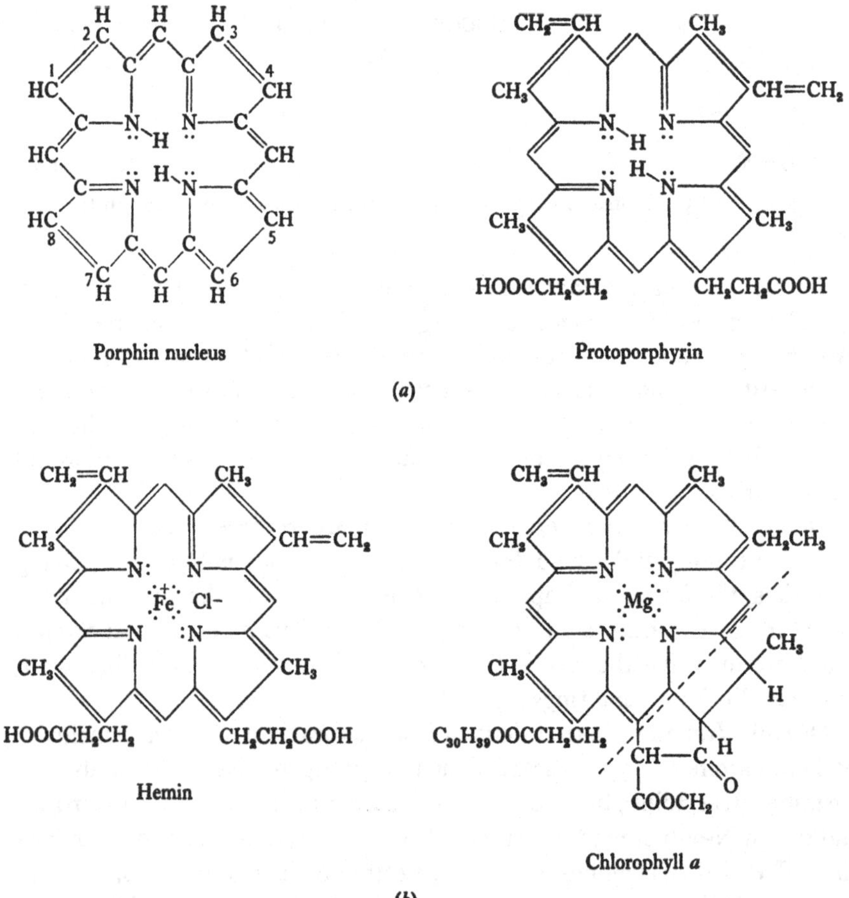

Porphin nucleus Protoporphyrin

(*a*)

Hemin Chlorophyll *a*

(*b*)

FIGURE 17 Heterocyclic compounds: tetrapyrrole pigments.

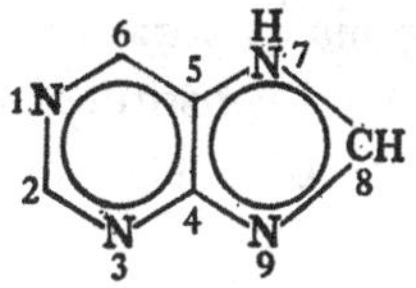

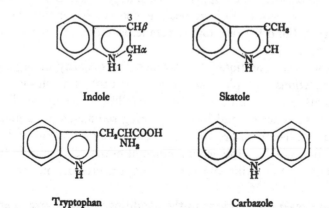

Purine 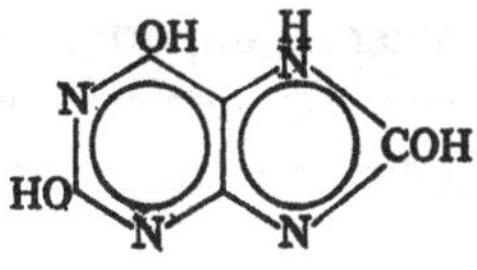   **Uric acid** 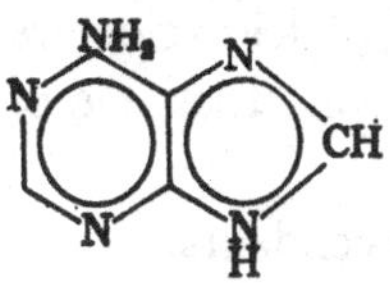   **Adenine**

FIGURE 18   Heterocyclic compounds: N-heteroaromatics; purine and adenine are constituents of nucleic acids, and uric acid is present in blood and urine.

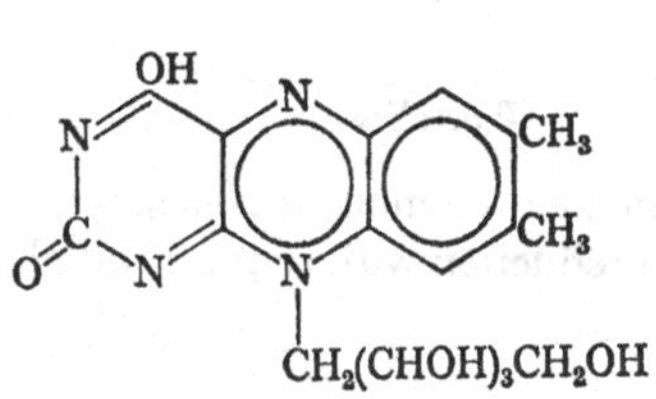

Riboflavin (vitamin B₂)   Flavonol (yellow)

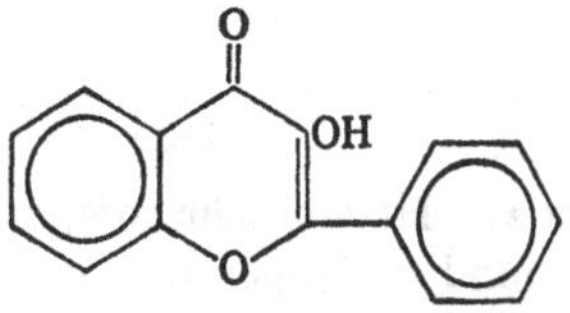

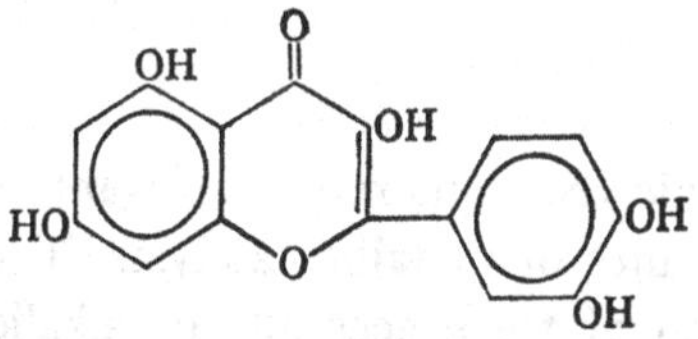

Quercitin

FIGURE 19   Heterocyclic compounds: heteroaromatic pigments.

Indole   Skatole

Tryptophan   Carbazole

FIGURE 20   Heterocyclic compounds: N-heteroaromatics; indole and skatole are formed during putrefaction of protein; tryptophan is an essential amino acid that generally is lost during acid hydrolysis of proteins in geochemical studies; carbazole is an important constituent of coal tar.

31

Nucleic acids are of fundamental importance in organic evolution as it appears that one of them, deoxyribonucleic acid (DNA) a constituent of the gene, transmits the hereditary characteristics of the organism to its descendants.

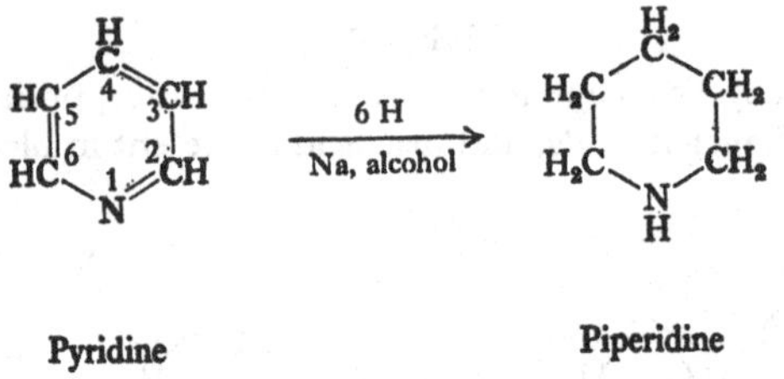

Pyridine   Piperidine

FIGURE 21 Heterocyclic compounds: pyridine, a 6-membered N-heteroaromatic compound, and piperidine, its derivative on reduction with sodium and alcohol: constituents of coal tars.

I.4 IMPORTANT ORGANIC REACTIONS IN ORGANIC GEOCHEMISTRY

I.4.I *Introduction*

Many significant organic reactions require elevated temperatures and the presence of certain alkalies or metallic catalysts. Taking into account the widespread distribution of trace accounts of alkalies and metals in the sedimentary environment and the substitution of the geologic time factor for elevated temperature, all or nearly all known organic reactions might conceivably take place in sediments. In the present discussion, however, only those reactions which appear most likely to take place in the sedimentary environment or to have possible geochemical importance, are to be considered. For other reactions the reader is referred to standard organic chemistry textbooks.

The geological conditions under which the majority of organic geochemical reactions take place are those of marine or freshwater aquatic environments under mildly oxidizing to mildly reducing conditions, in relatively shallow water, and in the resulting sediments in various stages of lithification. Temperatures in the accumulating sediments and in their lithified products fall in the general range of 0–200 °C. Probably for most of their history the sedimentary rock temperatures are in the range of 20–50 °C.

There are several exceptions to the conditions noted above. Terrestrial plants, the remains of which are later incorporated in aquatic sediments, undergo partial oxidative degradation prior to burial, and in some cases are deposited with eolian or tuffaceous sediments under more or less severe conditions of oxidation. Organic materials trapped in glacial sediments or ice, on the other hand, may be preserved intact until melting and consequent rapid degradation occur.

Evaporite conditions in the sea or in lakes are characterized by strongly negative redox potentials and consequent reducing conditions in the accumulating sediments. Such conditions favor rather intensive microbial humification and alteration of the organic matter but not necessarily destruction of it. In such environments, however, the organic productivity in the first place may be quite low.

Aquatic environments having poor circulation of the bottom waters owing to thermal or chemical stratification may also develop strongly negative redox potentials that favor preservation of the organic matter but that also encourage its microbial humification.

Temperatures above normal for most sedimentary organic matter such as those of hot springs are generally unfavorable to organic preservation but where thermal springs occur in deep water, such as in the Red Sea, production and preservation of organic sedimentary material may not be hampered.

Although most of the metallic catalysts necessary for organic reactions are found in sediments, they are, other than iron, so scarce as to be of limited importance in organic geochemistry except where local concentrations of copper and other sedimentary ores occur.

The following organic reactions, briefly described, are arranged for the most part according to the principal products. This arrangement is probably not the most desirable from a chemical point of view, but the geologist may find it more convenient to locate a particular reaction.

1.4.2 *Aliphatic hydrocarbons*

In the Fischer–Tropsch synthesis, which possibly occurs in volcanic regions, carbon monoxide formed by the water–gas reaction may be reduced catalytically to yield a mixture of alkanes.

$$C + H_2O \rightarrow CO + H_2 \tag{15}$$

$$2\,Co + CO + H_2 \rightarrow Co_2C + H_2O \tag{16}$$

$$Co_2C + H_2 \rightarrow 2\,Co + [CH_2] \tag{17}$$

$$x\,[CH_2] \rightarrow (CH_2)x \tag{18}$$

In the electrolysis of the water-soluble salt of carboxylic acid, carboxylate ions are discharged at the anode and lose CO_2; the resulting alkyl radicals couple to form a saturated hydrocarbon.

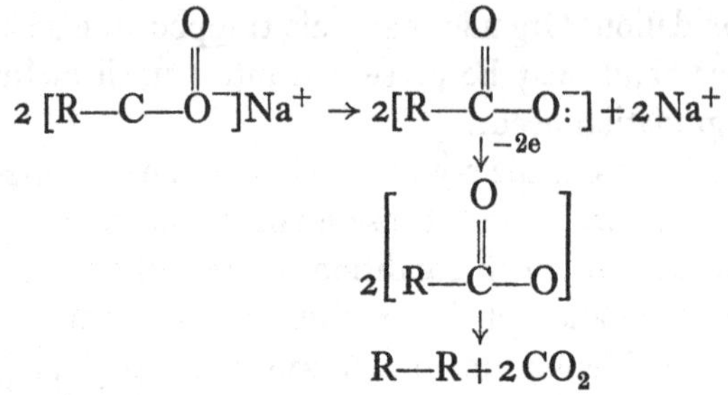

$$2\,[R-\!\!\underset{\underset{\displaystyle O}{\|}}{C}\!\!-O^-]Na^+ \rightarrow 2[R-\!\!\underset{\underset{\displaystyle O}{\|}}{C}\!\!-O^{\cdot}] + 2\,Na^+$$

$$\downarrow -2e$$

$$2\left[R-\!\!\underset{\underset{\displaystyle O}{\|}}{C}\!\!-O \right] \tag{19}$$

$$\downarrow$$

$$R-R + 2\,CO_2$$

Reactions of this type might occur at subsurface formational contacts across which large electrical self potentials occur, but which have not been studied from this viewpoint.

The formation of acetylene is possible in nature but naturally occurring carbides are not known.

$$CaC_2 + 2H_2O \rightarrow Ca(OH)_2 + HC \equiv CH \tag{20}$$
$$\text{Acetylene}$$

Formation of organic compounds from iron carbide (cementite) also is theoretically possible but the carbides have not been reported in natural occurrences.

$$FeC + H_2O \rightarrow \text{methane, ethane, ethylene and other hydrocarbons.} \tag{21}$$

Condensation between paraffins and olefins at high temperature (500 °C) and pressure (5000 p.s.i.) forms new compounds.

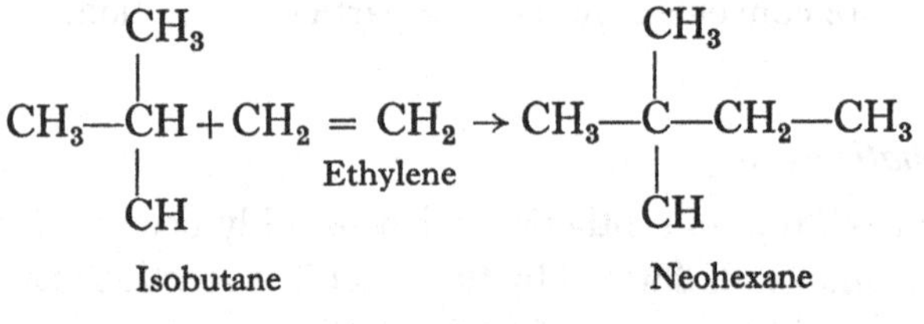

$$CH_3-\!\!\underset{\underset{\displaystyle CH}{|}}{\overset{\overset{\displaystyle CH_3}{|}}{CH}}\!\!+CH_2 = CH_2 \rightarrow CH_3-\!\!\underset{\underset{\displaystyle CH}{|}}{\overset{\overset{\displaystyle CH_3}{|}}{C}}\!\!-CH_2-CH_3 \tag{22}$$

$$\text{Isobutane} \qquad\qquad \text{Ethylene} \qquad\qquad \text{Neohexane}$$

Such reactions requiring higher temperatures and pressures than are normally found in sediments and sedimentary rocks may nevertheless occur very slowly in nature taking into account the geologic time factor.

Methane and ketene are formed by thermal decomposition of acetone (Schmidlin Ketene synthesis)

$$CH_3COCH_3 \xrightarrow{\text{therm. decomp.}} H_2CCO + CH_4 \tag{23}$$

When acetone is used as an extraction solvent in organic sedimentary studies the possibility of this reaction must be taken into account. The process probably does not occur in the sedimentary environment except perhaps by a biochemical route.

Cyanoacetylene can be added to hydrogen cyanide and cyanogen as a starting material in synthesis of biologically important compounds (Sanchez *et al.* 1966). Cyanoacetylene was formed in electric discharge chambers containing hydrogen cyanide, cyanogen and acetylene, hydrogen cyanide and acetylene and methane and nitrogen, but not in methane and ammonia mixtures. A mixture of cyanoacetylene, ammonia and hydrogen cyanide yielded aspartic acid and asparagine.

1.4.3 *Organic acids and fats*

Acetic acid is formed by the enzyme-catalyzed oxidation of ethyl alcohol; the process may be effective in early diagenesis of sedimentary organic matter.

$$CH_3CH_2OH + O_2 \text{ (air)} \xrightarrow{Acetobacter} CH_3COOH + H_2O. \qquad (24)$$

Saponification of esters is accomplished by alkaline hydrolysis; biochemical processes might result in such products in alkaline aquatic environments, on a small scale

$$RCOOR' + NaOH \rightarrow [RCOO^-]\,Na^+ + R'OH \qquad (25)$$

Hydrolysis of esters of fats in presence of acids, enzymes or alkali to yield glycerol, free fat acids, or their salts

$$
\begin{array}{l}
RCOOCH_2 \\
\quad | \\
R'COOCH + 3H_2O \xrightarrow[\text{enzyme}]{[H^+]\text{ or}} \\
\quad | \\
R''COOCH_2
\end{array}
\begin{array}{l}
RCOOH \quad CH_2OH \\
\qquad\qquad\; | \\
R'COOH + CHOH \\
\qquad\qquad\; | \\
R''COOH \quad CH_2OH
\end{array}
\qquad (26)
$$

$$
+ 3\,M^+OH^- \rightarrow
\begin{array}{l}
[RCOO^-]M^+ \quad CH_2OH \\
\qquad\qquad\qquad | \\
[R'COO^-]M^+ + CHOH \\
\qquad\qquad\qquad | \\
[R''COO^-]M^+ + CH_2OH
\end{array}
\qquad (27)
$$

Early diagenetic degradation of organic matter might form products of this type on a small scale.

High molecular weight odd-numbered saturated carboxylic acids can be prepared by the Krafft method; the barium or calcium salts of acetic acid and the naturally occurring high molecular weight acid are distilled. The resulting mixed ketone is subsequently oxidized to form the acid.

$$(RCOO)_2Ca \xrightarrow{\text{distilled}} RCO \cdot + R \cdot \longrightarrow R_2CO$$
$$+ \qquad + \qquad + \quad +2CaCO_3 \quad (28, 29)$$
$$(R'COO)_2Ca \xrightarrow{\text{distilled}} R' \cdot \quad +R'CO \cdot \rightarrow R'_2CO.$$

It is not known whether these processes operate in nature.

Free radicals can form by electrolysis; the biogeochemical effects of natural electrolysis are probably small but a few suggestions have been made on their occurrence.

In the Crum-Brown and Walker Reaction the monoester of succinic acid is electrolyzed to form the diester of adipic acid:

$$2CH_2\,COOEt \rightarrow 2CH_2COOEt \rightarrow CH_2COOEt + 2CO_2 + 2Na$$
$$|\qquad\qquad\qquad |\qquad\qquad\qquad |$$
$$CH_2\,COONa \quad CH_2COO^- \quad (CH_2)_2 \qquad\qquad (30)$$
$$\qquad\qquad \text{At anode} \qquad\qquad |$$
$$\qquad\qquad\qquad\qquad CH_2COOEt$$

Succinic and adipic acids are among commonly occurring constituents of kerogen (Forsman, 1963).

Formation of anhydrides from *cis-* and *trans-*dicarboxylic acids:

(*a*) Maleic acid (cis-ethylene dicarboxylic acid) readily yields anhydride indicating carboxyl groups are on same side of double bond.

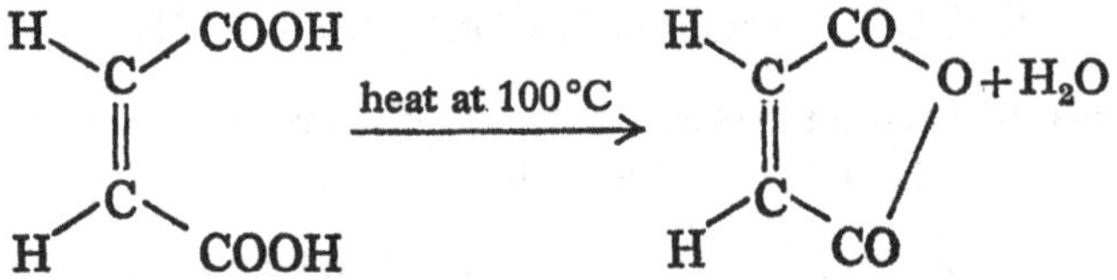

$$(31)$$

Maleic acid   Maleic anhydride

(*b*) Fumaric acid (trans-ethylene dicarboxylic acid), however, must be heated to 250–300 °C to yield maleic anhydride, and sublimes unchanged at lower temperatures.

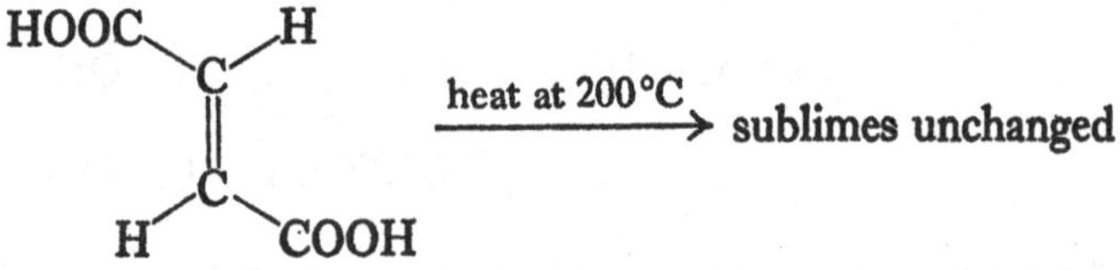

heat at 200 °C → sublimes unchanged   (32)

The first reaction may occur on a small scale in early diagenesis of sedimentary organic matter, given sufficient time.

Walden Inversion. Optically active compounds undergoing reactions involving the asymmetric atom may form inactive (racemized) products. In many instances, however, optical activity is retained after the reaction, and the sign of rotation may be either the same or the opposite. A change of configuration has taken place at one or more steps in the reaction and the phenomenon is called a Walden Inversion.

$$
\begin{array}{ccc}
\underset{\substack{|\\ \text{HOOCCH}_2\\ (-)\ \text{aspartic acid}}}{\text{HOOCCHNH}_2} & \xrightarrow[\text{+2HCl}]{\text{NaNO}_2} & \underset{\substack{|\\ \text{HOOCCH}_2\\ (-)\ \text{chlorosuccinic acid}}}{\text{HOOCCHCl}} + \text{NaCl} + \text{N}_2 + 2\text{H}_2\text{O}
\end{array}
$$

$$
\begin{array}{ccc}
\underset{\substack{|\\ \text{HOOCCH}_2\\ (-)\ \text{bromosuccinic acid}}}{\text{HOOCCH Br}} & \xrightarrow{\text{NaCl}} & \underset{\substack{|\\ \text{HOOCCH}_2 + \text{NaBr}\\ (+)\ \text{chlorosuccinic acid}}}{\text{HOOCCHCl}}
\end{array} \tag{33}
$$

This process could occur on a modest scale in the aquatic environment during diagenesis provided the necessary constituents (as in seaweed) occurred.

1.4.4   *Aldehydes, aldols and acetals*

Aldol condensation; aldehydes and ketones, having at least one α-hydrogen atom undergo condensation reactions in the presence of dilute aqueous alkalies and acids.

$$
\underset{\substack{|\\ \text{H}}}{\text{RCH} = \text{O} + \text{R}'\!-\!\text{CHCHO}} \underset{}{\overset{[\text{OH}^-]\ \text{or}\ [\text{H}^+]}{\rightleftharpoons}} \overset{\text{R}'}{\underset{\text{An aldol}}{\text{RCHOHCHCHO}}} \tag{34}
$$

Both this and the following process should occur (Breger, 1963) in aquatic environments but have been given little attention; the scarcity of potential starting materials may be unfavorable.

In acid-catalyzed aldol condensation the carbonyl group is attacked first and a proton is removed from the α-carbon atom; the product is the *enol* form of the carboxyl compound, and the process is enolization

$$
\underset{\text{H}}{\overset{\text{H}}{\text{RCH}_2\text{C}}} = \text{O} \underset{\text{Br}^-}{\overset{\text{HBr}}{\rightleftharpoons}} [\text{RCH}_2\overset{\text{H}}{\text{C}} = \overset{+}{\text{OH}}] \underset{\text{HBr}}{\overset{\text{Br}^-}{\rightleftharpoons}} \overset{\text{H}}{\text{RCH}} = \text{COH} \tag{35}
$$

Aldehydes form hemiacetals in presence of acid catalysts

$$
\text{RCHO} + \text{R}'\text{OH} \underset{}{\overset{[\text{H}^+]\ \text{or}\ [\text{Br}^-]}{\rightleftharpoons}} \underset{\text{a hemiacetal}}{\text{RCH}\!\!\begin{array}{c} \nearrow \text{OH}\\ \searrow \text{OR}' \end{array}} \tag{36}
$$

37

Hemiacetals form acetals with an excess of alcohol and an acidic catalyst; water is eliminated

$$\text{RCH}\overset{\displaystyle OH}{\underset{\displaystyle OR'}{\big<}} + HOR' \overset{[H^+]}{\rightleftharpoons} \text{RCH}\overset{\displaystyle OR'}{\underset{\displaystyle OR'}{\big<}} + H_2O \qquad (37)$$

Breger (1963) cited these reactions as possible in the sedimentary environment, but they have not been investigated.

1.4.5 *Organic nitrogen, amino compounds*

In the Kjeldahl method for determination of organic nitrogen in its non-oxidized form, the compound is treated with sulfuric acid which converts the organic nitrogen into ammonium sulfate. Copper is commonly used as a catalyst. The ammonium sulfate is treated with alkali (i.e. BaOH), ammonia is liberated, distilled into a known volume of standard acid, and the excess acid is titrated. The procedure is commonly used for determination of ammonia and amino-nitrogen content of geologic materials.

A soil organism of unusual type obtained from soil at Harlech, Wales (Siegel *et al.* 1967) was found to compare favorably with a species described from the Precambrian of Ontario, *Kakabekia umbellata* Barghoorn. Ammonia is required for growth. The occurrences raise the possibility that the Gunflint Chert in which *K. umbellata* occurs is of freshwater origin.

The thermal degradation of alanine was studied by Abelson (1954, 1955);

$$\underset{\displaystyle H}{\overset{\displaystyle NH_2}{CH_3-\underset{|}{\overset{|}{C}}-COOH}} \rightarrow \overset{\displaystyle NH_2}{CH_3-\overset{|}{CH_2}} + CCO_2 \quad (E = 44,000 \text{ cal/mol}) \quad (38)$$

He found that the reaction is of first-order type (one component reacting by itself), following the Arrhenius equation:

$$k = A\,e^{-E/RT}, \qquad (39)$$

where $k = -c\,dc/dt$, c is concentration, A is frequency factor $\sim 10^{13}/\text{sec}$, E is activation energy, T is absolute temperature and R is gas constant. His studies showed that near room temperatures alanine might persist in solution for 10^9 years. Other amino acids such as phenylalanine were found to be less stable.

Hydrolytic degradation of aspartic acid to malic acid takes place readily (Abelson, 1968) but aspartic acid is commonly found in geochemical materials.

$$\underset{\text{Aspartic acid}}{\begin{array}{c} CH_2-COOH \\ | \\ CHNH_2 \\ | \\ COOH \end{array}} + H_2O \rightarrow \underset{\text{Malic acid}}{\begin{array}{c} CH_2COOH \\ | \\ CHOH \\ | \\ COOH \end{array}} + NH_3 \qquad (40)$$

Strecker Synthesis of amino acids is a reaction of an aldehyde or ketone with a mixture of ammonium chloride and sodium cyanide, followed by acid hydrolysis of the amino nitrile. The process has been used by Miller (1957) in modified form to synthesize amino acids and possibly was responsible for prebiotic synthesis on the primordial earth.

$$NH_4Cl + NaCN \rightarrow NH_4CN + NaCl \qquad (41)$$

$$NH_4CN \rightleftharpoons NH_3 + HCN \qquad (42)$$

$$RCHO \xrightleftharpoons{NH_3} \underset{NH_2}{RCHOH} \xrightleftharpoons[H_2O]{HCN} \underset{NH_2}{RCHCN} \xrightarrow{H_2O[H^+]} \underset{NH_2}{RCHCOOH} \qquad (43)$$

Hydrogen cyanide and aldehydes were produced in spark-discharge chambers containing methane, ammonia, hydrogen and water. From these reaction products, amino acids and other substances form by a Strecker Synthesis. It is doubtful whether the processes operate in the normal sedimentary environment, except by some biochemical pathway.

Fischer peptide synthesis; amino acids react with α-halogenated acid halide, and the α-halogen is subsequently replaced by an amino group.

$$\underset{X}{\overset{R}{|}}CHCOX + H_2N\underset{}{\overset{R'}{|}}CHCOOH \rightarrow X\overset{R}{\underset{}{|}}CHCONH\overset{R'}{\underset{}{|}}CHCOOH$$

$$\xrightarrow{NH_3} \underset{\text{Dipeptide}}{H_2N\overset{R}{\underset{}{|}}CHCONH\overset{R'}{\underset{}{|}}CHCOOH} \qquad (44)$$

As such, this process probably is inoperative in the usual sedimentary environment, but biochemical processes produce the same results, i.e. a peptide bond linking the α-amino group of one amino acid with the α-carboxyl group of another.

The Schiff Reaction involves the condensation of aromatic aldehydes with primary amines in the presence of alkali, to give an anil. Aliphatic compounds may condense in the same way.

$$C_6H_5CHO + H_2NC_6H_5 \xrightarrow{\text{alkali}} C_6H_5CHNC_6H_5 + H_2O \tag{45}$$

It is not known whether this process is effective in sedimentary geochemistry but it probably occurs only on a small scale, if at all.

In the Biuret Synthesis and Reaction, two moles of urea on heating unite to yield biuret, through loss of one mole of ammonia. Cyanuric acid may be formed in the reaction owing to loss of the three moles of ammonia from three moles of urea. The biuret may be detected by treatment with a trace of cupric ion which, perhaps through development of a chelate ring, gives a violet coloration.

$$\underset{\text{Urea}}{2H_2NCONH_2} \xrightarrow{\text{heat}} \underset{\text{Biuret}}{H_2NCONHCONH_2} + NH_3 \tag{46}$$

Protein material is known to yield a positive test (violet color) for biuret on heating with sodium hydroxide solution and a trace of $CuSO_4$.

Wohler Reaction (Mueller, 1963): reaction of aldehydes and ketones with hydrogen cyanide or ammonia at low temperature and pressure.

$$NH_3 + HCHO \rightarrow \underset{\text{Urea}}{H_2NCONH_2} \tag{47}$$

Despite the fact that the reaction takes place at low temperature and pressure, it probably is uncommon in sedimentary geochemistry because of scarcity of reacting substances.

Synthesis of polyglycine by electrical discharge may occur as a possible primordial earth process (Fox, 1963):

$$HCHO + NH_3 + HCN \rightarrow H_2NCH_2CN + H_2O \tag{48}$$

$$X(H_2NCH_2CN) \rightarrow -(-NHCH_2-\underset{\underset{NH}{\|}}{C}-)x \xrightarrow{-H_2O}$$
$$-(-NHCH_2CO-)x- + XNH_3 \tag{49}$$

Polyglycine may be substituted by reaction with formaldehyde or acetaldehyde to yield serine and threonine residues on kaolinite substrate in electrical spark experiments.

$$-(-NHCH_2CO-)x- + HCHO \rightarrow -(-\underset{\underset{CH_2OH}{|}}{NHCHCO})y-$$

$$-(-NHCH_2CO-)z- \xrightarrow{[H^+]+H_2O} HOCH_2CHNH_2COOH + H_2NCH_2COOH \tag{50}$$

The process has also been postulated for the primordial earth as a mechanism for protein synthesis (Fox, 1963).

Formation of peptide bond at elevated temperatures to produce proteins:

$$H_2NCHRCOOH + H_2NCHR'COOH = H_2NCHRCONHCHR'COOH + H_2O \qquad (51)$$

The energy needed to overcome the energy barrier for the reaction to proceed in the synthetic direction is thought to be found in perivolcanic areas (Fox, 1963).

Production of ureidosuccinic acid as an intermediate in synthesis of pyrimidines, is suggested to have occurred at elevated temperatures that possibly existed on the primitive earth (Fox, 1963)

$$NH_3 + CO_2 + \text{L-aspartic acid} \rightarrow \text{ureidosuccinic acid} \rightarrow \text{orotic acid} \rightarrow \text{pyrimidines} \qquad (52)$$

Breakdown of proteins by proteolytic enzymes is accomplished as follows:

$$-\overset{\overset{\displaystyle O}{\|}}{C}-\underset{\underset{\displaystyle R}{|}}{\overset{\overset{\displaystyle H}{|}}{C}}-NH-\overset{\overset{\displaystyle O}{\|}}{C}-\underset{\underset{\displaystyle R'}{|}}{\overset{\overset{\displaystyle H}{|}}{C}}-NH- + H_2O \xrightarrow{\text{enzyme}} HOOC-\underset{\underset{\displaystyle R}{|}}{\overset{\overset{\displaystyle H}{|}}{C}}-NH_2 + HOOC-\underset{\underset{\displaystyle R'}{|}}{\overset{\overset{\displaystyle H}{|}}{C}}-NH_2 \qquad (53)$$

This process occurs early in the sedimentary environment and accounts for absence of most proteins, except fibrous types, in sediments (Breger, 1963).

Stable nitrogen bases may be formed from amino acids by bacterial putrefaction. These substances and their degradation products are found in sediments and acids, but the products have generally not been found in rocks.

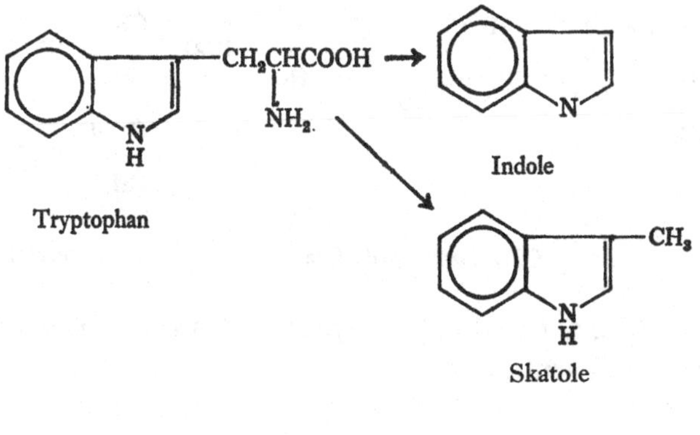
CH₂CHCOOH
NH₂
N
H
Tryptophan
N
Indole
CH₃
N
H
Skatole

(54)

$$\underset{\substack{|\\ NH_2}}{CH_2}CH_2CH_2CH_2\underset{\substack{|\\ NH_2}}{CHCOOH} \rightarrow \underset{\substack{|\\ NH_2}}{CH_2}CH_2CH_2CH_2\underset{\substack{|\\ NH_2}}{CH_2} \qquad (55)$$

Lysine   Cadaverine

$$HN = \underset{\substack{|\\ NH_2}}{C}-NH(CH_2)_3CHCOOH \rightarrow \underset{\substack{|\\ NH_2}}{CH_2}CH_2CH_2\underset{\substack{|\\ NH_2}}{CH_2} \qquad (56)$$

Arginine   Putrescine

1.4.6 *Carbohydrates*

Acetals (glycosides) can be formed from glucose, methyl alcohol and hydrogen chloride; the aldehyde group of the sugar first reacts with the alcohol on the fifth carbon to form a cyclic hemiacetal; the hemiacetal reacts with methyl alcohol to form diasteriomers called α- and β-glycosides (Fig. 13), also known as anomers. The presence of glycosides in geochemical materials remains an uncertainty.

Action of strong acids on carbohydrates

(*a*) Pentoses

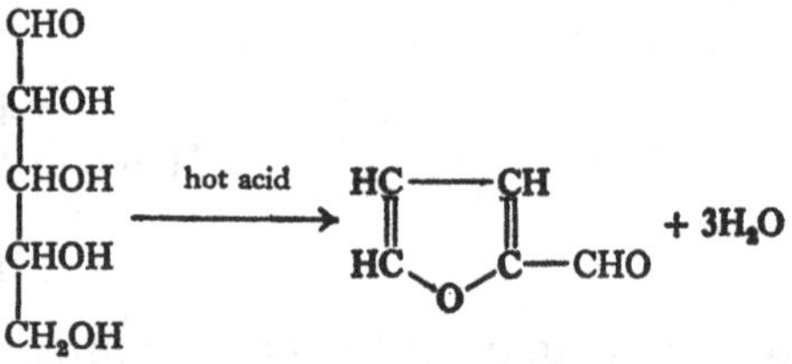

$$ \qquad (57)$$

Furfural

(*b*) Hexoses

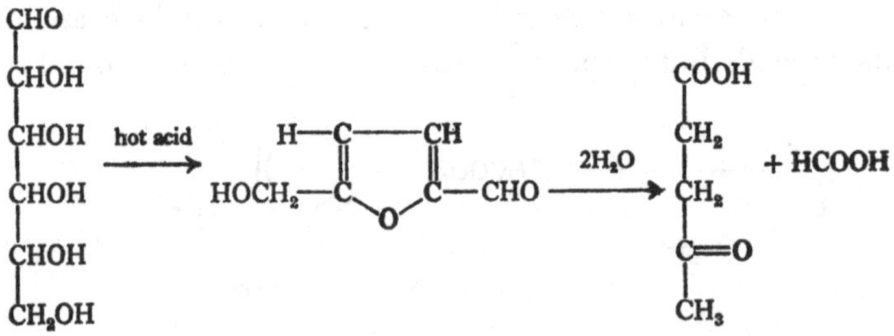

$$ \qquad (58)$$

5-Hydroxymethylfurfural   Levulinic acid

Both furfural and 5-hydroxymethylfurfural are widely distributed in sediments and sedimentary rocks.

Lobry de Bruyn and van Eckenstein Transformation. A solution of glucose and dilute $Ca(OH)_2$ standing for several days at room temperature results in isomerization of the glucose to produce 31 % fructose, 2·5 % mannose, and 3 % other substances in addition to glucose; the reaction is only partly reversible. The process illustrates the degradative action of bases on carbohydrates and must be given careful consideration in laboratory processing as well as in the natural environment.

Alkali degradation of monosaccharides involves the following products:

$$\text{Hexose} \xrightarrow{\text{strong alkali}} \text{formaldehyde} + \text{pentoses}$$

$$\xrightarrow{} \text{hydroxyacetaldehyde and tetroses}$$

$$\xrightarrow{} \text{trioses}$$

plus other products resulting from the above due to isomerization and condensation reactions.

Enzymatic conversion of glucose to ethyl alcohol (Breger, 1963 a)

$$\underset{\text{Glucose}}{C_6H_{12}O_6} \xrightarrow{\text{enzyme}} \underset{\text{Ethyl alcohol}}{2\,C_2H_5OH} + 2CO_2 \qquad (59)$$

Enzymatic conversion of starch to maltose

$$\underset{\text{Starch}}{2(C_6H_{10}O_5)_n} + nH_2O \xrightarrow[\text{in malt}]{\text{diastase}} \underset{\text{Maltose}}{nC_{12}H_{22}O_{11}} \qquad (60a)$$

$$\underset{\text{Maltose}}{C_{12}H_{22}O_{11}} + H_2O \xrightarrow{\text{maltase}} \underset{\text{Glucose}}{2C_6H_{12}O_6} \qquad (60b)$$

These reactions can be used in the laboratory to identify polysaccharides, but must also be considered as possibly having affected the geochemical samples studied in the natural environment.

Thermal degradation of glucose. Heating glucose in absence of, or in presence of, a catalyst produces dimer or polymer substances (Pictet & Costan, 1920; Cramer & Cox, 1922; Hurd & Edwards, 1949; O'calla & Lee, 1956; Mora & Wood, 1958; O'calla *et al.* 1962; Sugisawa & Edo, 1964). Glucose samples in aqueous solution, sealed in Carius tubes in a helium atmosphere, and heated in the range 180–250 °C, were found to degrade with an activation energy of about 22 kcal/mole (Swain, Bratt, Kirkwood & Tobback, 1969). In association with montmorillonite glucose degradated with an activation energy of about 21·2 kcal/mole, and in association with Devonian black shale, about 25·4 kcal/mole. These activation energies are overall values for the possible polymer formation referred to above, during the heating of glucose.

Thermal degradation of oligosaccharides and polysaccharides. Oligosaccharides (cellobiose and dextrose), on heating to 210–240 °C, degraded first by dehydration with loss of two moles of water, and second by the production of CO and CO_2. The activation energy of the second step in cellobiose is about 40 kcal/mole; that of dextrose in which initial water loss is less than 5% (compared to 15% for cellobiose), is about 29 kcal/mole; and that of maltose in which water loss is gradual is about 35 kcal/mole (Puddington, 1948). Potato starch degraded at about 29 kcal/mole over the range of 1–4% water given off.

1.4.7 *Organic sulfur compounds*

The reduction of sulfate resulting from activation by adenosine triphosphate (ZoBell, 1963):

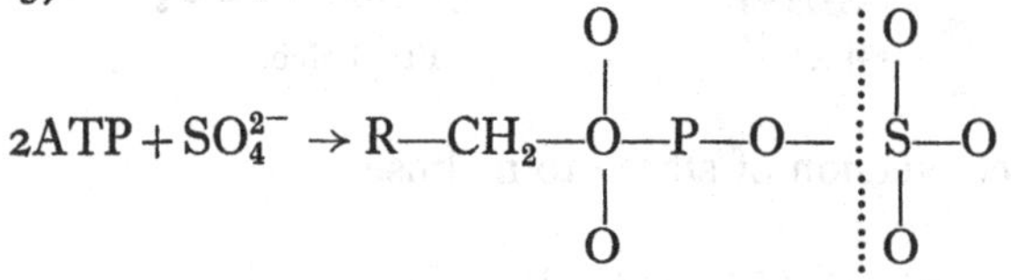

$$2ATP + SO_4^{2-} \rightarrow R\text{---}CH_2\text{---}O\text{---}\overset{\displaystyle O}{\underset{\displaystyle O}{\overset{|}{\underset{|}{P}}}}\text{---}O\text{---} \vdots \overset{\displaystyle O}{\underset{\displaystyle O}{\overset{|}{\underset{|}{S}}}}\text{---}O \tag{61}$$

is followed by breaking the bond at the position indicated. This may be an effective process in some high-sulfate sedimentary environments.

Reduction of sulfate for nutritional purposes is accomplished by *Escherichia coli*, by successive reductions to sulfite, thiosulfite and sulfide through a series of organic complexes shown by ZoBell (1963):

$$R'SO_4 \rightarrow R''SO_3 \rightarrow R'''S_2O_3 \rightarrow R^nSH \tag{62}$$

The next to last reaction produces sulfur that, with serine, forms the amino acid cystine (or cysteine).

Oxidation of carbon and hydrogen compounds with sulfate is carried out by *Desulfovibrio* (ZoBell, 1963):

$$R:CH_2 + CaSO_4 \rightarrow H_2S + CaCO_3 + R:O \tag{63}$$

in which $R:CH_2$ is an organic compound and $R:O$ is an oxidation product such as H_2O or CO_2.

Formation of elemental sulfur is accomplished by purple and green sulfur bacteria of families Thiorhodacae and Chlorobacteraceae (ZoBell, 1963):

$$CO_2 + 2H_2S + h\nu \rightarrow (CH_2O) + H_2O + 2S \tag{64}$$

where CH_2O is a component of organic matter in the bacterial cell.

Stepwise conversion of sulfate to cystine is done by *Aspergillus nidulans* (ZoBell, 1963)

> Sulfate $\rightarrow$ sulfite $\rightarrow$ sulfoxylate $\rightarrow$ thiosulfate $\rightarrow$ serine
>
> thiosulfate (cysteine-S-sulfonate) $\rightarrow$ cysteine $\rightarrow$ cystine. $\qquad$ (65)

Synthesis of dimethyl sulfide and methanethiol or methyl mercaptan by successive reduction and methylation of sulfate may be carried out by wood-rotting fungus *Schizophyllum commune* (ZoBell, 1963):

$$SO_4^{2-} \rightarrow SO_3^{2-} \rightarrow CH_3SO_3^- \rightarrow CH_3SO_2^- \rightarrow (CH_3)_2SO_2$$
$$\rightarrow (CH_3)_2S \rightarrow CH_3SH \tag{66}$$

1.4.8 Cyclic compounds

The Diels–Alder Reaction is one in which an adduct may be formed at normal temperatures from dienes and dienophiles

$$\tag{67}$$

Diene Dienophile   Adduct

Cyclic products in this type of reaction may lead to many different products found in petroleum and coal (Breger, 1963*a*).

Formation of hemiacetals by reaction between carbonyl groups and alcohols is a process that can occur at normal temperatures

$$\tag{68}$$

Carbonyl-   Hydroxyl-   Hemiacetal
bearing   bearing
component   component

An example of the production of a hemiacetal that may further break down to yield compounds of geochemical importance is the conversion of cumene hydroperoxide by rearrangement to phenol and acetone (Morrison & Boyd, 1966).

Introduction to study of organic sediments

Aromatic hydrocarbons react with acyl halides to yield ketones by Friedel–Crafts Reaction:

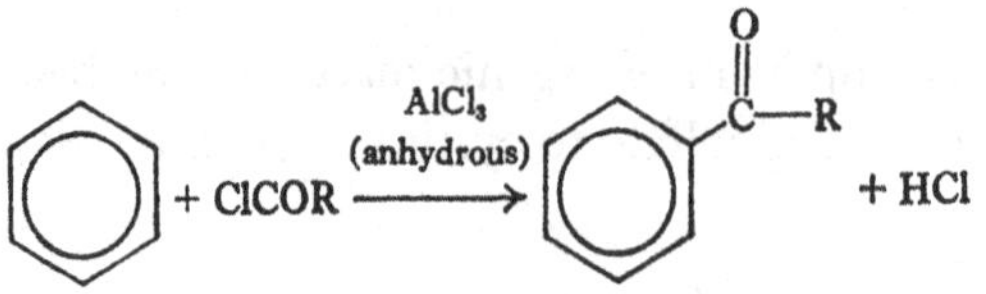

$$(69)$$

Geochemical materials that can participate in aldol condensations might in rare cases be made available by this process.

Bucherer Reaction:

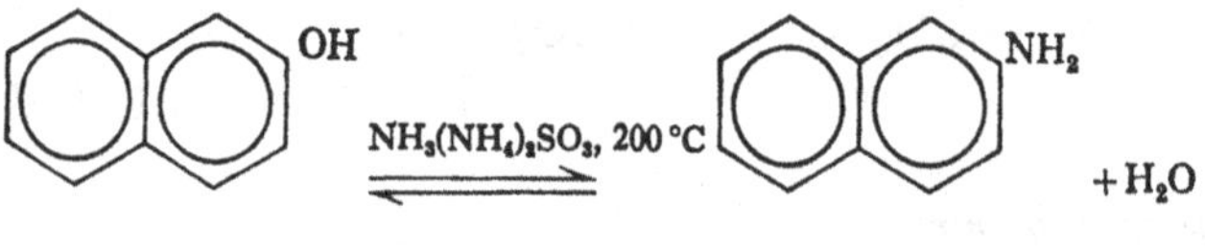

$$(70)$$

β-Naphthol β-Naphthylamine

The process has been cited as a possible source of naphthols in certain peat bogs from enzymatic and proteinaceous sources (Swain, 1967 b).

Oxidation of benzene yields biphenyl at high temperatures and pressures:

$$2C_6H_6 + (O) \rightarrow C_6H_5-C_6H_5 + H_2O \qquad (71)$$

This reaction has been suggested for the extraterrestrial realm and primordial earth, but is not likely to have occurred in subsequent earth history to any extent unless by biochemical processes.

In the Scholl Condensation, condensation of aromatic nuclei with loss of hydrogen can be brought about at 100 °C with the use of $AlCl_3$ as a catalyst:

$$2C_{10}H_8 \xrightarrow{AlCl_3 \text{ at } 100\,°C} C_{10}H_7.C_{10}H_7 + H_2 \qquad (72)$$

with continued treatment $C_{10}H_6.C_{10}H_6$ is formed. The process is not likely to have operated to any extent, possibly except by biochemical processes in humic material.

Wagner–Meerwein rearrangement: internal molecular rearrangements as a result of which, for example, α-Pinene undergoes stepwise changes to camphor (see (73) on next page).

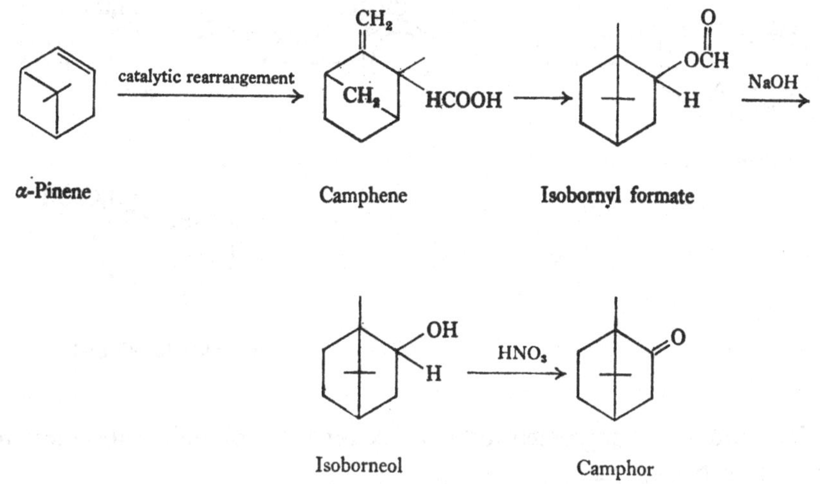
catalytic rearrangement
HCOOH
NaOH
α-Pinene
Camphene
Isobornyl formate
HNO₃
Isoborneol
Camphor

(73)

It is entirely possible that reactions of this sort have been important in richly humic accumulations, but have not been studied by geochemists.

Thermal decomposition of β-carotene may occur as follows:

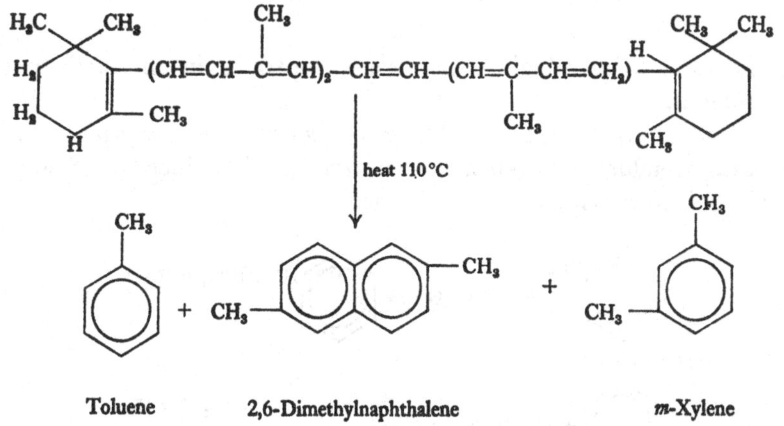
heat 110 °C
Toluene
2,6-Dimethylnaphthalene
m-Xylene

(74)

Carotenes are common constituents of modern aquatic sediments and may slowly degrade in this way. Similar degradation may occur in laboratory extraction of samples and must be anticipated.

Formation of α-naphthylacetic acid as a constituent of plant growth accelerators (auxins) may be of geochemical importance (Swain, 1967*b*). Synthesis of α-naphthylacetic acid involves the chloromethylation of naphthalene, conversion to the nitrile, and hydrolysis (Noller, 1951).

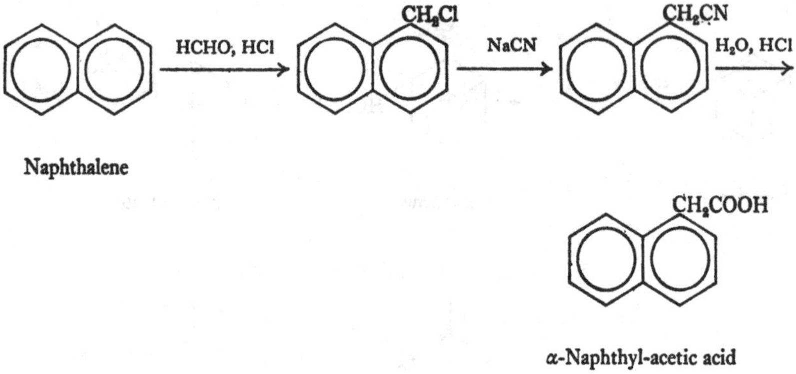

Naphthalene

α-Naphthyl-acetic acid

$$(75)$$

The process is suggested to help account for naphtholic substances in some peat bogs (Swain, 1967 b).

In the Claisen Rearrangement, unsaturated phenolic ethers rearrange when the solution is boiled, to yield O-substituted phenols:

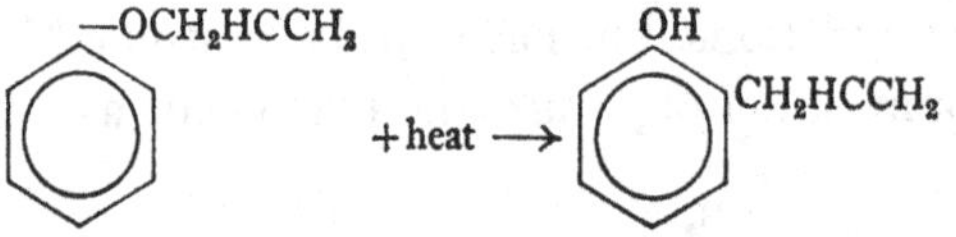

$$(76)$$

It is not known whether the process operates in the sedimentary environment.

The Ladenburgh Synthesis refers to the condensation of α-methylpyridine with acetaldehyde. When the intermediate is reduced with sodium and alcohol α-propylpiperidine is formed:

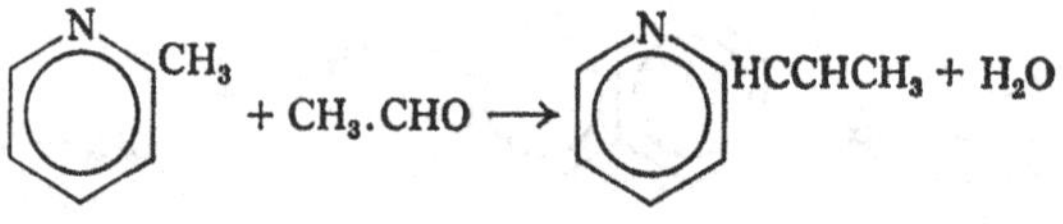

$$(77)$$

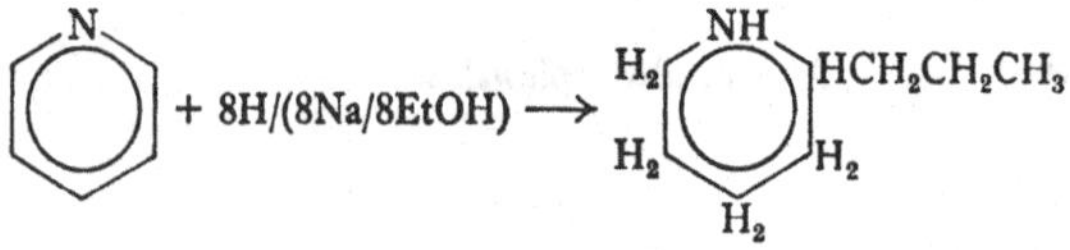

$$(78)$$

1,4-Diketones react with ammonia or primary amines to produce pyrrole derivative (Paal–Knorr Reaction):

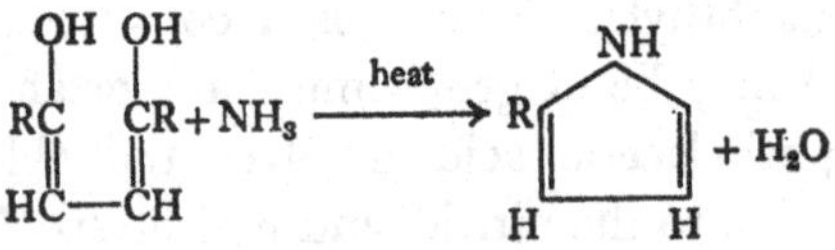

$$(79)$$

Such processes as the preceding two may have occurred biochemically in sediments but have not been investigated.

Alkaline degradation of 'Russell lignin' with demethylation may form as follows:

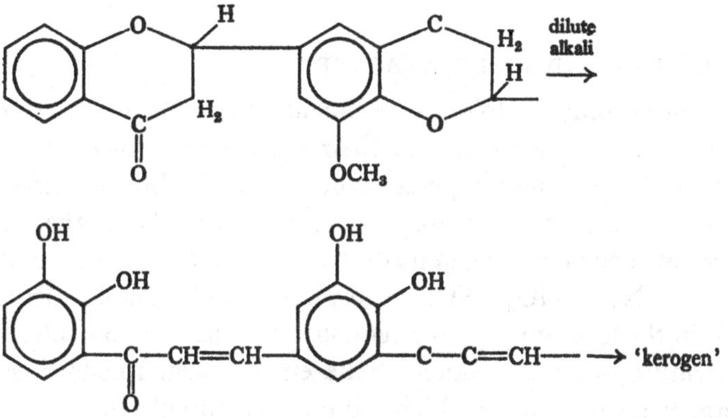

(80)

Several workers have postulated this and the following two possible mechanisms for formation of humic substances from lignin and quinones (Breger, 1963 a; Laatsch, 1944).

Formation of 'lignohumic acid' from coniferyl-*p*-aldehyde may take place:

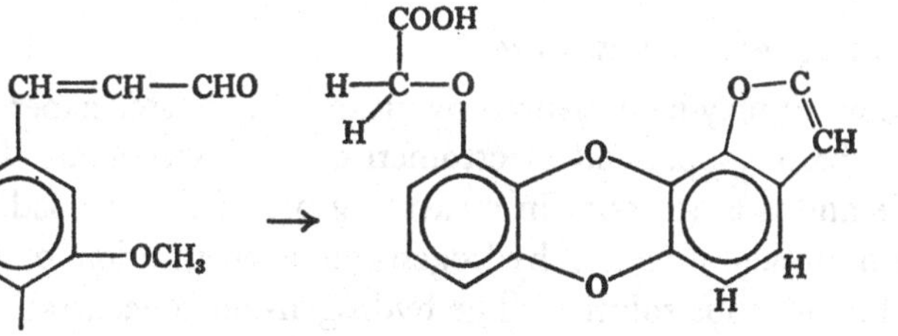

(81)

Benzenoid origin of humic acids is suggested to have proceeded as follows (Thiele & Kettner, 1953):

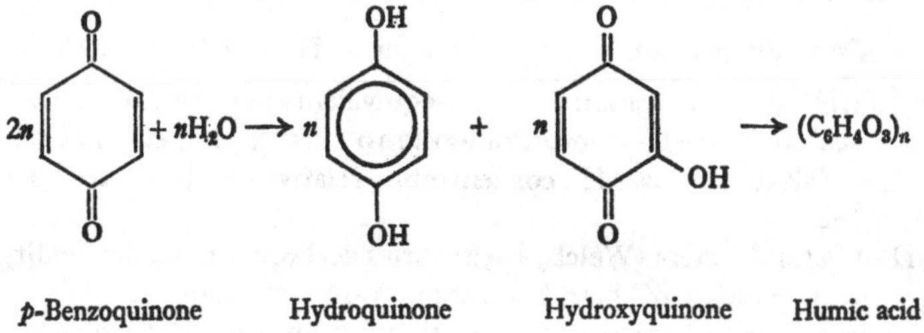

p-Benzoquinone   Hydroquinone   Hydroxyquinone   Humic acid

(82)

2 Biogeochemical field and laboratory analyses

Biogeochemical study of Holocene sediments and living organisms necessitates that certain measurements of their properties be made in the field. These include depth and temperature of water, pH, Eh, and oxygen content. Other measurements that may be made in the field or in the laboratory soon after collections have been made are: total alkalinity, Cl^-, Na^+, Ca^{2+}, Mg^{2+}, SO_4^-, NH_3, NO_3, NO_2. Phosphorus and organic nitrogen are measured in the laboratory soon after collections have been made. All such data contribute to an evaluation of the environmental conditions under which organisms lived or in which sediments accumulated.

Holocene aquatic sediments are collected by means of various dredges and coring devices. Freshly collected samples should be placed in dry ice for transmission to the laboratory and then stored in a deep-freezer. Methods for these procedures are described by Trask (1932) and Wright *et al.* (1965) and are summarized briefly here (table 6).

2.1.1 *Hydrogen-ion concentration*

pH of the water may be measured by means of indicator papers or powders or with a battery-operated electrometric pH instrument. The latter is preferable and is a necessity in measuring pH of the wet sediments.

pH is a measure of the hydrogen-ion concentration of the sample, measured in aqueous solution. The hydrogen-ion concentration of natural waters is a useful parameter of the amount and kinds of dissolved solids they contain, and bears a limiting relationship to organic productivity of the waters. Hydrogen-ion concentration is measured with reference to its concentration in distilled water and is expressed as pH:

$$Kw = \text{dissociation constant of water} = H^+ \times OH^- = 10^{-14} \quad (83)$$

pH $= \log_{10}(1/H^+)$; concentration in gram-equivalents per liter pH of water $= 7$. Addition of acid increases H^+ concentration to 10^{-6}, 10^{-5}, 10^{-4}, etc., i.e. pH 6, 5, 4, etc. Addition of alkali decreases H^+ concentration relative to OH^- to 10^{-8}, 10^{-9}, etc., i.e. pH 8, 9. etc.

The pH of natural waters (Welch, 1952) varies according to its net acidity or net alkalinity or lack of either of these from about 3–10·5 as follows:

Seawater 8·1–8·3 where water is in equilibrium with CO_2 in atmosphere.

Deep seawater with much dissolved CO_2, about 7·5. Shallow surface seawater,

TABLE 6   *Collection of sediment samples for organic geochemical study*

Sampling device	Type of material to be sampled	Description of method	Remarks
Pipe sampler or 'Mann' sampler (Trask, 1932)	Sandy bottom	10–15 cm diameter iron pipe with cloth bag on one end and handle on other; dragged across bottom; sample stored in plastic jars or bags and frozen	May recover hard-bottom sample where dredges and coring devices fail
Ekman Dredge	Soft bottom	10–30 cm square brass dredge with spring jaws; samples stored in plastic jars or bag and frozen	Excellent for soft bottom but not for sand or gravel bottom
Free-fall corer of type of Ekman (Trask, 1932), Hvorslev & Stetson (1946) or Phleger (1951)	Soft bottom	A weighted steel tube 1–2 m long with or without plastic tube liner that is dropped into sediment; cores retained in plastic tube	Phleger type weighing 30–50 lb plastic liner 1·5 m long is best for lake sediments; may compress sediment 10% or more
Davis Sampler (Trask, 1932; Wright *et al.* 1965)	Non-sandy peat and marsh sediments	Brass or steel tube 2–4 cm × 30 cm fitted with piston and set of rods each 1 or 2 m long; cores extruded into plastic or glass test-tubes and refrigerated	Most widely used for sampling peat; can not be used in coarse mineral sediments; may compress sediment
Hiller Sampler (Wright *et al.* 1965)	Non-sandy peat	Steel tube 3 cm × 1 cm with longitudinal slot to receive sample, a sliding cover for slot and set of rods; core extracted, cut in short pieces and wrapped in plastic	Difficult to avoid contamination, but does not compress sample vertically
Livingstone Piston Corer Kullenberg Deep Sea Corer (Wright *et al.* 1965)	Soft sediment	Steel tube having piston fixed at bottom of tube until sampling depth is reached in sediment; core extruded and wrapped in plastic	Other modifications for larger diameter core (Cushing & Wright, 1963) or deep water (Mackareth, 1958)

evaporation rate high, about 8·4. Seawater near bays with H_2S, about 7 or lower. Lake water, 3·2–10·5, averages 6·5–8·5. River water, 3–9, Mississippi drainage basin, 6·4–8·2. Sediments, range from 4 to 8·5 or higher.

Some of the substances affecting pH are:

(a) Inorganic; strong acids, HCl, H_2SO_4, HNO_3; weak acids, H_2CO_3; acid salts, ionized acid potassium sulfate; salts of weak bases with strong acids, aluminum chloride, ammonium sulfate, hydrolyzed and yielding H^+ ions.

(b) Organic; strong acids, highly ionized, oxalic HOOC-COOH; weak acids, slightly ionized, acetic, CH_3COOH; salts of weak acid with strong bases, as aluminum citrate; amino acids, as aspartic acid.

pH readings are made in the field with an electrometric instrument using a glass electrode and a calomel electrode. The electrodes are inserted directly into the freshly collected sample and the pH read. Well-buffered

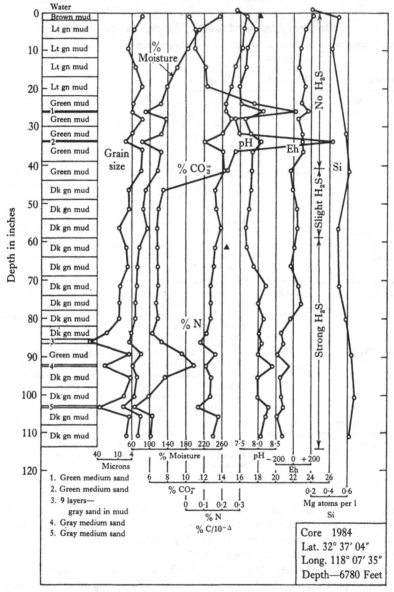

FIGURE 22 Analytical data for Core 7, representing marine sediments from the San Clemente Basin, California (Emery & Rittenberg, 1952, *American Association of Petroleum Geologists Bulletin*, **36**, no. 5, pp. 735–902, figs. 2 and 11).

aqueous systems have steady pH readings, whereas poorly buffered solutions are unsteady in this respect and undergo drift until equilibrium with the atmosphere is reached. Emery & Rittenberg (1952) (fig. 22) and Swain (1956) have discussed field measurements of pH.

2.1.2   *Oxidation-reduction potentials*

The intensity with which natural solutions will oxidize or reduce a redox system (i.e. Fe^{2+}–Fe^{3+}) is of value in understanding the general chemical properties of the water and the effect of the redox condition on accumulating organic matter (Emery & Rittenberg, 1952 (fig. 22); Swain, 1956). The general equation for the oxidation-reduction potential (Eh) of a solution containing a redox system is (ZoBell, 1946):

$$Eh = E_0 + 0.03 \log_e \frac{Ox.}{Red.}, \tag{84}$$

where e = Briggsian logarithm 2·3025,

Eh = oxidation-reduction potential in mV,

E_0 = a constant for the particular system in question, i.e.
Fe^{2+} (50%), Fe^{3+} (50%); an E_0 of +0·1 V (100 mV) will oxidize a system of +0·2 V.

0·03 = a factor dependent on the gas constant ($R = 1·99$ cal/degree), the absolute temperature T, n = number of equivalents of electricity transferred and F = a Faraday of electricity, $RT/nF = 0·03$.

Ox. = concentration of oxidized form of redox substance,

Red. = concentration of reduced form of redox substance.

Redox systems may be:
ions, Fe^{2+}, Fe^{3+},
atoms and ions, Cl_2, Cl^-;
compounds and radicals, MnO_2, MnO_4^-;
compounds, H_3AsO_4, H_3AsO_3.

Oxidation (giving up electrons) and reduction (gaining electrons) may or may not involve oxygen in natural systems.

Examples in which Eh = 1:

$$(O_2 - H_2O), \quad O_2 + 4H + 4e = 2H_2O, \quad E_0 = 123 \text{ mV}; \tag{85}$$

$$(H^+ - H_2), \quad 2H + 2e = H_2, \quad E_0 = 0 \text{ mV}; \tag{86}$$

$$(S - S^{-2}), \quad S + 2e = S^{-2}, \quad E_0 = 70 \text{ mV}; \tag{87}$$

Other systems in natural waters (Eh = 1):

$$(MnO_4 - MnO_2), \; MnO_4^- + 4H^+ + 4e = MnO_2 + 2H_2O, \; E_0 = 159 \text{ mV}; \tag{88}$$

$$(MnO_4^- - Mn^{2+}), \; MnO_4 + 8H^+ + 5e = Mn^{2+} + 4H_2O, \; E_0 = 150 \text{ mV}; \tag{89}$$

$$(Cl_2 - Cl^-), \; Cl_2 + 2e = 2Cl^-, \; E_0 = 140 \text{ mV}; \tag{90}$$

TABLE 7   *Correction of Eh readings for E_0 of calomel reference electrode*

Uncorrected Eh of sample (mV)	E_0 of ref. electrode (mV)	Corrected Eh of sample (mV)	Interpretation
+520	+245	+765	Strongly oxidizing
+360	+245	+605	Oxidizing
+102	+245	+347	Weakly oxidizing
+000	+245	+245	Null
−220	+245	+25	Weakly reducing
−450	+245	−205	Strongly reducing

$$(MnO_2 - Mn^{2+}), \quad MnO_2 + 4H + 2e = Mn^{2+} + 2H_2O, \quad E_0 = 135 \text{ mV}; \qquad (91)$$

$$(Fe^{3+} - Fe^{2+}), \quad Fe^{3+} + e + Fe^{2+}, \quad E_0 = 76 \text{ mV}. \qquad (92)$$

Measurement of the Eh is made in the field electrometrically using a platinum electrode and a calomel reference electrode. If possible, the electrodes are inserted directly into the sediment in place; otherwise a core of the sediment is taken and the electrodes are inserted in the freshly collected core. Measurement may be made in the laboratory several hours after the sediment is collected if the sample is refrigerated. The E_0 of the reference electrode is +245 mV which must be added algebraically to the Eh readings of the samples (table 7).

Using oxidizing conditions the organic matter will tend to be destroyed by bacterial action rather rapidly. Under reducing conditions the organic matter will be preserved under anaerobic conditions for much longer periods of time.

The analogue of buffer capacity, or lack of it, in redox systems is poise. Most negative Eh natural systems are poorly poised. Such systems drift badly in Eh and may not permit meaningful redox values to be obtained. In such cases there typically is rapid drift at first as carbon dioxide or hydrogen sulfide escape from the system, followed by slow drift as chemical or bacterial action takes place in the solution. Eh readings of sediment and water samples are generally made when the first rapid drift has stopped.

At best, Eh readings are only a rough indication of the oxidizing or reducing of a sedimentary environment. This information, however, is valuable in many situations because in several instances noted by the writer Eh values were unexpectedly strongly negative or positive based on what would have been deduced from other measurements.

2.1.3 *Oxygen content*

The oxidizing or reducing conditions in natural waters and in the associated sediments are further demonstrated by the content of oxygen as
measured electrometrically in sediments of the Delta–Mendota Canal,
California (Swain and Prokopovich 1969). In deep ocean waters the decrease
in temperature and lack of light result in an increase in O_2 content below
the photic zone (fig. 23).

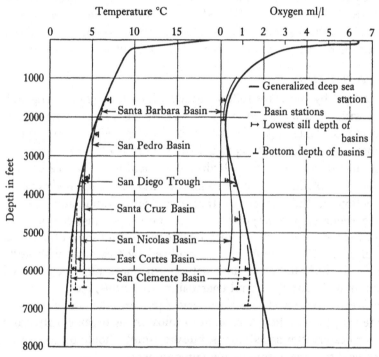

FIGURE 23   Depth distribution of temperature and oxygen in basins and in
the open sea off California (Emery & Rittenberg, 1952 *loc. cit.*).

2.1.4 *Rock samples*

Except for safeguards against contamination from living organisms, and
weathering, the principal care that must be taken in the collection of rock
samples for biochemical study is that of artificial contamination in handling
the samples. The collection of samples follows a more or less typical sched-
ule that applies to regular stratigraphic work.

Biogeochemical field and laboratory analyses

2.2 ORGANIC GEOCHEMICAL ANALYSES

Laboratory studies of organic constituents of sediments and rocks include analyses for total organic matter, organic carbon, organic nitrogen, protein amino acids, amines, carbohydrates, organic acids, hydrocarbons, chlorinoid pigments, carotenoid pigments, heteroaromatic and other carbocyclic compounds. The purpose of such analyses is to determine the source organisms, their diagenetic history and their significance in the interpretation of the sedimentary environment, as well as their possible role in biogeochemical evolution. Some of the methods used for laboratory analysis of organic materials are described below. For other methods and details of these methods refer to the chapters in which each type of organic substance is discussed.

2.2.1 *Organic matter*

It is determined by one of the following methods: loss on ignition (fig. 24), treatment with 30% hydrogen peroxide, treatment with chromic acid or other strong oxidants, analysis for organic carbon and calculation of organic matter from the resulting value. In determining loss in ignition the sample is first treated with 0·5 N hydrochloric acid to remove carbonate, then ignited in a platinum crucible. The accuracy of the method is about 1%. Objections to the method are that presence of sulfide and of more insoluble carbonates (siderite) or other substances may cause errors.

Most of the living plant and animal organic matter and some of the dead material of a sediment sample can be oxidized by treatment with 30% H_2O_2. This method, however, does not remove more highly condensed 'pyrobituminous' matter, mineral charcoal, resins, gums, etc.

A mixture of concentrated sulfuric acid (400 ml) and potassium permanganate (15 g) is an effective oxidant for most living organic matter and much of the dead matter of sediments, but the sample should be pretreated with weak acid to remove soluble inorganic minerals.

Organic carbon is measured by the Pregl combustion train. Total organic matter is obtained by $C \times 1·72$ for Holocene sediments and $C \times 2·0$ for rocks. The accuracy of this method is about 0·1%.

Organic nitrogen is determined by the Kjeldahl process or by a combustion train process. For quantities less than 1% the Kjeldahl method is preferable because of its greater sensitivity. These processes, measure the amino acids, glycoproteins, etc. Inorganic nitrogen is excluded from the measurements. Total organic matter of sediments is calculated by $N \times 16$; in rocks, owing to the gradual degradation of nitrogen with geologic time, organic matter is estimated by $N \times 24$. The accuracy of the Kjeldahl method is about 0·002%.

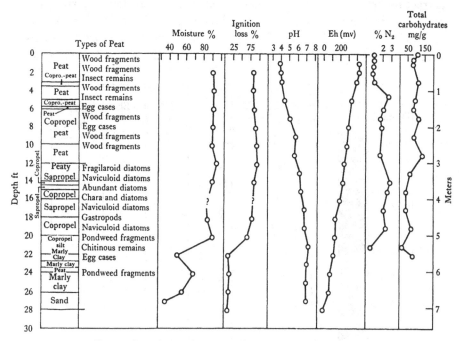

FIGURE 24   Properties of peat from Rossburg Bog, Minnesota (Swain, 1967*b*).

2.2.2   *Lipids and bitumens*

The oils, fats and waxes of living organisms collectively referred to as 'lipids', comprise from about 1% in most organisms to over 90% in *Botryococcus* and other boghead-type algae. Many lower organisms and probably some higher ones greatly increase their production of lipids during times of danger or under other adverse conditions. Ether, benzene, chloroform and alcohol are generally used to extract lipids. When organic matter accumulates in a sedimentary environment the original lipids become altered and mixed together from different organisms and the resulting ether– or benzene–methanol extractable material is referred to as 'lipoid' or 'bitumens'.

The following kinds of substances are included in the lipoids (fig. 25) and bitumen extracts: hydrocarbons; chlorinoid pigments; fatty acids and esters; carotenoid pigments; resins, alkaloids, carbohydrates, heterocyclic compounds.

The method used to study the lipids is (Smith, 1954): (1) extract 2–5 g of wet sample with 80% benzene + 20% methyl alcohol for 8 h in Soxhlet extractors; (2) separate water from extract in separatory funnel; (3) dry extract to constant weight in vacuum dessicator under flow of nitrogen to prevent oxidation; (4) weigh and record nature, color, texture, fluorescence and odor of extract; (5) run infrared absorption spectrum of extract interpretations of organic groups of which include those listed in table 8;

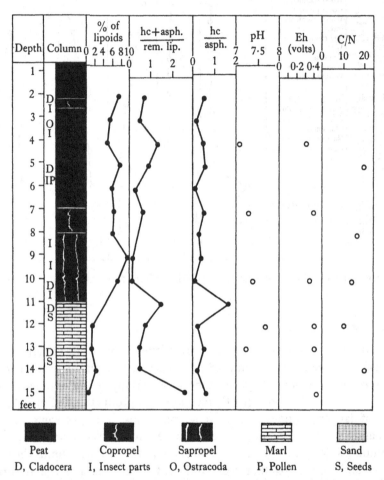

Peat   Copropel   Sapropel   Marl   Sand

D, Cladocera   I, Insect parts   O, Ostracoda   P, Pollen   S, Seeds

FIGURE 25 Comparison of sedimentary types to total lipoids, hydrocarbon fractions and other properties, Cedar Creek Bog, Minnesota (Swain & Prokopovich, 1954).

(6) separate extract on a 1 cm × 10 cm column of activated alumina (Alcoa A-20) into eluted fractions: the first eluted with *n*-heptane (10 ml), followed by benzene (10 ml), followed by pyridine (8 ml) and by methanol (8 ml), the last two elutions are combined. The *heptane-eluted* fraction is dried, described under the microscope, and is taken to represent the *saturated*

TABLE 8   *Interpretation of absorption bands in the infrared spectra of lipids (Jeffrey et al. 1964)*

Wavelength of maximum absorption (μm)	Most probable assignment
2·90–3·0	OH stretching of C—OH (also NH group)
3·3–3·5	C—H stretching (CH$_3$, CH$_2$)
5·72–5·80	C=O stretching (acids, esters)
6·1	C=O stretching, phospholipids, free acids
6·80–6·90	CH bending
7·22–7·40	CH bending
7·90–8·40	Band progression in solid fatty acids
8·2	C—O—C linkage in phospholipids
8·40–8·60	C—O stretching (COOR)
8·9–9·2	C—O stretching (complex glycerides)
9·5–9·6	Glycerides containing OH groups (mono and di)
9·2–9·4	P—O—C linkage of phospholipids
9·5	Steroid absorption band
10·3	P—O—C linkage
10·3–10·4	CH bending about a *trans* C=C group of complex glycerides
10·6–11·1	OH deformation of carboxylic acids and alcohols
13·8–14·1	Long carbon chain
14·7	*cis* Double bond

hydrocarbon fraction. Its nature is ascertained by obtaining an infrared spectrum which should show no aromatic absorption peaks. The saturated hydrocarbons may be separated further by gas chromatography (fig. 26), preceded by partition of the straight chain from branched chain hydrocarbons on 5 Å molecular sieve which will adsorb the straight chain compounds which can then be recovered and analyzed by dissolving the sieve in HF. Further analysis of the straight chain and branched chain aliphatic hydrocarbons can be made by scanning the individual hydrocarbon fractions in a mass spectrometer of high resolution type. The *benzene-eluted* fraction represents the *aromatic hydrocarbons*; a u.v. spectrum should show presence of aromatic nuclei if any occur (figs. 27–29); this fraction frequently contains the carotenoid pigments if any are present. The *pyridine +methanol-eluted* fraction is taken as the *asphaltic* material of the sample and consists of tars and resins as well as the chlorinoid pigments (chlorophyll, pheophytin and other chlorophyll-derived pigments) in modern materials; the pigments can be detected by means of a visible absorption spectrum. A fourth, non-eluted fraction is that material retained on the alumina column; it may contain resins, carbohydrates, some pigments and other high-molecular weight substances.

The column chromatographic method of analysis of hydrocarbons can be used to detect quantities in amounts as small as 10^{-6}–10^{-8} g/g of sediment of rock.

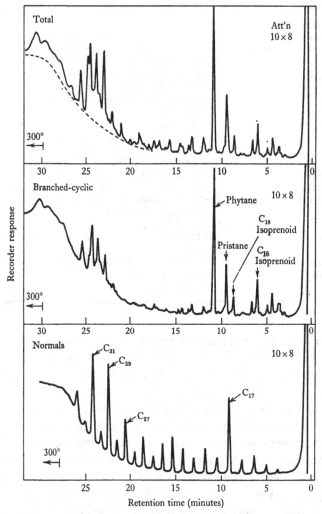

FIGURE 26 Gas–liquid chromatograms of alkane fractions from the Green River Shale, Colorado; (A) total, (B) branched cyclic, (C) normal alkanes (Eglinton, Geoffrey *et al.* (1966 *c*) In *Adv. in Org. Geochem.*, 1964. Oxford, Pergamon Press. All rights reserved).

Gas chromatographic analysis of hydrocarbons can detect quantities as small as 10^{-8}–10^{-9} g/g of rock or sediment.

Eglinton *et al.* (1966 *b*) used column chromatography, gas chromatography and mass spectrometry to separate hydrocarbons from geological

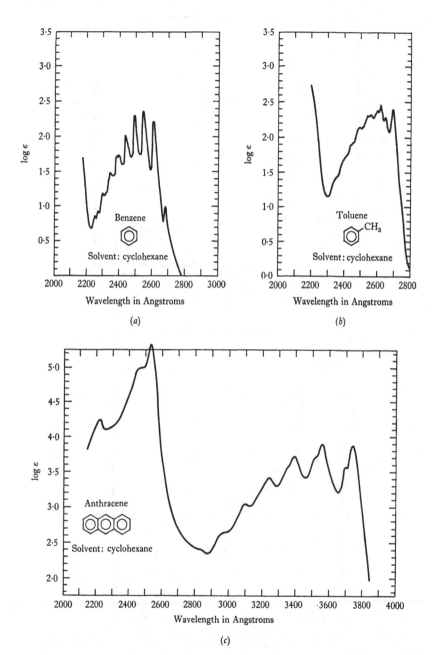

FIGURE 27   Absorption spectra of typical aromatic hydrocarbons: (*a*) benzene; (*b*) toluene; (*c*) anthracene (Friedel & Orchin, 1952. *Ultraviolet Spectra of Aromatic Compounds*, Wiley & Sons, New York).

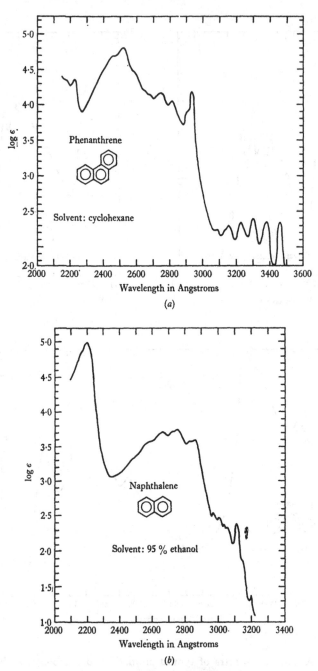

FIGURE 28 Absorption spectra of typical aromatic hydrocarbons: (*a*) phenan-threne; (*b*) naphthalene (Friedel & Orchin, 1952. *Ultraviolet Spectra of Aromatic Compounds*, Wiley & Sons, New York).

materials. The cleaned and powdered sample (a few grams to several hundred grams) is extracted in Soxhlet extractors or under ultrasonic conditions for several hours using benzene + methanol or other solvents. Separation of the saturated alkanes was on columns of freshly activated alumina, with _n_-hexane or _n_-heptane the eluting agent. The eluate was

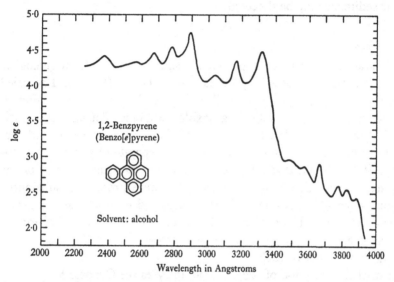

FIGURE 29   Absorption spectra of typical aromatic hydrocarbons: 1,2-Benzpyrene (Friedel & Orchin, 1952. _Ultraviolet Spectra of Aromatic Compounds_, Wiley & Sons, New York).

dried, dissolved in dry benzene and refluxed with 5 Å molecular sieve in pellet form for 1–3 days. The solution, containing branched alkanes, was dried and stored for future analysis. The sieve, containing normal alkanes, was placed in 24 % HF and benzene until the sieve dissolved; the benzene solution with the normal alkanes was filtered through Na_2CO_3 and the solvent removed. The extract was then separated by gas chromatography. The eluted fractions were in part caught in small tubes by use of a stream splitter and further analyzed by ultraviolet, and infrared spectrophotometry, and by mass spectrophotometry.

Other studies of methods of extraction of hydrocarbons and lipids in geological samples have been made by Bray & Evans (1961); Baker (1962); Dunton & Hunt (1962); Erdman _et al._ (1958_a, b_); Evans _et al._ (1957); Forsman & Hunt (1958); Hunt & Jamieson (1956); Orr & Emery (1956); and Oró (1967).

Thin-layer chromatography of lipids and hydrocarbons may be useful as a first approximation of what compounds are present (Eglinton _et al._

1966c). The chromatographic plate, coated with a thin layer of activated alumina or silica gel, is spotted with μl amounts of solutions containing the hydrocarbons and developed in benzene or other solvent. Detection of the hydrocarbon spots is by means of iodine vapor, charring with sulfuric acid or other method. Quantities as small as 10^{-9} g of hydrocarbons per g of rock or sediment may be detected.

2.2.3 *Organic acids*

The separation of fatty acids, principally saturated types from geological samples has been discussed by Abelson *et al.* (1964) (fig. 30). The dried samples are powdered in a ball mill, treated with aqueous HCl, filtered, dried and extracted with water in a Soxhlet extractor. The extract is purified from tarry materials by further extraction and the organic acids are converted into methyl esters which are then separated by gas chromatography. It was found that fatty acids comprised as little as 0·1% of the crude extract and that only saturated acids were present in substantial amounts. Urea adduction has also been employed in geochemical analysis of organic acids as well as of hydrocarbons. The smallest amounts of organic acids resolvable by gas chromatographic analysis are in the range 10^{-6}–10^{-8} g/g of sediment or rock.

For further discussion of organic acid analyses see Chapter 8.

2.2.4 *Carbohydrates*

Total carbohydrates are determined by a phenol-sulfuric method (see p. 222). The monosaccharides are separated in the following way (Rogers, 1965; Swain, 1966a). The quantity of sulfuric acid necessary to neutralize the alkaline constituents of the sample is determined by preliminary acid extraction and subsequent back-titration with NaOH in presence of phenolphthalein. This amount of 0·5 N-H_2SO_4 is added in excess of that needed to treat 1–20 g of the sample. Pretreatment of the sample with cold 70% H_2SO_4 for 8 hours may aid in extraction of humified samples. The sample and 0·5 N acid mixture is centrifuged and the solutions neutralized with $CaCO_3$ or $BaCO_3$. Desalting is done first by ethanolic precipitation, reduction to 25–30 ml in a flash evaporator, followed by passage through ion-exchange resins in three superposed columns: (1) Dowex 50 cation resin 8% cross-linked 50–100 mesh; (2) Duolite A-4 weak anion resin; (3) Dowex 50 or Amberlite Ir-120 (H^+) cation resin. The lower column is used to insure that the effluent from the anion column is neutral or at least not alkaline. Capacity of resins should be at least 5 × the anticipated volume of salts.

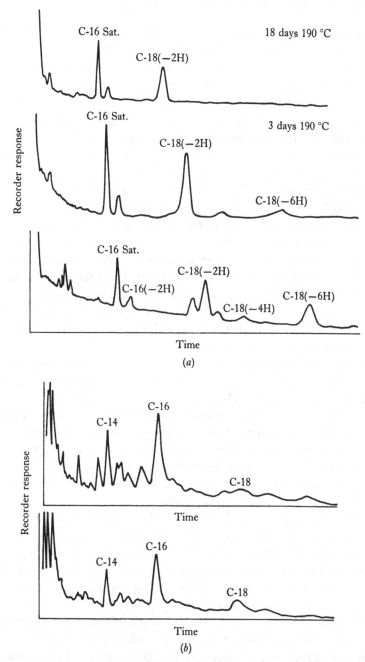

FIGURE 30 (a) Gas–liquid chromatograms of methyl esters of fatty acids extracted from *Chlorella pyrenoidosa*, that had been exposed to heat, compared to an unheated specimen; (b) chromatograms of esters of fatty acids extracted from a dredged sediment sample (lower curve) from San Nicolas Basin and from a core from Pedernales, Venezuela (Abelson *et al.* 1964).

Biogeochemical field and laboratory analyses

Prior to use the ion-exchange resins are regenerated as follows. The cation resin is converted to the Na^+ form with 2 N-NaOH, followed by rinsing with distilled water and is then converted to the H^+ form with 2 N-HCl, again followed by thorough rinsing with distilled water. The anion resin is similarly regenerated with HCl and NaOH. In the deionization of the carbohydrate solution, 25–30 μl of solution were passed through the successive columns; the columns were flushed with 10 volumes of distilled water. The effluent from the columns was reduced to dryness in a flash evaporator and taken up in 10 m of distilled water. Desalting may be done in an electric desalter but the process is more time consuming.

The carbohydrates in the samples are then separated by one-dimensional paper chromatography. The chromatograms are developed in a solvent system of butanol:acetic acid:water (4:1:5) or pyridine:ethyl acetate: water; from 1 to 80 μl of sample is spotted on the chromatogram and development time is 24–48 h. The spray used to detect the monosaccharides is aniline:phthalic acid:water saturated butanol (0·9 g:1·6 ml:100 ml) or ammoniacal silver nitrate. After development the chromatograms are air-dried in a hood for 2–3 h; then are sprayed quickly and evenly and placed in an oven at 85–95 °C for 4–5 min.

The unknown monosaccharides are identified by comparing the developed spots with those of a mixture of known monosaccharides placed on the chromatogram and by observing their color reactions; the hexoses (galactose, glucose, mannose) stain grayish brown with this spray while the pentoses (arabinose, xylose, ribose, rhamnose) stain reddish brown. Semiquantitative determinations of monosaccharides were made by scanning the chromatograms in a recording densitometer (Photovolt, Corp., N.Y.).

The paper chromatographic method is useful for estimating quantities of sugars down to 10^{-6}–10^{-8} g/g of sample.

Gas chromatography of silyl ethers of carbohydrate-bearing extracts can be carried out. Trimethylsilyl ethers of glycose-containing water extracts or acid hydrolysates are prepared (Swain *et al.* 1967*a*) by treating 10 mg of carbohydrate with 1 ml of anhydrous pyridine, 0·2 ml of hexamethyldisilazane, and 0·1 ml of trimethylchlorosilane (Applied Sci. Lab., State College, Pa.) for 5 min or longer (fig. 31). From 0·1 to 0·5 μl of the resulting-reaction mixture is injected on a gas chromatographic column using a 5-ft × $\frac{1}{8}$ in stainless steel column packed with QF-1 on Aeropak 30 (Varian-Aerograph Co., Walnut Creek, Calif.) using a flame-ionization detector.

Chloromethyldimethylsilyl ethers of glycose mixtures are prepared (Swain, Bratt & Kirkwood, 1968) by reacting 0·1 ml of extract with 0·09 ml of pyridine, 0·03 ml of dichloromethyltetramethyl-disilazane and 0·01 ml

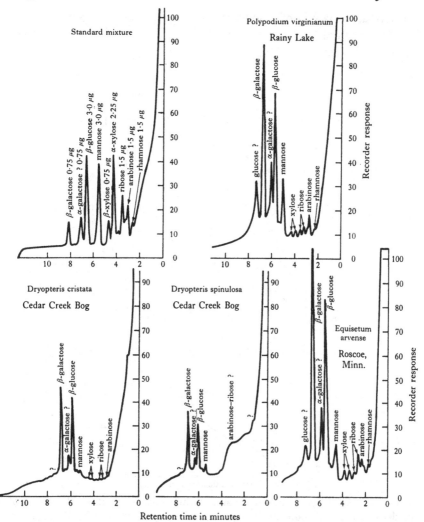

FIGURE 31   Gas–liquid chromatograms of trimethylsilyl ethers of monosaccharide components of a sulfuric acid extracts of living plants compared with a standard mixture (Swain, Bratt & Kirkwood, 1967, *Journal of Paleontology*, **41**, pp. 1549–54, fig. 1).

of chloromethyldimethylchlorosilane (Applied Sci. Lab., State College, Pa.) for 30 min at room temperature (fig. 32). This chloromethyldimethylsilyl (CMDMS) solution in 0·1–2·0 μl quantities is injected on the gas chromatographic column as above, but for this purpose an electron capture detector is used. Standard mixtures of sugars are chromatographed in conjunction with the unknown samples. These methods can detect carbohydrates in amounts of 10^{-7}–10^{-9} g/g of rock or sediment sample.

Enzymatic analysis of D-glucose and D-galactose in amounts down to 10^{-9} g/g of sample is made by use of enzyme-oxidase preparations. These are available from Worthington Biochemical Corp., Freehold, N.J. The

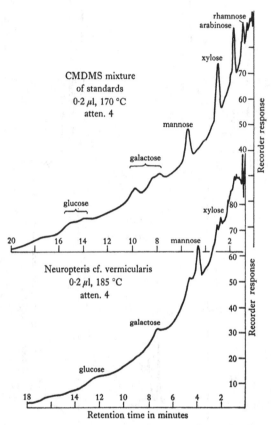

FIGURE 32 Gas–liquid chromatograms of chloromethyldimethyl silyl ethers of a standard mixture of monosaccharides and of an acid extract of a Pennsylvanian fern (Swain, Bratt & Kirkwood, 1968, *Journal of Paleontology*, **42**, 1078–82, fig. 1).

glucose-oxidase technique is a coupled enzyme system based on the following scheme of reactions:

$$\text{D-glucose} + O_2 + H_2O \xrightarrow{\text{glucose oxidase}} H_2O_2 + \text{gluconic acid} \quad (93)$$

$$H_2O_2 + \text{reduced chromogen} \xrightarrow{\text{peroxidase}} \text{oxidized chromogen} \quad (94)$$
$$(\textit{o}\text{-dianisidine})$$

The absorbance of the oxidized chromogen solution at 400 nm is read and compared to a standard curve for quantitative determination. The method

is specific for D-glucose except that 2-deoxy-D-glucose is oxidized at 12%
of the rate for glucose. The galactose-oxidase determinations are based on
the following schematic reactions:

$$\text{D-galactose} + O_2 \xrightarrow{\text{galactose oxidase}} \text{D-galacto-hexodialdose} + H_2O_2 \quad (95)$$

$$H_2O_2 + \text{reduced chromogen} \xrightarrow{\text{peroxidase}} \text{oxidized chromogen} \quad (96)$$

The method is less specific than the glucose-oxidase reaction, since the
galactose-oxidase system also attacks D-galactosamine and other monosaccharides, galactosides and oligosaccharides.

Another method for separation and identification of monosaccharides
involves the preparation of alditol acetates of monosaccharide mixtures and
their separation by gas chromatography (Swain *et al.* 1967 *a*). The glycose
mixture is reduced with sodium borohydride for 3 h; the excess borohydride is neutralized with acetic acid and the solution evaporated to dryness.
The dry mixture is refluxed for 4 h with a mixture of equal amounts of
acetic anhydride and pyridine (*ca.* 1 ml/100 mg of sugars). The solution is
cooled and injected directly in a gas chromatograph equipped with a flame
ionization detector.

2.2.5 *Protein amino acids*

The protein amino acids have been found to be preserved as residues in
many different deposits of Holocene sediment. The amino acids are preserved either as nitrogen-bonded peptides or are complexed with humic
materials in the sediments (Swain *et al.* 1959). The amino acids differ
measurably in their thermal stability as shown by Abelson (1957) and by
Jones & Vallentyne (1960) and Vallentyne (1964). The individual acids also
differ in their relative abundance and stability in aquatic sediments (Swain,
Paulsen & Ting, 1964). There is fair agreement between the thermal studies
of Vallentyne and the data from natural materials.

Extraction. Free amino acids are extracted with water; protein amino acids
are extracted with acid, generally 6 N-HCl, to break the peptide bonds of
the protein and liberate the individual amino acids. A 1- to 25-g sample of
organism or sediment is treated with concentrated HCl to neutralize the
carbonates in the sample, 6 N-HCl is added and hydrolysis under reflux is
carried on for 24 h. The mixture is centrifuged, the supernatant decanted
and saved, the precipitate is washed twice with distilled water and the
washings combined with the supernatant. The combined solution is reduced to small volume in a flash evaporator and then transferred to a porce-

lain dish in a vacuum dessicator over silica gel dessicant. The precipitate is dissolved in 10–30 ml distilled water, centrifuged, poured off, and the supernatant saved; the precipitate is further washed with distilled water which is added to the supernatant and is then dried to eliminate any remaining HCl. The residue is taken up in distilled water and passed through a column of Dowex 50 ion-exchange resin. Water is added until effluent pH is neutral and a test for iron is negative (amino acids are now absorbed on the column and inorganic salts have been washed off the column); 2 N-NH$_4$OH is added to the column to elute amino acids using 4–5 times as much ammonia solution as column is high. The eluate is reduced to dryness at 50 °C, the residue is taken up in exactly 5 ml of 10% isopropyl alcohol and placed in a labelled bottle. Alternatively, for automatic amino acid analyzer the residue is taken up in 5 ml of pH 3·28 sodium citrate buffer solution.

Paper chromatography. A known amount of solution containing a mixture of amino acids as standards, and known amounts of the unknown solutions are spotted on a sheet of Whatman No. 1 filter paper; typically a total of 1–5 μl (λ) of each solution is spotted in successive 1 λ quantities and allowed to dry to avoid spreading of the spot. The chromatogram is placed in a chromatographic chamber with a suitable solvent such as butanol: acetic acid:water (4:1:5) for 24 h after which it is air dried. For better separation of amino acids by 2-dimensional chromatography the chromatogram is turned 90° and rerun in a second solvent system such as phenol : water. For other amino acid solvent systems see the Appendix. The chromatogram is stained by dipping it in a solution of 0·25% ninhydrin in acetone, dried in an oven for 2 min at 50 °C and stored in the dark. The spots which have characteristic colors come out fully in several hours.

For semi-quantitative estimation the chromatogram is cut into strips and scanned in a recording densitometer. Amounts of amino acids down to 10^{-6} g/g of sample can be detected by this method.

Automatic amino acid analyzer. For more quantitative determination and better separation the amino acid hydrolyzates are analyzed in an automatic analyzer. From 0·5 to 0·2 ml of amino acid-bearing solution is analyzed. The apparatus consists of ion-exchange resin columns in which circulate buffers of differing pH. The amino acids placed on the columns are continuously eluted and the effluent is mixed with a ninhydrin solution. The resulting colored solution is monitored photoelectrically and the effluent peaks representing individual amino acids are recorded on a chart.

The unknown chromatograms are compared quantitatively with a standard mixture of amino acids. The method is useful for analysis of down to 10^{-6} g/ g of geologic sample.

2.2.6 *Organic pigments*

Chlorinoid pigments. The chlorinoid pigments were extracted in the following way. Samples ranging from 0·2 to 1·7 g are extracted in the dark in ultrasonic tanks with 90% aqueous acetone until no additional color could be obtained with addition of fresh solvent. A visible spectrum in the 660–670 nm (nanometers = millimicrons) range was obtained with a Beckman DU Spectrophotometer and a spectrum in the range 280–700 nm was obtained in a Bausch and Lomb 505 Spectrophotometer. Chlorophyll *a* and pheophytin *a* both have absorption maxima at 661–667 nm (fig. 115). Chlorophyll *a* has another max. at 432 nm, while the comparable max. of pheophytin *a* is at 409–412 nm.

Similar extractions and absorption spectra were obtained of fresh spinach in which all the chlorinoid pigment occurs as chlorophyll *a*. A portion of the spinach extract was allowed to degrade to pheophytin. The chlorinoid pigments are tabulated as SCDP (sedimentary chlorophyll degradation product) using the method of Vallentyne (1955) where:

$$\text{SCDP} = \frac{(\text{absorbance at 665 nm})\,(\text{vol. of extract})}{\text{dry weight of sample}}. \tag{97}$$

The quantity of chlorophyll *a* in acetone extracts of sediments may be calculated also in the following way:

$$Ca = \frac{Au\,Ws/As}{Ds}\frac{Vu}{Vs}, \tag{98}$$

where

Ca = chlorophyll *a* in mg/g of dry sample;

Au, As = absorbance (log I_0/I) of unknown and standard samples, respectively above a base line, that of the solvent, at maximum of 665–670 nm;

Ws = weight of standard chlorophyll sample in mg;

Vu, Vs = volume of solution of 90% acetone containing unknown and known chlorophyll samples, respectively, in ml;

Ds = dry weight of sample.

The quantities of chlorinoid pigments that can be detected by these methods range down to 10^{-6}–10^{-8} g/g of rock or sediment sample.

Biogeochemical field and laboratory analyses

Flavoproteins. Yellow-fluorescing pigments of the flavinoid type are detected in recent sediments in the following way (Swain & Venteris, 1964): the wet sediment is extracted with dilute HCl; the extract is neutralized with $BaCO_3$, desalted on cation-exchange resins, reduced to dryness and taken up in 10% isopropyl alcohol; the extract containing the flavinoid pigments and amino acids is separated by paper chromatography; using butanol:acetic acid:water (4:1:5) as developer. The fluorescent flavinoid pigments are cut out of the chromatogram and rechromatographed; the developed spots are again cut out of the chromatogram, eluted with *n*-hexane and the solution scanned in the ultraviolet and visible ranges in a spectrophotometer. Riboflavin has absorption maxima at 262, 370 and 440 nm. In geological samples the interaction of amino acids with flavinoids reduces or eliminates the absorption at the longer wavelengths 370 and 440 nm (Swain, Paulsen & Ting, 1964). The quantities of pigments detectable by this method are as low as 10^{-6}–10^{-8} g/g of sample.

Carotenoid pigments. The extraction of carotenoid pigments from recent sediment samples was accomplished by Vallentyne (1956) by extracting the wet sediment with methanol and petroleum ether. The carotenes and related terpenoid hydrocarbons are driven into the ether by adding water to the mixture leaving the associated xanthophylls in the hypophase. The epiphasic carotenes in the ether are chromatographed over activated alumina using petroleum ether as the developer. The colored zones are cut out and rechromatographed on $Ca(OH)_2$. The redeveloped fractions produce a bluish color with $SbCl_3$ in chloroform (Carr–Price reaction); and each fraction is epiphasic to both ether and 95% methanol. The spectra of the fractions are determined with a spectrophotometer, which can be used for detection of 10^{-6}–10^{-8} g/g of geologic sample.

3 Characteristics of non-marine sediments and sedimentary rocks

3.1 CLASSIFICATION OF NON-MARINE SEDIMENTS

Most classifications of the bottom deposits of lakes and bogs have been by limnologists whose viewpoints and purposes are somewhat different from those of the geologist. Commonly, such deposits are referred to as 'soils'. Some works in limnology do not present a separate discussion of the bottom deposits, although these may be taken up in connexion with organic productivity and other subjects. The reader is referred to Lundquist (1927), Naumann (1930), Troels-Smith (1955), Welch (1952), Ruttner (1953), Deevey (1939), Lindeman (1941), Roelofs (1944) and Swain (1956). The classification of lake bottom and bog sediments used here attempts to systematize discussion of lacustrine deposits from the point of view of the geologist.

One of the most widespread organic lake sediments in temperate and tropical regions is brownish and grayish pulpy coprogenic ooze in which the organic materials comprise settled phytoplankton, fragmentary aquatic vegetation of other types and animal contributions much if not all of which has been worked over by benthonic organisms of various kinds. This includes the 'gyttja' of European limnologists, but that term has also been applied at least informally to black, more completely anaerobic sapropel, as well as to settled humus colloids (dy), and lake peat (förna). To avoid this confusion the term *copropel* was introduced (Swain & Prokopovich, 1954) in place of gyttja.

Fox *et al.* (1952) proposed the term *leptopel* for suspended finely particulate sludge of organic and inorganic composition in ocean water. It includes particles ranging in size from finely colloidal micelles to tiny visible particles with intimately associated micro-organisms. The material may become concentrated at the surface, creating slicks which gradually sink to the bottom. Adsorption of marine colloidal matter to solid surfaces of the sea bottom forms a muddy slime, *pelogloea*.

Both leptopel and pelogloea are thought to be important sources of food for certain marine animals, and they may also be considered as possible sources of petroleum. Pelogloea is comparable to the copropel of lake sediments.

In an examination of the vertical variation in organic carbon, hydrogen, and nitrogen in lake sediments Koyama (1966) found that the microbial degradation of these three components was of the order

$$\text{org C} > \text{org N} > \text{org H}.$$

Terrestrial opal phytoliths have been found 500 km at sea, east of Cape Verde Island during a dust storm (Folger *et al.* 1967), thus explaining anomalous occurrences of presumably freshwater diatoms on the Middle Atlantic Ridge.

A variety of opal called tabashir, found in bamboo, was found to contain more water but less alkali and alkaline earths than most opal (Jones *et al.* 1966). It resembles other phylolith opal in consisting of particles about 100 Å in diameter linked together in clumps.

In the following classification of lake and bog deposits, the lithified equivalent is given in parentheses.

Classification of non-marine sediments (after Swain, 1956)

(A) Organic detritus

(1) Peat, förna; phytogenic (lignite and coal; vitrinite, durinite, clarinite); coarse (caustopsephite); fine (caustopsammite, caustopelite).

(2) Copropel, gyttja; brown and greenish gray phytogenic and zoogenic, the latter commonly includes chitinous arthropod exoskeletons (copropelite, some boghead coal, oil shale, etc.).

(3) Sapropel, gyttja; black, mainly phytogenic involving bacterial decay by both aerobes and anaerobes, but in an essentially anaerobic environment (sapropelite, some boghead coal).

(B) Biogenic and authigenic mineral substance

(1) Marl (limestone, dolomite or dolostone) may be subdivided on the basis of: (*a*) texture (calcirudite, calcarenite, calcilutite); (*b*) composition (algal, *Potamogeton, Chara*, etc.); (*c*) fabric (öolite, chalk).

(2) Diatoms (diatomite).

(3) Agglutinated Protista; sarcodinids—*Difflugia, Centropyxis*, etc.; ciliates—tintinnids.

(4) Others; iron and manganese oxides and carbonates of partly bacterial origin (ironstones, etc.), sodium chloride (rock salt), calcium sulfate (anhydrite, gypsum), magnesium carbonate (magnesite), potassium chloride, borax, sodium carbonate (trona), phosphates (vivianite, etc.), sulfide (melnikovite), etc.

(C) Terrigenous substances

(1) Gravel, sand, silt, and clay (conglomerate, sandstone, silt-stone, shale, mudstone, etc.).

Triangular diagrams of: (1) variation in particle size of organic matter, (2) a 3-component organic sediment, and (3) a combined organic, terrigenous and authigenic sediment, are given in fig. 33.

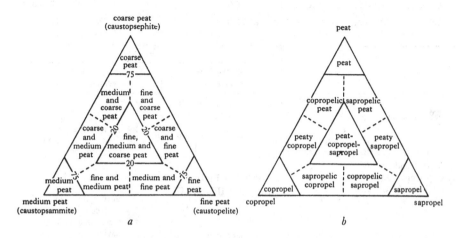

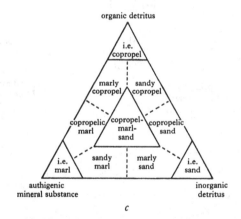

FIGURE 33 Nomenclature of lake sediments: (*a*) variation in particle size of organic matter; (*b*) 3-component organic sediment; (*c*) combined organic, terrigenous and authigenic sediment (Swain, 1956, *AAPG Bull.* **40**, 4, pp. 600–53, fig. 16).

The sediments of some lakes and bogs representing diverse geologic conditions are briefly described in the remainder of the chapter. Those of unglaciated regions will be discussed first and those of glaciated regions second. The descriptions and discussion will provide a background for consideration of the organic materials to be dealt with in later chapters.

3.2 LAKE SEDIMENTS OF UNGLACIATED REGIONS

The recent lake sediments of unglaciated regions vary with the character of source material, relief surrounding the lake basin, and climatic conditions. Lake basins in those areas are: of tectonic origin, are formed on river

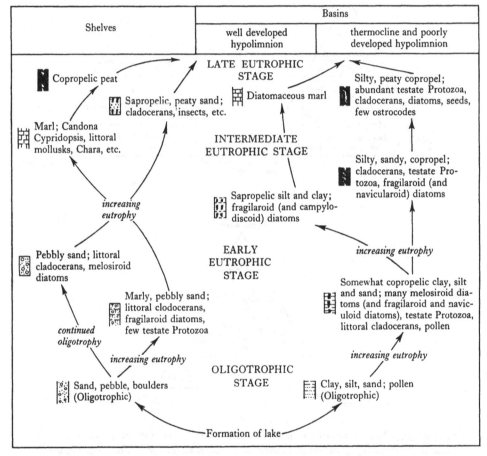

FIGURE 34   Make-up of coarse fractions of various kinds of bottom deposits of lakes; organic remains of fine fractions added in parentheses (Swain, 1956).

floodplains, or are marine basins separated from the sea by sand bars, or perhaps are earlier depressions. Lakes that form in such depressions rather rapidly assume the conditions of eutrophication that reflect the prevailing climate and sources of nutrients. Once the stage of nutritional development for the region has been established in the lake the accumulating sediments acquire the geochemical and detrital products which exemplify the environment. The resulting sedimentary history is relatively uniform and stable, but if climatic changes or other factors intervene, corresponding changes in stratigraphy of the lake sediments will occur (fig. 34).

A few examples of sedimentary sequences in such lake basins are given below.

3.2.1 *Copropelic and peaty sediments of humid temperate regions*

The deposits of Lake Drummond, Dismal Swamp, Virginia and North Carolina consist of two kinds of peat (Swain *et al.* 1959). Samples obtained about 500 ft north of the mouth of Dismal Swamp Canal feeder, locality near Osbon's (1919, p. 48) locality D where he found 10 ft of peat which at a depth of 4 ft had the following dry-weight composition: C 30-66%, volatile matter 50·02%, ash 26·32%, N 1·84%, S 0·81%. The peat is referred to as 'black gum' peat because of the abundance of *Nyssa biflora* in this region; in addition, the following plants are found in the area (Osbon, 1919, p. 43): *Fraximis caroliniana* (water ash), *Berchemia scandens* (rattan), *Gelsemium sempervirens* (yellow jessamine), *Bigonea capreolata* (cross vine), *Taxodium distichum* (bald cypress), *Acer rubrum* (red maple), and *Nyssa aquatica* (cotton gum).

The sediments at the locality sampled consist of 7 ft of peat, the upper 4 ft consisting of reddish brown, porous, punky, woody peat, the lower 3 ft are dark brown, copropelic fibrous peat, very sandy in the lower part and rest upon sand. The pH of the peat waters ranges between 4·5 and 5. The sands underlying the peat are either Pleistocene (Osbon, 1919, p. 50) or Pliocene (Shaler, 1890, p. 315). There is no evidence for more than one cycle of peat development. The lower copropelic peat layers represent eutrophic lake deposits that were succeeded by reddish-brown forest peat. Thicker peat deposits in which bog development may have been more complex have been reported in other parts of Dismal Swamp. The amino acids and some pigments of Dismal Swamp sediments are discussed below.

The bottom sediments of the 'bay lakes' of Bladen County, North Carolina Coastal Plain (Frey, 1949, 1953), comprise *dy* and *copropel* up to 12 ft thick overlying sand which was derived from the underlying late Miocene Duplin Formation (fig. 35). The lakes became eutrophic soon after formation of the lake basins and the resulting organic sedimentation persisted more or less unchanged until the present time.

A transitional marine to freshwater sedimentary organic sequence is represented in south-western Florida. The Everglades and fringing mangrove swamps of south-western Florida contain 6–15 ft of peat (fig. 36), that overlies a porous marine limestone, the Miami Oölite, and sandy

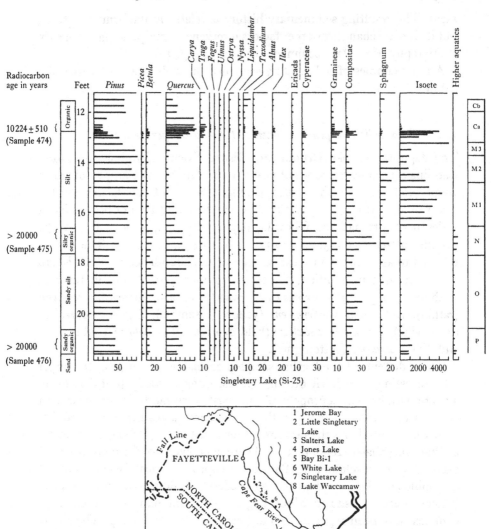

FIGURE 35 Pollen diagram of sediment core from Singletary Lake, North Carolina (Frey, 1953, *Ecological Monographs* **23**, pp. 289–313, figs 1 and 2).

limestone and calcareous sandstone of the Tamiami Formation (Spackman *et al.* 1966) along Shark River a variety of sedimentary environments includes coastal mangrove swamp, intermediate mixed hardwood environments to freshwater saw grass marsh. The distribution of uranium in the surface samples shows a decreasing trend up Shark River (fig. 37). Manganese does not show this relationship in Shark River, but inland sites of brackish nature have about three times more manganese than the non-marine Shark River localities.

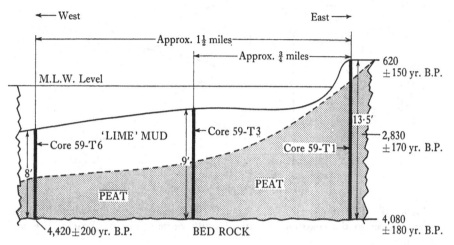

FIGURE 36   Sediment sequence in Everglades Swamp of south-western Florida (Spackman *et al.* 1966).

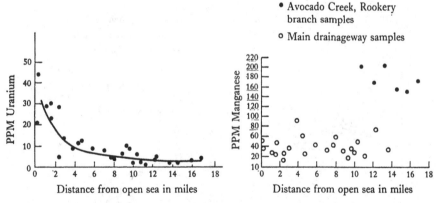

FIGURE 37   Distribution of uranium and manganese in sediments of part of Everglades Swamp, south-western Florida (Spackman *et al.* 1966).

The distribution in surface samples of red mangrove (*Rhizophora*) pollen and of black mangrove (*Avicennia*) pollen shows a decreasing trend away from the ocean while that of chenopods increases in that direction (fig. 38). Core samples show that at the coastline and to distances of 1·5 miles offshore, an autochthonous peat layer which is continuous with the

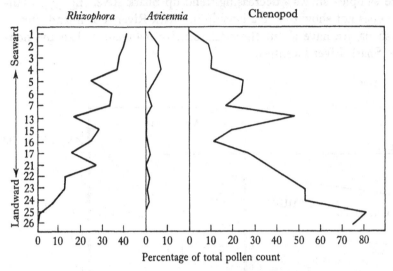

FIGURE 38 Distribution of red mangrove, black mangrove and chenopod pollen with respect to the coastline in part of south-western Florida (Spackman *et al.* 1966).

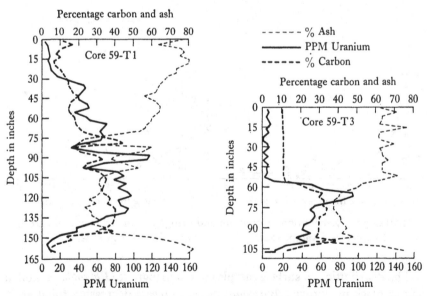

FIGURE 39 Vertical distribution of ash, uranium and carbon contents of cores from Everglades Swamp, Florida (Spackman *et al.* 1966).

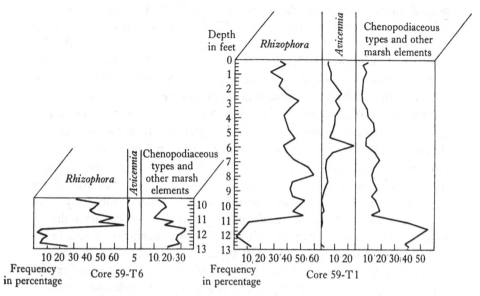

FIGURE 40   Vertical distribution of pollen types in cores from Everglades Swamp,
Florida (Spackman *et al.* 1966).

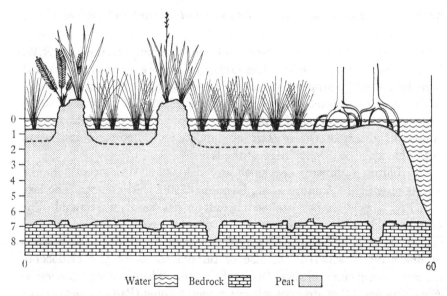

FIGURE 41   Steep-sided peat hummocks formed by erosion accompanying a rise
in sea level, Everglades Swamp, Florida (Spackman *et al.* 1966).

peat of the mainland extends out beneath the Gulf of Mexico floor where it is overlain by modern marine marls. This evidence plus data from cores (figs. 39, 40) show a former low sea level stage 4000–5000 years ago that has since risen to its present level. The rise in sea level has also resulted in erosional effects on the peat deposit that produces steep-sided hummocks of peat stabilized by herbaceous vegetation and by red mangrove (fig. 41).

Catahoula Lake, Winn and Morehouse Parishes, Central Louisiana, is an overflow basin of Black River that is subjected to rather severe seasonal flooding. The lake level fluctuates 10 m or more in level from the spring and summer high water stage to the late fall and winter period of near dryness. The bottom sediments of Catahoula Lake (Swain, 1961 *a*) are represented by light-gray and reddish-brown organic silty clay. The gray clay is prevalent in the middle of the basin where there are reducing conditions as shown by negative Eh values. The reddish-clay occurs in the peripheral part of the basin, subject to the most pronounced water-level fluctuations and represented by high positive Eh values in the sediments. The lake clays lie on Neogene (Catahoula Formation) sands and except for this sand intercalation the lacustrine stratigraphy is relatively simple. The bitumen contents of sediments of Catahoula Lake are described in Chapter 4.

3.2.2 *Calcareous, organic, clastic lake sediments of temperate high-carbonate regions*

The bottom sediments of Bultsee, Holstein, Germany near Kosel, N.W. Eckernforde were shown by Ohle (1965) to comprise organic oozes that had the following properties: the pH of the surface lake water changed from 6·9 in 1939 to 8·6 in 1961, the pH varied in 1961 from 8·63 at the surface to 7·09 at 11·5 ms of depth. A rich growth of pondweeds and plants contribute much of the organic accumulations. These also indicate a eutrophic stage of development of the lake.

The bottom sediments of a small artificial lake in Ordovician dolomite are represented in Zumbro Lake, Goodhue County, Minnesota. The lake is fed by a small stream and sedimentation has been rather rapid. The bottom sediments are gray and brown silty somewhat diatomaceous and calcareous clays and clayey-silts, typically becoming sandy, or with layers of fine sand, downward; laminae of carbonized leaves and cladoceran carapaces occur (unpublished data by the writer). The presence of *Melosira*, *Fragilaria* and other diatoms suggest a mesotrophic (Patrick, 1954) stage of development of the lake.

Neutral to slightly alkaline pH values characterize the lake water and upper sediments while slightly acidic pH is represented in the deeper

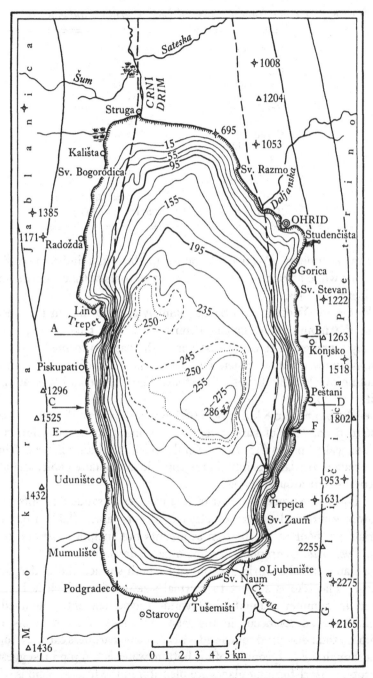

FIGURE 42   Lake Ohrid, Yugoslavia and Albania (Stankovic, 1960).

TABLE 9   *Composition of water of Lake Ohrid (Stankovic,* 1960)

Loss on ignition	19·92 %
Ash	80·08 %
CO_2	8·8 %
Carbonates	30·52 %
Organic matter	11·12 %
Xanthophyll	0·01 % × 10^{-2}
Mineral P	0·059 %
Organic P	0·011 %
HCl soluble	
Ca^{2+}	9·14 %
Mg^{2+}	1·57 %
Fe (Fe^{2+}, Fe^{3+})	2·80 %
Al	0·023 %
Total N 0·280 %; total P 0·071 % (as (P_2O_5)	

(0·5–1 m) sediments. Positive Eh values in the water indicate oxidizing conditions while negative values in the sediments show that they are under reducing conditions.

Lake Ohrid, Yugoslavia–Albania (Stankovic, 1960) lies in the western part of the mountainous Balkan Peninsula. It is a tectonic depression, probably a graben. The eastern border of the basin is formed of Triassic limestone in mountainous relief. The western part of the basin is in Jurassic serpentine overlain by Triassic limestone. The climate and vegetation are Mediterranean.

The lake basin is 30·8 km long, 14·8 km wide, covers 348 km², and is up to 286 m deep. It is of karst origin, representing coalesced solution depressions, called poljes (fig. 42). Several detached solution blocks of limestone occur in the lake. Sandy limnogenic, deltaic potamogenic, and abrasion coast occur around the lake.

The bottom sediments consist of fine calcareous sands and silts in the shallower waters and calcareous clays in deeper areas (table 10). Fossil materials are poor in the bottom sediments but include *Bosmina*, ostracodes, *Cyclotella*, gastropods and fine organic detritus.

The water of Lake Ohrid is supplied by both surface and underground drainage. The lake is subject to thermal stratification which is best developed in summer. Longitudinal wind-derived currents, modified by Coriolis-force effects occur in the lake. The lake water contains 125–130 p.p.m. total dissolved solids and is a carbonate-hardness early-eutrophic lake, although the CO_2 content of the bottom water is 5–9 p.p.m. suggesting oligotrophy. Marl-forming plants are plentiful in the littoral regions of the lake, *Chara* and *Potamogeton* both are responsible for marl deposition.

TABLE 10   *Composition of bottom sediments of Lake Ohrid*
(Stankovic, 1960)

Depth of sample (m)	Fine sand (%) 0·25– 0·05 mm	Silt (%) 0·01– 0·005 mm	Fine silt (%) < 0·005 mm	Colloidal clay 0·001 mm	Silt and clay (%)	CaCO₃ (%)	Humus (%)
6	94·88	1·44	1·12	2·56	5·12	44·23	—
25	88·32	2·08	2·76	6·84	11·68	60·93	—
71	72·75	13·12	6·28	7·85	27·25	30·04	—
130	59·30	12·82	6·88	21·00	40·70	—	2·91
220	46·73	8·82	16·44	26·11	53·27	24·07	2·94

Chara ceratophylla is reported to form prairies in the littoral parts of the lake. An analysis of the deep mud below 200 m gave the results shown in table 10 (Stankovic, 1960).

The lake is characterized as a typical oligotrophic water (Stankovic, 1960) (table 9).

3.2.3 *Copropelic and clastic lake sediments of tropical and subtropical regions*

The Pleistocene and Holocene lake sediments of the Mexico City Basin were studied by Foreman (1955) and Clisby & Sears (1955), (fig. 43). The sediments have accumulated up to 80 m or more thick and consist of diatomaceous volcanic ash, ashy clay, sand, and molluscan, ostracodal coquinas. There were intermittent intervals of volcanic activity and of drier and more moist climate that resulted in several mineralogic and paleontologic zones.

The sediments of Lakes Nicaragua and Managua, Nicaragua, Central America, large tropical lakes of tectonic origin, are characterized by uniform, profundal, volcanic copropelic silts to depths of 3 m or more (Swain, 1961b; 1966a), (figs. 44, 45; table 11). There is a rich population of diatoms and boghead algae (*Botryococcus*) (Swain & Gilby, 1965). Active volcanism in and near the lakes, together with active erosion of the cultivated volcanic soil of the area result in a rapid sedimentation rate in the lakes.

Lake Titicaca, Peru and Bolivia, an oligotrophic lake (Gilson, 1964) is also a tectonic basin due to block-faulting that took place in late Pliocene time (Newell, 1949) (fig. 46). An earlier-stage, Plio-Pleistocene stage of the lake has been called Lake Ballivian and was up to 80 m above the present

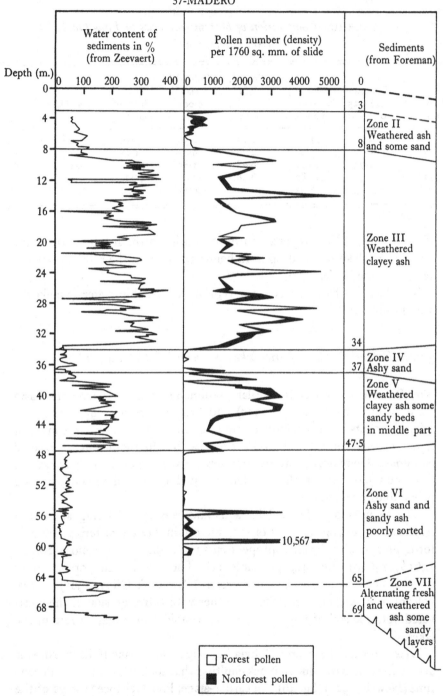

FIGURE 43 Properties of Pleistocene and Holocene lake sediments

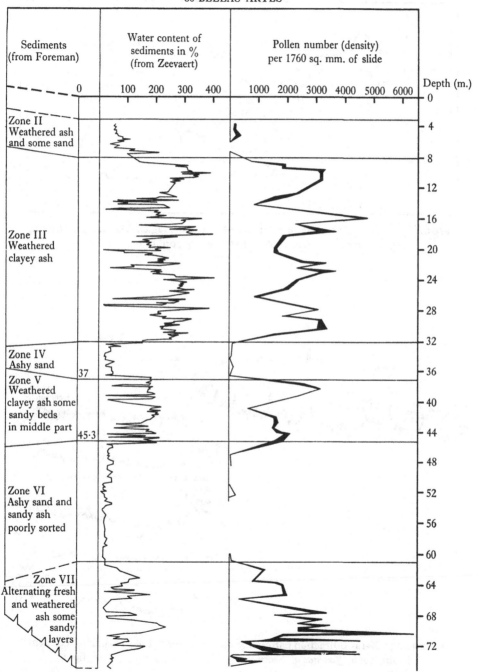

of Mexico City Basin (Foreman, 1955; Clisby & Sears, 1955).

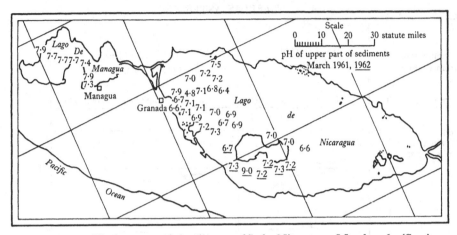

FIGURE 44   pH of upper cm of sediment of Lake Nicaragua, March 1961 (Swain, 1966, *Journal of Sedimentary Petrology*, **36**, pp. 522–40, fig. 5).

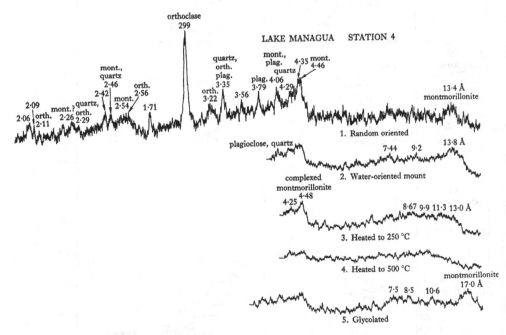

FIGURE 45   X-ray diffraction patterns of clay minerals of Lake Managua sediments (Swain, 1966, *Journal of Sedimentary Petrology*, **36**, pp. 522–40, fig. 9).

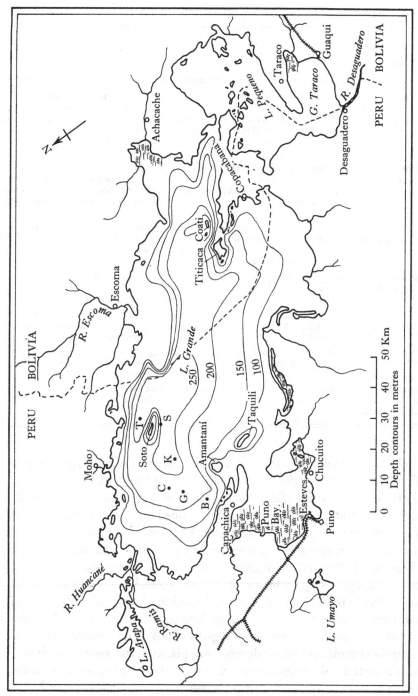

FIGURE 46   Lake Titicaca, Peru and Bolivia, showing bathymetry (Gilson, 1964).

TABLE 11   *X-ray analyses of bottom sediments from Lakes Nicaragua and Managua, Nicaragua, compared with marine sediments from a bay of the nearby Pacific coast, and two bedrock formations of the area (Swain, Paulsen & Ting, 1964)*

Locality	Mineral composition	Description of sediments
Lake Nicaragua Sta. 4, depth 11 m	Quartz, feldspar (albite); organic complexed dioctahedral montmorillonite; trace kaolinite and chloritoid	Pale grayish brown, silty, ashy, clayey copropel, fecal pellets, melosiroid and naviculoid diatoms, gastropods (Tyronia?)
Lake Nicaragua Sta. 13, depth 10 m	Feldspar (albite?), quartz, organic complexed dioctahedral montmorillonite, trace kaolinite and chloritoid	Pale grayish brown ashy, silty, copropelic fine sand and sandy silt, abundant melosiroid, fragilaroid and naviculoid diatoms, cladocerans, few ostracodes
Lake Nicaragua Sta. 15, depth 16 m	Feldspar (albite), quartz, organic complexed dioctahedral montmorillonite, kaolinite and/or chlorite, trace mica	Pale brownish gray waxy, clayey copropel, fecal pellets, melosiroid and naviculoid diatoms
Lake Managua Sta. 4, depth 26 m	Quartz, feldspar, dioctahedral montomorillonite, organic complexed	Brown copropel
Corinto Bay, Nicaragua Sta. 4	Feldspar (albite), quartz, calcite, small amount of kaolinite, chlorite,? trace mixed-layer clay	Tan argillaceous sand, feldspar and volcanic glass fragments; shell fragments common
Rivas Formation, Upper Cretaceous, Rivas, Nicaragua	Calcite, 7Å kaolinite and a 9·3Å group clay mineral	Light greenish gray finely sandy graywacke shale, planktonic Foraminifera
Brito Formation Eocene, San Juan Del Sur, Nicaragua	Albite, quartz, calcite, pyrite, chlorite and kaolinite	Pale brownish gray, calcareous, sandy siltstone

deep lake. The raised deposits of Lake Ballivian are laminated light-gray to buff clays up to 100 m thick, apparently low in organic matter.

Lake Victoria, Africa (Beauchamp, 1964) (fig. 47) receives most of its water as rain (table 12). It is thermally stratified but has a regular annual cycle of overturn as in temperate-region lakes, despite its location on the equator, its large size and relatively shallow depths ($\sim$ 100 m). According to Beauchamp, thermal stratification develops at the beginning of the rainy season, perhaps owing to inflow of cool sediment-laden rainwater which sinks to the bottom. The eulittoral and sublittoral parts of the bottom of Lake Victoria down to 60 m have deep deposits of black to greenish sapropel; below 60 m the bottom is principally sandy. The organic matter of the copropel comprises partly decomposed plankton and vegetal detritus in which a bacteria-like micro-organism is present in large quantities. In the lake the sapropel undergoes little or no apparent decomposition but boiled samples in which the micro-organisms have been killed will decompose. The reduced state in which the copropelic mud is held in the lake helps to account for its stability.

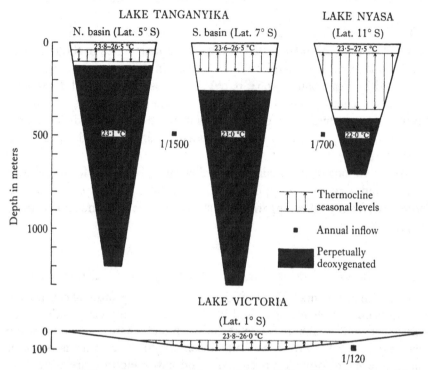

FIGURE 47   Transverse sections of some African Lakes (Beauchamp, 1964).

TABLE 12   *African Rift Valley Lakes (Beauchamp, 1964)*

(Expressed as mm over Lake Area (from Gillman, 1933).)

	Victoria	Tanganyika	Nyasa
Rainfall	1260	900	1000
Inflows	330	530	490
Total intake	1590	1430	1490
Evaporation	1310	1350	1300

The Rift Valley lakes of Africa (Beauchamp, 1964) are characterized limnologically by a small supply of water and thermal stability which combine to restrict the decomposition of organic matter in the lakes. Lake Tanganyika is an example (fig. 47). The water budget of this lake and Lake Nyasa, another Rift Valley lake, and of Lake Victoria, a non-Rift

Valley lake, are shown in table 12 (from Beauchamp, 1964, after Gillman, 1933). Evaporation accounts for most of the water lost from the lakes; to some extent the evaporating water is returned to the lakes as dew (they are likened by Beauchamp to enormous dewponds) or from local storms that develop over the lakes.

The bottom sediments of Lake Tanganyika vary from black, fine grained sapropel in the profundal parts to gray, less organic oozes in the region of the thermocline. The benthonic conditions in Victoria and Nyasa are suggested by Beauchamp (1964, p. 97) to be those of perpetual deoxygenation, rich accumulations of anaerobic bacteria in more or less static equilibrium and an abundance of mineral substances in the reduced state: metallic sulfides, hydrogen sulfide, methane, carbon dioxide and other mineral salts.

Lake Kivu, one of the smaller Rift Valley lakes, is notable for its content of CO_2 and CH_4 in the hypolimnion (Beauchamp, 1964, p. 97) but the larger Rift Valley lakes apparently lack these large concentrations (*ibid.*, p. 99).

3.2.4 *Calcareous, diatomaceous lake sediments of arid and semi-arid regions*

The bottom sediments of Pyramid Lake, Nevada are profundal calcareous, diatomaceous, ostracodal gray and black sapropelic silty clay grading into deltaic silts and nearshore sands at the south end where a river enters the lake (Swain & Meader, 1958; Hutchinson, 1937). The water is brackish and the levels of organic productivity and eutrophication are restricted. Hot springs in the lake issue along faults which are also responsible for the tectonic origin of the lake basin. The hot springs which issue from limestones provide the high bicarbonate content of the lake and form tufa deposits around the lake margin. An apatotrophic limnetic regimen has been imposed on the lake as a result of which the sediment types now forming will continue until the geologic setting changes.

Lake Niriz, Iran, is a large shallow swampy area in an intermontane basin (Loeffler, 1959). Cretaceous limestones, Eocene limestones and breccias of Holocene tectonic origin surround the lake basin. Some of the characteristics of the lake are shown in fig. 48.

The bottom sediments and sediment stratigraphy of Lake Patzcuaro, Michoacan, Mexico were studied by Hutchinson *et al.* (1956) (fig. 49): The lake basin is in a late Tertiary and Pleistocene lava plateau. Na- and K-bicarbonates are high, Ca and Mg are low and Cl and SO_3 are moderate in amount in the lake waters (table 13). The lake is about 15 m in maximum depth and fluctuates about 6 m annually. Pollen dating, sedimentation

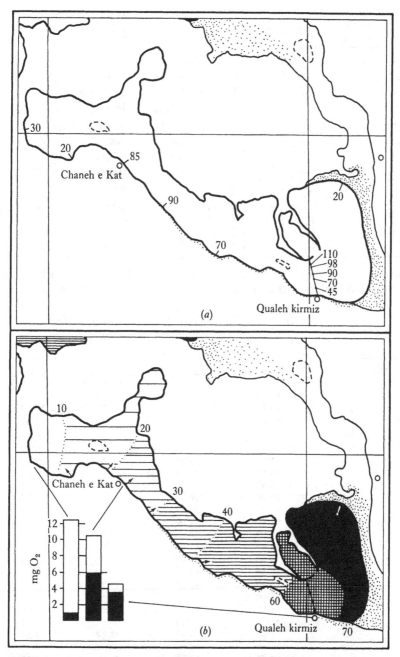

FIGURE 48   (a) Depth measurements in Niriz Lake, Iran, in cm (Loeffler, 1959).
(b) The chloride contents in Niriz Lake (in g/l) and the oxygen fluctuation
(black=night measurement; white=day measurement) (Loeffler, 1959).

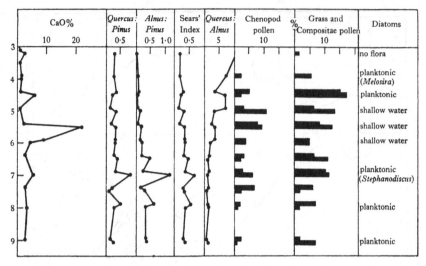

FIGURE 49   Calcium, pollen and diatom stratigraphy of core P-3, Lake Patzcuaro, Mexico (Hutchinson *et al.* 1956, *American Journal of Science*). Depth in m.

TABLE 13   *Analyses of cores from Lake Patzcuaro, Mexico (Hutchinson* et al. *1956, American Journal of Science)*

Depth (m)		Ignition loss	SiO_2	$(Al, Fe)_2O_2$	CaO
P-1	4·70	11·63	40·75	39·09	2·38
	5·60	16·84	40·71	12·51	18·37
	6·80	25·40	48·34	14·33	5·47
P-3	3·0	18·90	40·60	—	0·59
	3·1	16·90	43·00	—	0·92
	3·2	12·45	43·51	39·85	2·49
	3·4	20·00	42·07	—	0·33
	3·9	14·91	40·58	35·75	1·51
	4·0	28·70	36·66	—	1·50
	4·5	30·07	39·30	17·33	5·72
	4·9	23·90	54·96	—	9·93
	5·0	16·33	50·55	24·19	1·22
	5·4	27·40	52·03	—	2·64
	5·5	19·00	34·43	12·94	21·63
	5·9	28·40	44·59	—	8·70
	6·0	19·38	55·51	13·30	4·55
	6·4	22·90	57·64	—	2·76
	6·5	18·60	59·05	—	3·90
	6·9	21·50	52·42	—	3·04
	7·0	11·93	53·58	22·42	5·30
	7·9	26·20	50·15	—	2·15
	8·0	20·80	50·88	19·07	3·10
	9·0	25·60	54·67	—	2·97
	9·1	17·85	54·32	18·64	2·72

rates and C^{14} dates of the silty and clayey sediments from 3 m to 9·1 m below the lake floor suggest that a moist interval began about 3,300 years ago and continued to 2,300 years ago after which a drier interval with a distinct rise in CaO content set in. The low-water, high CaO stage is indicated by several species of diatoms that frequently occur in alkaline water: *Navicula oblonga* (Kutzing), *Anomoeonis sphaerophora* (K.), *A. polygramma* (Ehrenberg) and *Amphora ovalis affinis* (K.).

3.3 LAKE SEDIMENTS OF GLACIATED REGIONS

The sediments of many lakes in glacial-drift regions have been studied. A characteristic oligotrophic–eutrophic–dystrophic cycle is recorded in lakes typical of such regions (Swain, 1965).

3.3.1 *Low-organic, clastic sediments of oligotrophic lakes*

Examples of *oligotrophic* lakes that will be discussed are: Lake Superior, North Central United States; Lake Baikal, Siberia; Great Slave Lake, Canada; and Rainy Lake, Minnesota–Ontario.

The bottom sediments of a small area of Lake Superior near Silver Bay, Minnesota were studied by Swain & Prokopovich (1957). This lake lies in late Precambrian basalt lavas and sediments (Fig. 50). The sediments were collected from depths of 140–300 m (table 14), and consist principally of olive-gray to reddish-brown slightly sandy, silty clay. Near the Beaver River effluent the deposits become coarser and lenses of fine- and coarse-grained angular, feldspathic sand occur. Authigenic pyrite, vivianite $(Fe_3P_2O.8H_2O)$, coscinodiscoid and fragilaroid diatoms, cladocerans, thecamoebid protozoans, fish bones, land plants and fecal pellets are present in the sediments. The reddish sediments extend to depths of one foot or more near the shore but thin to a feather's edge a few miles offshore where the sediments are all gray. Oxygen brought in by the river and introduced by eddy diffusion in the shallower waters evidently keeps the iron oxides in the nearshore sediments in an oxidized state, while offshore the iron oxides remain in a reduced state. Other properties of the Lake Superior sediments are shown in table 15.

Lake Baikal (Kozhov, 1963), Siberia, the world's deepest lake, is a crescentic depression 636 km long, up to 80 km wide and is 31,500 km² in area. Its maximum depth is 1620 m and its bottom lies 1164 m below sea level. It lies in an area of glaciated mountains in which there are several active glaciers (fig. 51). The bedrocks in the mountain ranges are Archean gneisses, slates, amphibolites and marbles, and Proterozoic schists; granites

TABLE 14 *Descriptions of core samples of bottom sediments, Silver Bay area, Lake Superior (Swain & Prokopovich, 1957)*

Station	Type of sample	Depth of lake (ft)	Description of core
1	Core lost, only few cc recovered	888	10 YR 6/2. Slightly sandy, silty diatomaceous clay; coscinodiscoids, fragilaroids
2	Core (27″)	892	5 Y 6/1. Sandy, silty clay
3	Grab	900	Near 5 Y 6/1. Slightly sandy, silty diatomaceous clay; coscinodiscoids, few plant fragments, carbonized
4	Core (27″)	892	Top 10 YR 6/2. Slightly sandy, silty, diatomaceous clay; coscinodiscoids Bottom. 10 YR 6/2 to 5 Y 6/1. Slightly sandy, silty diatomaceous clay, coscinodiscoids
5	Grab	892	10 YR 6/2. Silty, slightly sandy diatomaceous clay; coscinodiscoids, plant fragments, patches of vivianite
6	Grab	876	10 Yr 6/2. Silty diatomaceous clay; coscinodiscoid diatoms; patches blue-green vivianite
7	Core (17″)	814	Bottom 10 in 10 YR 6/2. Slighty sandy, silty diatomaceous clay; fragilaroids, coscinodiscoids, chitinous exoskeletons; trace vivianite
8	Grab	795	10 YR 6/2. Only slightly sandy, silty clay; few diatoms, pollen, trace ostracodes, plant fragments, carbonized
9	Grab	807	10 YR 6/2. Slightly sandy, silty diatomaceous clay; fragilaroids; coscinodiscoids, chitinous exoskeletons, dark-brown egg case
10	Core (42″)	838	Top. Pale-yellowish brown (10 YR 6/2) slightly sandy silty diatomaceous clay; very abundant coscinodiscoids, few plant fragments, testate Protozoa; cylindrical aggregates organic rich silt, probably coprolites; trace vivianite near edge of core, possibly contamination Middle. 10 YR 6/2. Slightly sandy silty, clay, few diatoms, scattered red-brown feldspar grains and scattered small pebbles and granules of quartz Bottom. Light-olive-gray (5 Y 6/1) slightly silty clay; few carbonized plant fragments chitinous exoskeleton; small patches are pale red

TABLE 14 (*cont.*)

Station	Type of sample	Depth of lake (ft)	Description of core
11	Core (40″)	838	Top. 10 YR 6/2. Slightly sandy, silty clay; few diatoms, fragilaroids, carbonized plant fragments, pollen, common chitinous exoskeletons, dark-brown egg cases, fish bone fragments Bottom. Light-olive-gray, 5 Y 6/1, slightly sandy, silty clay; testate Protozoa, few plant fragments, chitinous exoskeletons
12	Grab	970	10 YR 6/2. Slightly sandy, silty diatomaceous clay; coscinodiscoids, fragilaroids, scattered red-brown feldspar grains
13	Grab	927	10 YR 6/2. Slightly sandy, silty clay; few diatoms, fragilaroids; red-brown fish teeth or bone fragments
14	Core (25″)	745	Bottom 10 in 10 YR 6/2. Sandy silty clay; coscinodiscoid diatoms, chitinous exoskeletons; scattered coarse quartz grains

TABLE 15   *Inorganic analyses of Silver Bay sediments, Lake Superior (Swain & Prokopovich, 1957)*

Station	3	8	9	10b	11t	11b	13	19t	19b	23t	23b
SiO_2	54·39	53·94	54·79	50·92	53·82	54·41	54·16	54·53	54·66	55·48	46·60
Al_2O_3	14·33	12·86	15·31	16·01	14·67	14·83	14·90	14·85	14·34	14·73	14·57
Fe_2O_3	9·92	9·46	9·43	10·09	12·20	10·09	10·32	11·89	9·75	13·09	8·94
FeO	2·57	2·78	2·98	3·51	2·14	2·69	2·37	1·63	2·67	1·64	3·40
MgO	4·00	—	1·77	5·08	3·15	3·86	3·63	3·69	4·02	4·10	5·09
CaO	2·39	—	2·74	3·28	2·75	2·65	2·67	3·30	3·04	3·82	7·06
Na_2O	1·55	1·74	1·82	1·61	—	1·72	1·68	1·69	1·76	1·73	1·57
K_2O	2·83	2·06	1·70	3·62	—	3·03	3·17	2·81	2·95	3·16	3·42
TiO_2	0·47	0·41	0·60	0·63	0·67	0·62	0·66	0·63	0·65	0·57	0·52
P_2O_5	0·49	0·33	0·37	0·27	0·33	0·36	0·48	0·23	0·16	0·26	0·30
MnO	0·16	0·10	0·12	0·18	0·33	0·15	0·16	0·77	0·14	0·23	0·15
Ign. loss	6·89	5·65	—	5·07	—	—	—	—	—	—	7·94
CO_2	(0·16)	(0·20)	0·24	(1·24)	—	—	(0·20)	—	(0·14)	0·28	(4·38)
S	(0·02)	(0·08)	0·02	(0·04)	—	—	(0·03)	—	—	—	(0·04)
H_2O^+	—	—	4·89	—	—	4·99	4·79	4·86	4·91	—	—
H_2O^-	—	—	3·50	—	—	—	—	—	—	—	—
Totals	99·99	—	100·28	100·27	—	99·15	99·24	100·88	99·05	99·09	99·56

Map showing location of Silver Bay Area

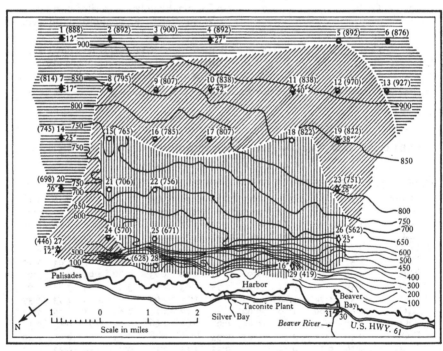

FIGURE 50   Silver Bay area, Lake Superior, location of sampling stations, type of sample, color of bottom sediments (Swain & Prokopovich, 1957). Circles and discs are core stations in gray and red sediments, respectively. Parallelograms and rhombohedrons are of Petersen dredge samples in gray and red sediments, respectively. Triangles are samples collected from Beaver River. Half discs and half rhombohedrons have less than 1 ft of red sediment overlying gray. Depths in parentheses are those determined at time of sampling and more often than not do not conform to echo soundings on which contours are based.; depths in feet.

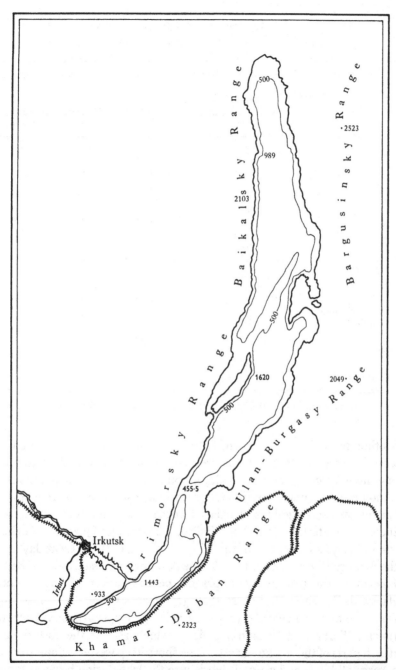

FIGURE 51   Outline map of Lake Baikal and its basin (Kozhov, 1963);
depths in meters.

TABLE 16 *Average chemical composition in part of Lake
Baikal and in affluent waters (Kozhov, 1963)*

	Open waters (mg/l)			Affluents (mg/l)	
	After Vereshchagin (1949) for South Baikal*	After Votintsev (1961)		After Votintsev (1961)	After Bochkaryov (1959)
HCO_3'	63·5	63·5		79·3	72·8
SO_4''	4·8	5·2		6·7	7·0
Cl'	0·7	0·6–1·4		1·8	1·3
Ca··	15·2	15·2		20·0	18
Mg··	4·1	3·1		4·3	3·6
Na·	3·9	3·8	5·1	4·6	
K·	2·3	2·0			
Si	SiO_2 2·5–5·5	Si 1·070		4·4	—
Al	Traces	Traces		—	—
Mn	—	0·0015		—	—
Fe, total	—	0·028		0·28	—
N	NO_3-0·19–0·62	N-0·045		—	—
P	PO_4-0·01–0·06	P-0·024		—	—
Oxidizing ability mg O_2/l	—	1·62		4·3	—
CO_2 free	0·44–5·28	1·49		—	—
O_2	14·4–9·6	11·64		—	—
N_2	22·4–16·8	—		—	—
Sum of ions	—	93·4		117·2	—

* First figure stands for the superficial layers, second figure for deep waters.

and other plutonic Precambrian, Palaeozoic and younger rocks occur in places. Tertiary basalts lie on the crests of the mountains and Quaternary basalts lie in some of the river valleys. Some older tectonic depressions in the mountains contain continental Mesozoic carbonaceous sandstones, conglomerates and argillites. Tertiary lacustrine deposits occur in terraces of the south-eastern coast of Lake Baikal. Tertiary and Quaternary deltaic sands and clays 2,000–3,000 m thick lie in the area of the present day delta of the Selenga River, eastern Baikal. The north-western side of Lake Baikal is formed of the precipitous slopes of the Baikal Range only recently deglaciated. Six rivers and more than 300 small streams drain into Lake Baikal. The more noteworthy rivers are Selenga, Upper Angona, Kichera, Barguzin, Turka and Snezhnaya, all of which rise in the eastern and northern parts of the catchment area. The short steep western-slope streams may carry all kinds and sizes of debris into the lake during heavy summer rains. The lake is subjected to summer stratification, winter overfreeze, seiches of up to 15 m and complex current and circulation patterns. The

TABLE 17   *Great Lakes of Canada (Larkin, 1964)*

	Area (km²)	Drainage basin (km²)	Total dissolved minerals (p.p.m.)	Maximum depth (m)	Mean depth (m)	Elevation (m)
Great Bear	28,490	141,400	99	> 365	> 100?	157
Great Slave	27,200	994,500	22 to 174 Average 150	614	62	162
Athabasca	7,770	274,500	52 to 130 Average 58	120	26	213
Winnipeg	24,300	984,200	50 to 560 Average 220	36·5	13	217

basin is divided into three parts by submerged elevations: North Baikal, Maloye More or Central Baikal and South Baikal. Near the river mouths are several extensive shallow bar-formed lagoons or 'sors'.

The surface-waters of Lake Baikal contains about 93 p.p.m. total solids and the profundable water about 117 p.p.m. It can be classified as a mesotrophic lake of low carbonate hardness (table 16).

The lake is noted for its endemic fauna of 87 animal genera and 11 families and subfamilies, and is believed to be of great age within the Pleistocene and Holocene Epochs.

The bottom sediments of Lake Baikal consist of profundal silty, diatomaceous clays with abundant *Melosira* and *Cyclotella* and littoral to sublittoral gravel sands and diatomaceous silts. The thickness of the lake sediments may be up to several thousand meters.

Great Slave Lake, north-west Territories is one of the glacial and glaciotectonic Great Lakes of north-western Canada. Its physical characteristics and those of three other Great Lakes of that region are shown in table 17. Although there does not seem to be a thermocline developed, except in the smaller areas and bays of Great Slave Lake, bottom temperatures of less than 4 °C occur stably in Christie Bay. Oxygen is at near-saturation levels in all these large lakes. Great Slave Lake is oligotrophic but is somewhat more productive than Great Bear Lake and is less productive than mesotrophic Lake Winnipeg (Larkin, 1964, p. 79) The bottom sediments are silts and clays more or less void of organic accumulations.

Rainy Lake, St. Louis and Kootchiching Counties, Minnesota and Ontario, Canada occupies an elongate depression carved by glacial ice in early Precambrian granite and graywacke (Swain, 1961 a). The narrow basins and intervening islands of Rainy Lake are structurally aligned (Swain, 1961 a). The lake waters when sampled in 1951 had 22·5 p.p.m. total alkalinity, 0·8 p.p.m. chlorides, and 0·6 p.p.m. sulfates, which indicate a soft water lake of oligotrophic character. The littoral bottom sediments are sand and boulders along the exposed coasts and lake peat and copropel in the bays. The profundal deposits are pale-reddish brown clay and silt to depths of about 40–60 ft and are light-gray clays and silts at greater depths, owing to the status of oxidation of iron oxides in the sediments. Varved clays occur in the profundal deposits below depths of 2 ft in the sediments. The stratification is due to silty and non-silty layers and in part to color variations related to the grain size. Assuming the varves to be seasonal, the average rate of deposition as observed in one core is 2·56 mm/year.

3.3.2 *Copropelic, peaty, clastic, calcareous, early-eutrophic lake sediments*

Several lakes can be cited as examples of *early-eutrophic* (or *mesotrophic*) and *alkalitrophic* lakes, all of which are in Minnesota (Swain, 1961 a): Clear Lake, Sherburne County; Leech Lake, Itasca County and Kabekona Lake Hubbard County. These lakes all are marl-forming and lie in carbonate-rich late Pleistocene glacial drift.

Clear Lake, Sherburne County, Minnesota, occupies a small basin in the sandy valley train of Mississippi River (Swain, 1961 a). The bottom deposits are gray, flaky copropelic and shelly marl having abundant ostracod and gastropod shells in addition to several species of testate protozoans and cladocerans. The water composition in 1954 was 147·5 p.p.m. total alkalinity, 15 p.p.m. sulfate, 0·005 p.p.m. total P, and 0·83 p.p.m. total N. The lake is in a eutrophic-alkalitrophic condition. It is evidently spring fed as neither surface inlet nor outlet occur.

Leech Lake, Cass County, Minnesota is a large mostly shallow lake in late Pleistocene ground moraine till; the western arm of this lake contains a fairly deep ice block depression. The bottom deposits consist of a mixture of marl, sand, silt and copropel. pH values range from 6·7–6·95 in the lake water, 6·65–7·65 in the surface sediments and 6·5–7·9 deeper in the sediment. A 1950 analysis of the lake water in p.p.m. was: SO_4 5·0, P 0·6, inorganic P 0·0051, Cl 0·2, NO_2 0·015, NO_3 0·04, total N 0·118, total alkalinity 155. Ostracods, mollusks, cladocerans and diatoms are abundant in the sediments.

Lake Kabekona, Hubbard County, Minnesota lies in a broad terminal moraine belt of one of the lobes of late Pleistocene (Wisconsin) ice from the Winnipeg, Canada region that extends across central Hubbard County. Kabekona Lake represents an ice-block depression. It reaches about 40 m of depth and has unusually clear water. The marly shelves support large stands of tall lake plants. The shelf deposits are molluscan, ostracodal sands and shelly, peaty coal, the latter consisting of flakes deposited around the stems of *Chara*. Remains of the latter have accumulated in windows along the beaches. The profundal deposits are pure fine-textured marl that probably was derived in large part from the shelf-marls by bottom current action.

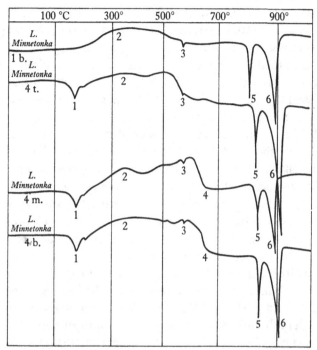

FIGURE 52   Differential thermograms of copropel samples from Lake Minnetonka (Swain, 1956). (1) Dehydration zone, (2) broad zone of oxidation of organic matter, (3) quartz endothermic reaction, (4) possible illite endothermic reaction, (5) dolomite, (6) calcite endothermic reactions. Slightly lower temperature of dolomite reaction in 1 b may be caused by iron content.

3.3.3   *Copropelic and peaty sediments of eutrophic lakes*

Lakes Minnetonka, Mille Lacs, Blue and Kandiyohi all in Minnesota are selected as examples of *eutrophic* lakes (Swain, 1956).

Lake Minnetonka, Hennepin County, east-central Minnesota (fig. 52) consists of a series of ice-block basins (Zumberge, 1952) of late Wisconsin (Mankato) age in a till facies of Mankato red and gray drift. From a limnologic viewpoint the lake is divided into three areas: (1) the lower lake which resembles more northerly Minnesota lakes such as Mille Lacs; (2) the upper lake which are like those of east-central Minnesota in having medium productivity; (3) the isolated western and northern bays having the most productive waters and resembling the prairie lakes of western Minnesota (Swain, 1956). The bottom sediment types are profundal copropelic silt and copropel (fig. 52) from less than 1–40 m or more thick, and littoral peat or sand and gravel (fig. 53). Other properties of the lake sediments and waters are given in table 32.

Mille Lacs Lake, Aitkin and Crow Wing Counties, is a large shallow lake in late Wisconsin ground moraine. The lake water in July 1954 had the following composition: total alkalinity 125 p.p.m.; SO_4 17·0 p.p.m.; total P 0·006 p.p.m.; total N 1·66 p.p.m. The littoral bottom sediments are peaty and shelly sand and the profundal sediments are silty, sandy copropel overlying sand. A rich population of planktonic algae and diatoms and nearshore pondweeds pervades the lake.

A small lake in an elongate ice-block depression, Blue Lake, Isanti County, lies on the Anoka Sand Plain, east-central Minnesota (Swain, 1961 *a*). The lake had 112 p.p.m. alkalinity in 1949 and is in a eutrophic-alkalitrophic condition. Other water properties are SO_4 4·5, P 0·002 and N 0·036 p.p.m. Although the lake has no thermocline the oxygen content is very low in the bottom waters and reducing conditions prevail in the profundal areas owing to high productivity of pondweeds and algae. The bottom sediments are copropel, marl and silt.

Lake Kandiyohi, is a shallow basin in late Wisconsin ground moraine in west-central Minnesota. The composition of the lake water in 1953 was: total alkalinity 192·5 p.p.m., SO_4 40 p.p.m., total P 0·032 p.p.m., total N 1·78 p.p.m. The bottom sediments are silty copropel and copropelic silt containing many mollusks, ostracods, Cladocera and diatoms. pH values of the bottom water ranged from 8·1–9·7, that of the top of the sediments 7·3–8·4 at 2·4 in depth 7·8–8·2. Eh values all were negative in the bottom sediments (Meader, 1956).

3.3.4 *Humic sediments of dystrophic lakes*

Lakes of dystrophic type passing into bog stages in which nitrogen is present in excess quantity, brown humic acid complexes color the water and productivity is lower than in the eutrophic stage are represented by Linsley Pond, Connecticut and Cedar Creek Lake, Anoka County, Minnesota.

The bottom deposits of small shallow lakes in recently deglaciated regions typically begin with one or two meters of sand and gravel overlain by silt and clay, representing the oligotrophic to mesotrophic stages (fig. 34). These deposits are overlain in limestone-rich areas by marl (alkalitrophic stage) which grades upward into pulpy copropel, sapropel and dy (eutro-

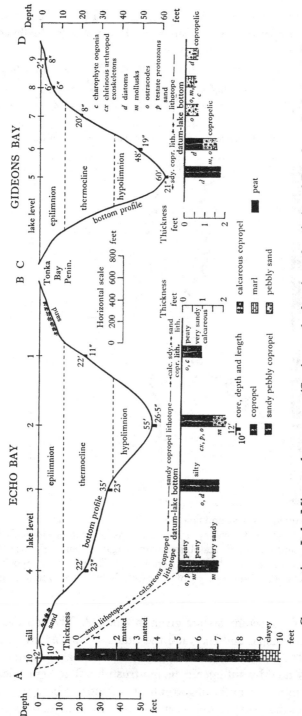

FIGURE 53   Cross-section, Lake Minnetonka traverses (Swain, 1956, *American Association of Petroleum Geologists Bulletin*, **40**, no. 4, pp. 600–653, fig. 3).

phic and dystophic stages). If the lake changes to a bog, lake peat or förna succeeds the sapropel and dy and sedge and forest peat gradually accumulate with advancing stages of bog development.

A typical sequence without marl is that of Linsley Pond, Connecticut (Hutchinson & Wollack, 1940; Vallentyne & Swabey, 1955). The post-glacial stratigraphy of this small lake is shown in fig. 54. The inorganic

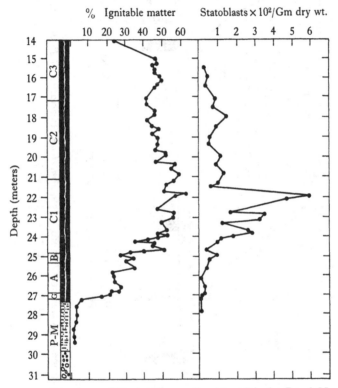

FIGURE 54   Sequence of lake and bog sediments at Linsley Pond, North Branford, Connecticut (Vallentyne & Swabey, 1955, *American Journal of Science*). Per cent ignitable matter and *Plumatella* statoblasts plotted against depth. Points at the right refer to sediment levels where 'wheels' were found. The numbers opposite each X refer to the number of wheels in 0·5 g of wet mud. Sediment zonations as in fig. 25.

constituents of the sediments are given in table 18. The silica as quartz, the alumina as feldspar and clay, and the titania as sphene are principally detrital from the surrounding source materials. A small quantity of silica may be represented by authigenic diatom frustules. The high silica values in the lower 3 m represent the oligotrophic to mesotrophic stages; the lower silica content of the 11–13 m unit results from the abundant organic matter

TABLE 18   *Analyses of sediments, Linsley Pond, Connecticut*
(Hutchinson & Wollack, 1940, American Journal of Science)

Constituent	Silty clay 40–43 ft	Clay and gyttja 34–40 ft	Coarse-detritus gyttja 3–34 ft	Silty gyttja 0–3 ft
SiO_2	47·3	28·6–37·8	50·0–50·2	63·6–73·0
Fe_2O_3	12·1	3·1–7·8	9·3	4·3–9·2
Al_2O_3	5·5	3·3–5·0	6·2–8·4	10·3–15·0
TiO_2	0·08	0·05–0·18	0·25	0·63–0·71
P_2O_5	0·254	0·096–0·140	0·055–0·087	0·078–0·105
MnO	0·23	0·08–0·15	0·30–0·55	0·11–1·03
CaO	1·2	1·0–1·5	2·1–2·7	1·2–1·9
MgO	1·0	0·27–0·68	0·48–0·62	1·2–1·77
SO_3	1·0	1·0–2·1	1·9–3·2	0·44–2·1
Loss on ignition	2·8	46·0–59·0	23·8–28·7	2·9–7·4

formed in the eutrophic stage; organic matter is still forming but is diluted by silt content that has risen as a result of cultivation of the surrounding land. The iron in the lake sediments has evidently been precipitated in large part as ferrous sulfide hydrate (melnikovite); some may occur as siderite, in diatom frustules or complexed with humic matter. The iron as well as manganese are suggested by Hutchinson & Wollack (1940) to reflect biochemical changes in the lake. Both Ca and Mg appear to occur mainly as detrital carbonates in the lake sediments. The sulphide probably occurs as iron sulfide. The phosphate maximum of the sediments is at 13 m. According to Drake & Owen (1950) phosphorus, with high pH in the presence of $CaCO_3$ tends to be precipitated as calcium orthophosphate. In addition, photosynthesis may result in precipitation of phosphates together with calcium carbonate. Studies by Moyle (1954) have shown that the concentration of P in stratified lakes is higher below than above the thermocline. The concentration of P in the lower part of the Linsley Pond sediments is the result of sorption with mineral matter according to Livingstone & Boykin (1962). The high ion-exchange capacity of the mud in early stages of lake development brought about the concentrations of P.

A dystrophic lake and associated bog sequence in which marl is abundant is represented by Cedar Creek Bog, Anoka and Isanti Counties, Minnesota (fig. 55) (Lindeman, 1941; Swain & Prokopovich, 1954). Partial chemical analyses of the peats and marls in this late-senescent bog are shown in table 19. In addition to the abrupt change at 3·5 m from calcareous copropel above to marl below, the Ca:Mg ratio also changes abruptly from about 8:1 to between 20:1 and 30:1. Dolomitization may be taking place in the

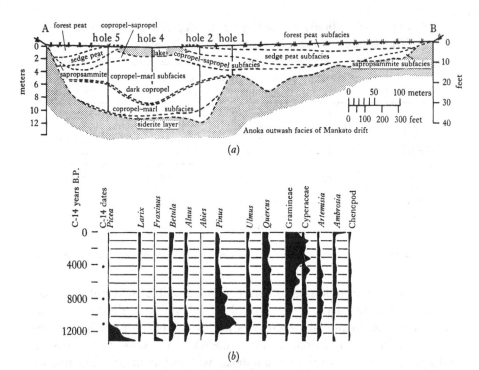

FIGURE 55 (*a*) North-west–south-east cross-section through Cedar Creek Forest, Anoka County, Minnesota (Swain, & Prokopovich, 1954, after Lindeman, 1941). Showing subfacies of the peat and marl accumulation. Modified after Lindeman (1941*a*). Locations and depths of stations sampled in this study are shown except station 3 which is directly back (NE) of station 2. (*b*) Pollen diagram and time scale for Cedar Creek Bog Lake (Cushing, 1963).

TABLE 19   *Analyses of sediments, Cedar Creek Bog, Minnesota*

Station	Depth (ft)	CaCO₃ (%)	MgCO₃ (%)	SO₄ (%)	Total S* (%)	Total Fe† (%)	Total P‡ (%)
1	4–5	7·99	0·96	Nil	0·21	2·27	—
1	7–8	6·88	0·90	Nil	0·32	1·59	—
1	9–10	7·11	0·88	Nil	0·80	1·72	—
1	11–12	74·80	2·37	Nil	0·37	1·04	0·021
1	13–14	40·75	2·05	Nil	1·46	2·01	0·036
2	35–36	38·60	1·13	Nil	0·43	15·63	0·255

* May be present as sulfide, native sulfur, or an organic compound.
† May occur as siderite, ankerite, marcasite, ferrous sulfide hydrate or some other form.
‡ May occur as inorganic apatite, vivianite ($Fe_3P_2O_8 \cdot 8H_2O$), bobierrite ($Mg_3P_2O_8 \cdot 8H_2O$), or organic phosphate ('collophane').

peat carbonates. The high S value at 4·5–4·8 m is matched by a slight increase in Fe and both seem to occur as ferrous sulfide at that depth as well as at shallower depths. At the 11·5–12 m level the Fe content is very high and the S content low, and iron carbonate is present at that depth. Total P is low in the shallow samples but is an order of magnitude higher at the 11·5–12 m level where it probably occurs as vivianite or similar bog-phosphate.

3.4 LACUSTRINE AND PALUDAL SEDIMENTARY ROCKS

There are several examples of well-developed lacustrine and paludal formations in the geologic time scale. These include paludal coal and lignite beds and associated lacustrine clays and freshwater limestones of the Upper Carboniferous, Triassic, Jurassic, Cretaceous and Cenozoic of many parts of the world; the interbedded fluvial and lacustrine Purbeck and Wealden beds of Jurassic–Cretaceous age of England and their equivalents in central Europe, Brazil, and Western United States; the Paleogene Green River Formation of the Western Interior United States; the Neogene Humboldt Formation of Nevada, and similar deposits in Australia and the U.S.S.R.

From a mineralogic, stratigraphic and geochemical point of view the rock deposits are mostly similar to one or another of the modern lake sediments described above. The Green River Formation of Colorado, Utah and Wyoming is different in several respects from most other lake sediment sequences. Of particular importance is the laminated carbonate oil-shale up to several thousand feet thick that forms much of the upper part of the formation and which was apparently deposited in a saline or alkaline phase of the lake. The Green River formation is discussed further in Chapters 5–8.

Other lake sediment and coal-swamp sequences are described in later portions of the book.

3.5 FLUVIAL SEDIMENTS

Fluvial sediments are perhaps characterized more by their variability in composition, grain size, sorting and color than by uniformity in any of these features. Sediments of old-age river floodplains are typified by numerous fine-grained sand and silty sand bodies of lenticular cross-section and arcuate plan embedded in clay and silty clay; colors of the river sediments of humid regions are various shades of gray; organic content may be up to 1 % but is likely to be concentrated in small lenses or laminae, and is typically well humified or carbonized; carbonate content and other salts are typically low in humid-region flood plain sediments unless the terrain is one of soft marly or chalky sedimentary rocks. Redox potentials are likely to be low-positive or slightly negative in the accumulating sediments although few data are available on redox values in river sediments.

Fluvial sediments of arid and semi-arid regions are generally lighter in color or are more varicolored than those of humid regions owing to lower organic content, and positive redox potentials which result in predominance of ferric oxides. The organic matter is in a highly humified state.

The river sediments and those of three reservoirs of Columbia River are similar to those of graywackes (Whetten, 1966). Rock fragments are abundant silica contents are low, and sodium exceeds potassium. The sediments are somewhat better sorted than those of graywackes. Ignition loss ranges, in the reservoir sediments from 0·65 to 5·99% but how much of this represents organic carbon is not known.

Very few studies have been made on the organic geochemistry of fluvial sediments. The field would appear to be one in which studies are rather badly needed as regards pollutional problems in larger river systems.

3.6 FLUVIAL SEDIMENTARY ROCKS

Fluvial deposits of the geologic past are frequently part of 'red bed' sequences which suggest high, seasonally-variable temperatures. Among such sequences the following can be mentioned.

Precambrian Erathem: Torridon Sandstone, Great Britain, Fond du Lac Group of northern U.S.A.; Uinta Group of Utah; Upper Precambrian of Baltic Shield; upper Vindhyan System of India.

Paleozoic Erathem: Queenston-Juniata Formations (U. Ordovician) of Appalachian region, U.S.A; Bloomsburg Formation (U. Silurian) of Appalachians; Catskill Group (U. Devonian) of Appalachians; Old Red Sandstone, (Devonian) of Great Britain, Mauch Chunk Formation (U. Mississippian) of Appalachian region; Coal Measures of parts of Europe and North America; Artinskian, Kungurian and Kazanian Stages (Permian) of Russia; Cimmaron Group (Permian) of Kansas and Oklahoma; Ochoa Series (U. Permian) of Texas and New Mexico; Karoo Series (Permian part) of South Africa.

Mesozoic Erathem: New Red Sandstone (Triassic) of Great Britain; Keuper Series (U. Triassic) of Europe; Spearfish, Chugwater, Chinle and underlying Moenkopi Formations (part) of Western U.S.A.; Hosston Formation (L. Cretaceous) of southern U.S.A., Difunta Formation (U.

Cretaceous) of Mexico; Mesa Verde Sandstone (U. Cretaceous) of western U.S.A.

Cenozoic Erathem: Fort Union and Wasatch Formations (Paleocene and Eocene) of west-central U.S.A.; Humboldt and related formations (Oligocene-Pliocene), western U.S.A.; White River Group (Oligocene) South Dakota, U.S.A.; Ogalalla Group (Pliocene) Kansas–Nebraska, U.S.A.; Siwalik Series (Plio-Pleistocene) Pakistan.

The writer is aware of only a few studies on the organic geochemistry of fluvial sedimentary rocks. Swain (1966*b*) presented data on some organic constituents of Ordovician, Silurian and Devonian fluvial deposits of central Pennsylvania (see Chapter 6). Hedberg (1968) summarizes the distribution of hydrocarbons in non-marine rocks, including several of fluvial type (Chapter 4).

3.7 OTHER NON-MARINE SEDIMENTS AND SEDIMENTARY ROCKS

Glacial, aeolian, mass-wasting, pyroclastic types of sediments and their lithified products under some circumstances may contain considerable organic material, but very little study has been made of it. Growth of moss and lichens on exposed rock surfaces probably should also be included in such studies. Glacial sediments include tills, glaciofluvial deposits and gumbotils. Glacial tills and tillites commonly contain carbonized wood and other plant fragments. Glaciofluvial sediments also may have carbonized plant remains especially in the finer-grained deposits. Gumbotils are very likely low in organic residues owing to their leached nature but there is little information about them.

Aeolian sediments include dune sands, lag gravels and wind-facetted pebble accumulations, and loess. The oxidized environments of deposition of the first two kinds of deposits probably eliminates all but the most resistant organic debris but mineral charcoal, spores, etc. might be expected to occur in small amounts. Loess frequently contains reworked humus and larger plant fragments, although probably in small total amount. The writer is not aware of specific organic geochemical studies of these materials. Organic matter in small temporary ponds in loess and in sand-dune blow outs also would be of interest.

Among mass-wasting types of sediments mud-flows and debris slides may effectively trap and preserve plant and animal remains of various kinds that might yield valuable climatological and paleoecological information, but investigations have not yet been made.

Plant and animal remains are trapped and partly preserved in volcanic ash falls and more rarely in lava flows. In these and in preceding cases

Characteristics of non-marine sediments

where oxidative conditions prevail, the use of electron spin resonance, controlled permanganate-oxidation, and laser–mass spectrometer analyses should be of value.

In summary, there are several kinds of non-marine sedimentary deposits other than those of lakes and bogs that may contain organic matter of value in the interpretation of the climate, relief, and other paleoecological conditions in which the sediments formed.

4 Bitumens of non-marine sediments and sedimentary rocks

4.1 LIPID MATERIALS IN NON-MARINE SOURCE ORGANISMS

The lipid or bitumen substances of terrestrial plants are predominantly protective waxes on trunks, leaves and needles. In this discussion, *lipid* refers to substances in living organisms that are soluble in benzene, chloroform, petroleum- or diethyl-ether or similar organic solvents and which include hydrocarbons as well as fats, waxes and some organic pigments. These high molecular weight substances of terpenoid nature are represented by such compounds as the triterpenoid betulin that is present in large amounts in white birch outer bark:

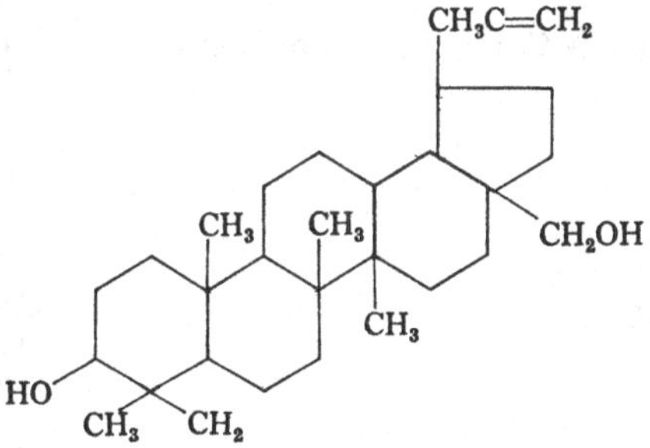

Betulin

Other terpenoid substances in terrestrial plants and animals are listed in table 20. Previous discussions of these materials and their geochemical aspects, for example, that of Bergmann (1963), although comprehensive for Holocene deposits of certain types have reported only a few occurrences in generalized kinds of non-marine sediments.

An interesting instance of methane occurrence in two pondweed species, *Elodea canadensis* and *Myriophyllum exalbescens* has been recorded (Hartman & Brown, 1966). The methane is present as a constituent of the internal atmosphere of the plants, and apparently enters the plant roots or rhizopodal portion from an anoxic substrate. High concentrations of methane occur in the plants during the winter, especially under ice cover. It attained 13% of the internal plant atmosphere in March in one specimen. It is also present at night and during early daylight hours during the summer. Methane was displaced rapidly from internal tissues in light-bottle experiments during periods of active photosynthesis but disappeared

TABLE 20 *Lipid substances in some modern organisms and soils*

Compound	Composition	Occurrence
Carnuba wax and related material, in ester form	A wax mixture containing acids and alcohols of order of C_{22}–C_{32}, with even number of C atoms in a straight chain; paraffinic hydrocarbons of order of C_{25}–C_{35} with odd number of C atoms (Piper *et al.* 1934)	In large amounts in Brazilian palm leaves; similar material in many other plants, exoskeletons of insects (Bergmann, 1963)
Dihydroxystearic acid	$CH_3(CH_2)_7CHOHCHOH(CH_2)_7COOH$	In soils of low fertility (Schreiner & Shorey, 1908)
α-Hydroxystearic acid	$CH_3(CH_2)_{15}CHOHCOOH$	Elkton silt loam, Maryland (Schreiner & Shorey, 1910)
'Lignoceric acid'	Fatty acid mixture?	Peaty soil (Schreiner & Shorey, 1910)
Fatty acids of high molecular weight	—	Water of Beloe Lake, Russia (Goryunova, 1954)
Solid hydrocarbons melting at 52 and 59·5 °C	Possibly $C_{25}H_{52}$, $C_{27}H_{56}$ and $C_{29}H_{60}$	Russian clayey silt (Petrova *et al.* 1956)
n-Heptane	C_7H_{16}	Oils of many species of pines; Jeffrey pine oil contains 98 % *n*-heptane (Bergmann, 1963)
n-Nonane	C_9H_{20}	*Sarothra gentianoides*, and in pine oil and *Pittosporum eugenioides* (Bogert & Marion, 1933; Carter & Heazelwood, 1949; Rock, 1951)
Undecane	$C_{11}H_{24}$	Ponderosa pine and other pine oils and in volatile oil of ant *Formica rufa* (Iloff & Mirov, 1954; Rock, 1951; Schall, 1892)
Pentadecane	$C_{15}H_{32}$	A component of Sanna oil (Nakao & Shibuye, 1924)
Phytol, an acyclic diterpenoid	$(CH_3)_2CHCH_2CH_2CH_2CHCH_2CH_2CH_2-$ $\qquad\qquad\qquad\qquad\qquad\mid$ $\qquad\qquad\qquad\qquad\quad CH_3$ $CHCH_2CH_2CH_2C{=}CHCH_2OH$ $\mid\qquad\qquad\qquad\quad\mid$ $CH_3\qquad\qquad\quad CH_3$	Derived from chlorophyll; it is produced in very large amounts and is relatively stable but evaporates readily
Squalene, a triterpenoid hydrocarbon	$\qquad\qquad\qquad\qquad CH_3$ $\qquad\qquad\qquad\qquad\mid$ $(CH_3)_2C{=}CHCH_2(CH_2C{=}CHCH_2)-$ $\qquad\quad CH_3$ $\qquad\quad\mid$ $(CH_2CH{=}CCH_2)_2CH_2CH{=}C(CH_3)_2$	Many animals and plants

TABLE 20 (*cont.*)

Compound	Composition	Occurrence
Caoutchouc, natural rubber, long chain of isoprene units	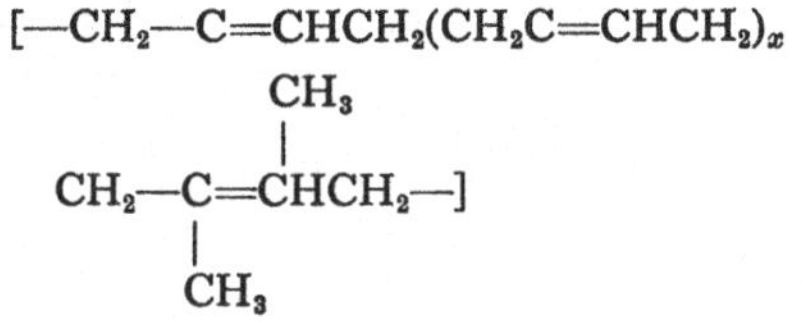 $[-CH_2-C=CHCH_2(CH_2C=CHCH_2)_x$ $\quad CH_3$ $CH_2-C=CHCH_2-]$ $\quad\quad CH_3$	Similar material ('monkey hair') found in lignites (Kindscher, 1924); *elaterite* is similar physically but different chemically; may include some resinous material in hard coal (Bergmann, 1963)
Terpene hydrate (flagstaffite)	$C_{10}H_{20}O_2 . H_2O$	Fossil woods, San Francisco Mountains, Ariz. (Guild, 1922)
4-isoPropylidene-cyclohexanone	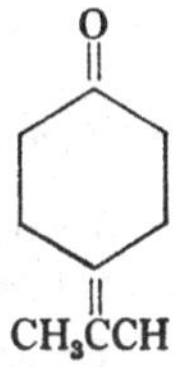	Pine needle oil, eucalyptus oil, lignite resins (Ruhemann & Rand, 1932)
1,4-dimethyl-7-isopropylazulene, one of the 'blue hydrocarbons'	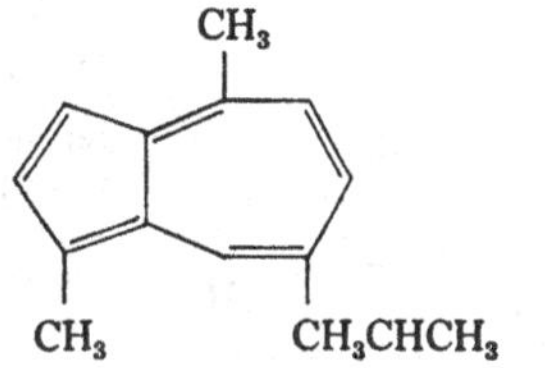	Derived from sesquiterpenoid alcohol guajol, by heating with sulfur, also found in fossil lignite resins from India (Varier, 1950)
Abietic acid, a diterpenoid resin acid	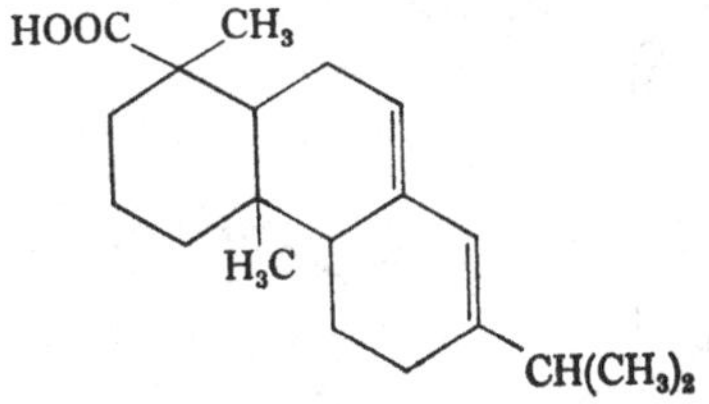	Component of pine resin; also found in bituminous coal (Golumbic *et al.* 1950)
Retene, an aromatic hydrocarbon	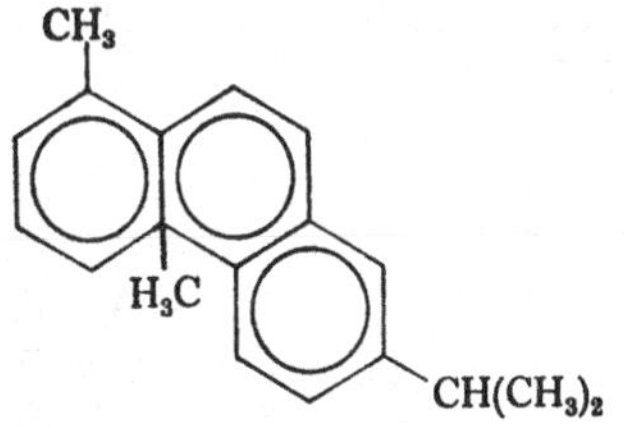	Dehydrogenation product of abietic acid; in fossil pine wood (Bergmann, 1963)

TABLE 20 (*cont.*)

Compound	Composition	Occurrence
Fichtelite, a non-aromatic hydrocarbon	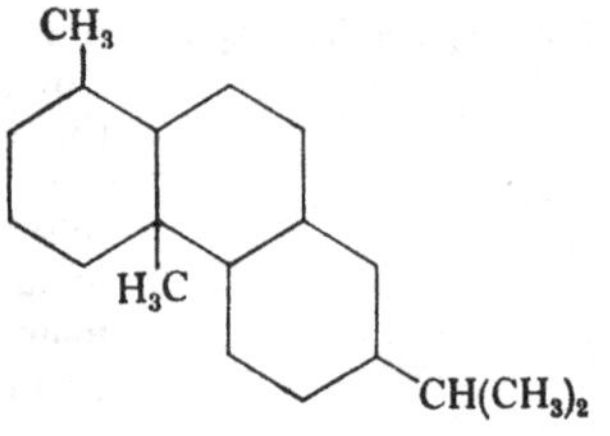 CH₃ ... H₃C ... CH(CH₃)₂	Derived from abietic acid, in fossil wood
Iosene, α-dihydro-phyllocladine (Briggs, 1937)	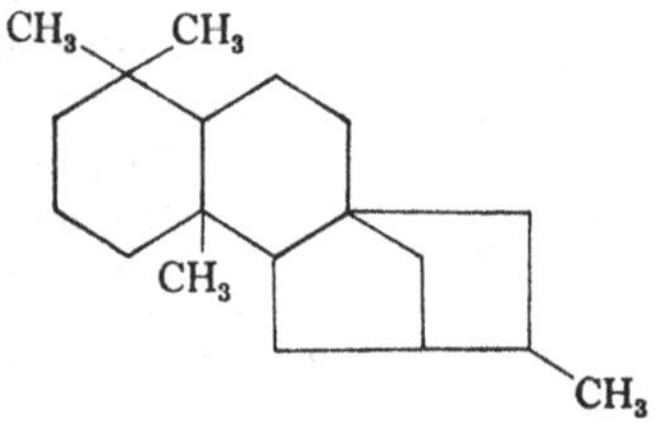 CH₃ CH₃ ... CH₃ ... CH₃	Found in essential oils of plants which yield retene on dehydrogenation; in Piberstein, Austria lignite (Brandt, 1952; Soltys, 1929); copal is similar; these resins may also contain agathene-dicarboxylic acid (Ryzicka & Hosking, 1929)
Agrosterol, m.p. 237°, a triterpenoid	Similar to betulin	Marshall Clay of North Dakota (Schreiner & Shorey, 1909)
Dimethylpicene	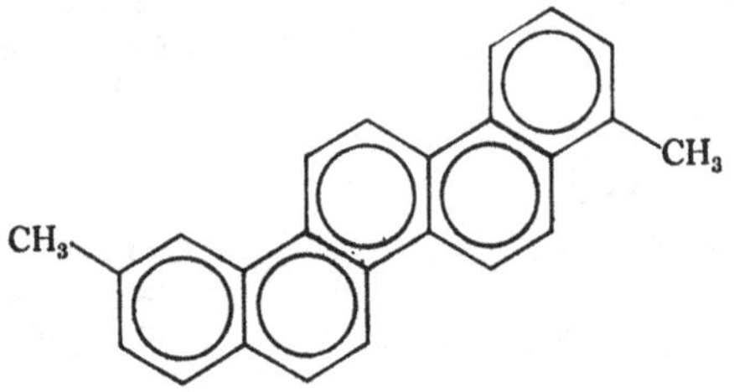 CH₃ ... CH₃	Derived from triterpenoid; found in Yuburi Coal, Japan (Sakabe & Sassa, 1952)
Cholesterol, an unsaponifiable fraction of animal and plant fats	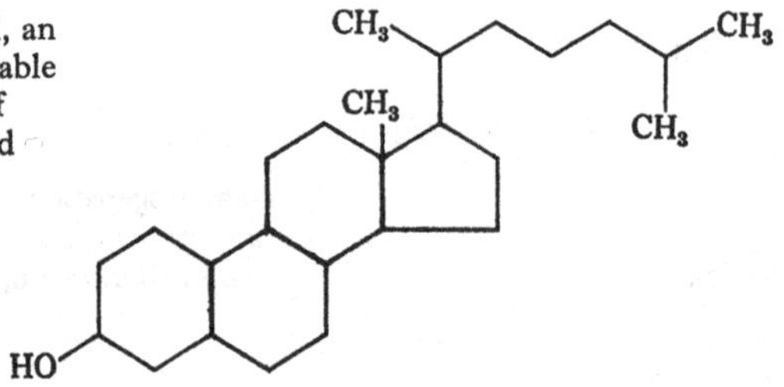 CH₃ CH₃ ... CH₃ CH₃ ... HO	Higher and lower plants and invertebrates
β-Sisosterol, a phytosterol	?	Associated with stigmastanol and triterpenoid friedelan-3 β-ol in peat (McLean *et al.* 1958)
Sterol mixtures	—	Peruvian guano and in *Andonta cygna* (mussel)

TABLE 20 (*cont.*)

Compound	Composition	Occurrence
Adipocire, Ca and Mg salts of animal fatty acids	Mainly palmitic acid; also stearic, oleic and hydroxystearic acids and Ca–Mg salts	Burial grounds, peat bogs, etc.; 'bog butter' is similar material (Bergmann, 1963)
Unsaturated hydrocarbons and related material	$C_{19}H_{30}$–$C_{30}H_{62}$	Obtained by vacuum distillation of montan wax (Kraft, 1907)
Cerotic acid	$C_{26}H_{52}O_2$	In Irish peat wax (Reilly & Emlyn, 1940), together with 'carboceric acid'
Behenic acid	$C_{22}H_{44}O_2$	In Irish peat wax (Reilly *et al.* 1943)
Russian peat wax (Tschischty *Sphagnum* peat)	Contains 'carboceric acid', unidentified oxy-acid, higher alcohol $C_{27}H_{56}O$, hydrocarbons $C_{33}H_{68}$ and $C_{35}H_{72}$, esters of cyclic alcohols and cyclic acids	Tschischty *Sphagnum* peat (Titov, 1932)
Saturated fatty acids	Hexanoic to stearic acids	Marine and freshwater algae without marked difference as to environment
Unsaturated fatty acids	Oleic and linoleic acid	Same as preceding (Bergmann, 1963)
Mono-unsaturated fatty acids	C_{16} acids and C_{18} with one double bond in the carbon chain; other acids in smaller amounts	Blue-green algae *Synectococcus cedrorum*, *Anacyctis nidulans* and *A. marina* (Holton *et al.* 1968; Parker *et al.* 1967); *Hapalosiphon laminosus* a hot spring form
Poly-unsaturated fatty acids	C_{16} acids, and C_{18} with two and three double bonds in carbon chain; other acids in smaller amounts	*Oscillatoria* sp., *Nostoc muscorum*
Fatty acids, saturated and branched aliphatic hydrocarbons	A complex mixture including isoprenoid types	*Elaeophyton* (=Botryococcus) *coorongiana* (boghead algae) (Eglinton *et al.* 1966*b*; Thiessen, 1925; Stadnikov & Weizmann, 1929

slowly in dark-bottle experiments. It appears likely that vascular plants serve as a significant pathway for release to the air of both methane and evolved oxygen.

Gas chromatograms of the *n*-heptane eluted chromatographic fractions of extracts of two freshwater plants are shown in fig. 56. The normal alkanes were separated on 5 Å molecular sieve. Branched alkanes and cyclic

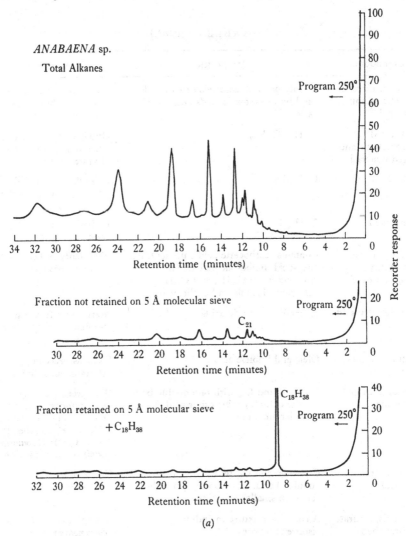

(a)

FIGURE 56 Gas chromatograms of saturated hydrocarbons of some freshwater plants (*a*) *Anabaena* sp.

saturated hydrocarbons are noticeably more abundant than normal alkanes in these specimens.

The olefin hydrocarbons of two species of non-marine algae (*Botryococcus braunii* and *Anasystis montana*) are shown in table 21 (Gelpi *et al.* 1968). Gas chromatographic analyses of the olefins were obtained (fig. 57). Both the algae show distribution of aliphatic hydrocarbons in the medium and high molecular weight ranges. The hydrocarbons seem to be mainly

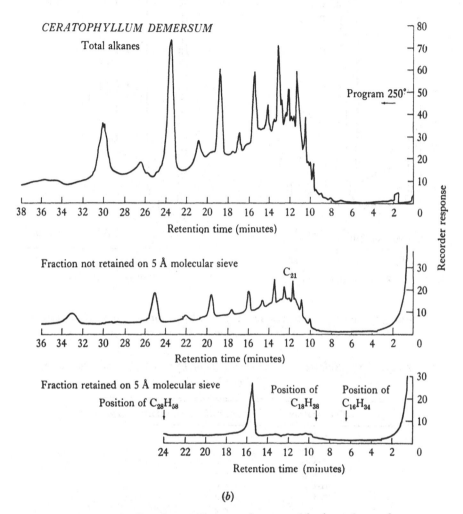

FIGURE 56   Gas chromatograms of saturated hydrocarbons of some freshwater plants (b) *Ceratophyllum demersum.*

monoenes, dienes or trienes as determined by mass spectrometric data (table 21). The organisms may have served as precursors of some of the hydrocarbons in continental sediments because the latter show similar distribution of *alkane* hydrocarbons with respect to carbon number. The scarcity of published data on the lipid components of possible non-marine source organisms indicates that a great deal more work must be done in this field. The non-hydrocarbon lipids of aquatic plants, on degradation,

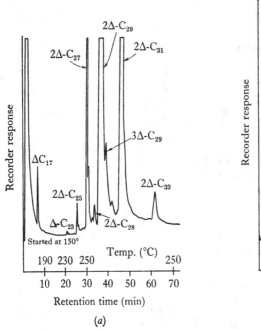

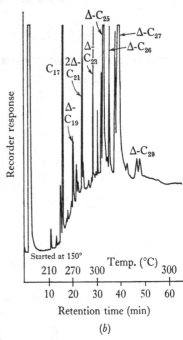

(a)

(b)

FIGURE 57   Gas chromatograms of aliphatic hydrocarbons of (a) *Botryococcus brauni* and (b) *Anacystis montana* (Gelpi *et al.* 1968).

TABLE 21   *Relative per cent content of hydrocarbons of two algae*
(Gelpi et al. *1968)*

Hydrocarbon	Anacystis	Botryococcus
$n\text{-}C_{17}$	11·5	—
$\Delta\text{-}Cr_{17}$	—	1·52
$\Delta\text{-}C_{19}$	0·2	—
$\Delta\text{-}C_{20}$	0·1	—
$2\Delta\text{-}C_{21}$	8·9	—
$\Delta\text{-}C_{23}$	8·0	0·14
$\Delta\text{-}C_{24}$	0·2	—
$\Delta\text{-}C_{25}$	14·6	—
$2\Delta\text{-}C_{25}$	0·2	0·65
$3\Delta\text{-}C_{25}$	—	0·10
$\Delta\text{-}C_{26}$	3·8	—
$\Delta\text{-}C_{27}$	34·7	—
$2\Delta\text{-}C_{27}$	2·8	11·10
$\Delta\text{-}C_{28}$	0·1	—
$2\Delta\text{-}C_{28}$	—	0·65
$\Delta\text{-}C_{29}$	0·2	—
$2\Delta\text{-}C_{29}$	—	50·40
$3\Delta\text{-}C_{29}$	—	5·54
$2\Delta\text{-}C_{31}$	—	27·90
$2\Delta\text{-}C_{33}$	—	2·00

TABLE 22   *Distribution patterns of fatty acids in alewife concretions and carcasses (Sondheimer et al. 1966)*

| Acid | Concentration (% by wt.) | |
	Concretions	Carcasses
Caprylic	Trace	—
Nonanoic	0·1	—
Caproic	0·4	—
Undecanoic	0·1	—
Lauric	3·0	0·2
Tridecanoic	0·3	—
Myristic	23·6	2·1
Pentadecanoic	1·7	0·7
Palmitic	55·4	32·6
Heptadecanoic	1·6	2·3
Stearic	4·3	15·3
Heneicosanoic	—	Trace
Docosanoic	—	5·5
Unknown 1*	—	1·7
Pentacosanoic	—	7·3
Dodecenoic	0·1	—
Tridecenoic	Trace	0·6
Tetradecenoic	1·2	0·2
Pentadecenoic	0·2	0·2
Hexadecenoic	2·5	4·7
Heptadecenoic	0·3	1·0
Octadecenoic	4·5	19·4
Octadecadienoic	0·6	3·1
Octadecatrienoic	—	2·0
Unknown 2†	—	0·5
Heneicosenoic	—	0·6

* Assumed molecular weight, 354.   † Assumed molecular weight, 278.

probably contribute principally to the humic complex through a series of condensation reactions. Several recent studies bearing on lipid source materials are briefly discussed below.

In pineapple plants naphthalene-acetic acid, applied as an auxin, stimulates the production of ethylene (Burg & Burg, 1966). The ethylene is suggested to bring about flowering of the plant. Previously it had not been considered that the auxin directly influences the flowering.

Many species of fungi metabolically produce ethylene (Ilag & Curtis, 1968). It is a regulator of growth in plants, especially higher forms. It was detected by gas chromatography of incubated fungi gaseous products. Of the 228 species studied 58 yielded ethylene, the amounts ranging from less than 1–514 p.p.m. in *Aspergillus clavatus*.

Bitumens of non-marine sediments

The isolation of water and soil bacteria of thermophilic character capable of growth on hydrocarbons was accomplished by Mateles *et al.* (1967). In the production of single-cell proteins (SCP) by culturing yeasts or bacteria on hydrocarbon substrates one of the technical problems has been the large amount of heat evolved per pound of cells produced, owing to the large amount of oxygen required for oxidation of hydrocarbons. The heat of fermentation evolved on hydrocarbon substrate is estimated to be double that on carbohydrate substrate, and expensive cooling measures must be introduced in the case of hydrocarbons. Thermophilic bacteria of the *Bacillus stearothermophilus* group from soil, ditch water and oil refinery water were found to grow on normal alkanes ranging from C_{12} to C_{20} at temperatures from 55 to 60 °C. Problems of fermentor cooling would appear to be mitigated by such bacterial types. Several geological possibilities for growth under these conditions also came to mind, and should be investigated.

Decomposition products of the alewife, *Alosa pseudoharengus* (Wilson) in an alkaline lake (Lake Onondaga, New York) have combined with calcium to form combustible concretions of calcium salts of fatty acids (Sondheimer *et al.* 1966). The alewife is normally a marine species that spawns in freshwater or coastal areas. It also occurs in lakes in upstate New York, perhaps as a marine relict fauna, or one that is able to exist because of the high chloride content of the water (about 400–600 p.p.m.). The authors did not mention the possibility that the chlorinity may derive from the Silurian salt beds which occur in the vicinity, but this seems to be worth considering. In Lake Onondaga, a polluted lake near Syracuse, dead alewives have been involved in the formation of combustible concretions of light-colored, crumbly, chalky substance. Much of the fine skeleton is missing, and the muscles are replaced by chalky material. The concretions are extracted by shaking with a mixture of 0·5 N aqueous hydrochloric acid and ether to yield an average of 85% ether-soluble material. The extracted substances were found to be fatty acids in concentrations high enough to bind the calcium present which in turn prevented their being extracted by ether alone. Normal alewife carcasses from Lake Ontario contained only 7–8% of ether extractables with or without hydrochloric acid. Neutral lipids, steroids and hydrocarbons were present only in trace amounts in the alewife concretions.

The ether extracts from the alewife carcasses and concretions were saponified and resulting fatty acids were converted to methyl esters with diazomethane. The esters were gas-chromatographed on Apiezon-L and on diethylene glycol succinate in order to define the chain length and structure of the acids (table 22). The concretions were dominated by myristic and palmitic acid and other lower molecular weight acids while the carcasses are relatively richer in high molecular weight acids. The concretions appear to be of the order of 10^3 years old but their age is somewhat uncertain and some may be still forming. In terms of C^{13}/C^{12} ratios the alewife concretions have δ values of about $-35\%_0$ whereas the carcasses are about $-20\%_0$ (fig. 58). The concretions apparently do not contain true adipocire, as this

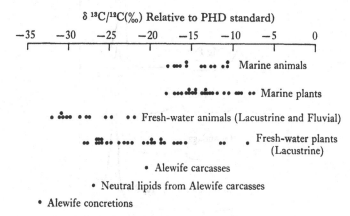

FIGURE 58   Carbon 13:carbon 12 ratios of samples of alewife and other aquatic organisms (Sondheimer *et al.* 1966).

residue of animal fats typically is high in hydroxystearic acid which was not found in the concretions. A concretion from Lake Barlewitzer, Germany, believed to be from an eel (Faber & Krejci-Graf, 1936) consisted of 48·4% lipids, and appeared to be composed of white, sectile, talc-like, crumbly material similar to the Onondaga Lake concretions. The best estimate as to origin of the concretions is that they formed in or on the bottom sediments by bacterial decomposition in an alkaline, calcium- and chloride-rich environment and are now being reworked out of the sediments and washed ashore.

A related study was made by Berner (1968) on the formation of calcium carbonate concretions from adipocire resulting from bacterial decomposition of butterfish and smelts in sealed jars of seawater. The decomposition resulted in an increase of dissolved bicarbonate, carbonate and ammonia, rise in pH, and precipitation of Ca^{2+} in the form of calcium fatty acid salts or soaps with 14–18 carbon atoms. X-ray diffractions for Ca and Mg palmitate, Lake Onondaga alewife concretions and butterfish samples are very similar (fig. 59) and suggest similar composition. The fossil-bearing carbonate concretions that are so common

throughout the geological column may have originated in part by conversion of fatty acid calcium salts to calcium carbonate, according to
Berner (*loc. cit.*).

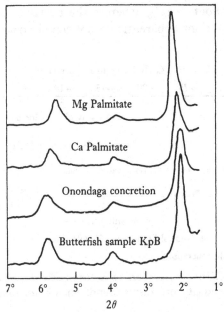

Mg Palmitate

Ca Palmitate

Onondaga concretion

Butterfish sample KpB

7°   6°   5°   4°   3°   2°   1°

2θ

FIGURE 59 X-ray diffraction patterns for pure Ca and Mg palmitate, adipocire
formed from decomposition of butterfish in the laboratory, and an alewife concretion from Lake Onondaga (Berner, 1968).

TABLE 23 (*a*) *Bituminous extracts of Pyramid Lake sediments; values
before correction for S are in parentheses.* (*b*) *Chromatographic analyses of
bituminous extracts from Pyramid Lake (Swain & Meader, 1958)*

(*a*)

Station	Position in core	Weight of extract (g)	p.p.m. of sample	Description of extract
5	Top 12 in	0·2167	(8200) 8188	Dark brown to black tar; tarry odor; bladed and fibrous pale yellow sulfur crystals 0·14 % S
7	Top 4 in	0·1231	(6800) 6120	Brown earthy wax; long needlelike yellow crystals of sulfur; slight reddish fluorescence 9·6 % S
10	General	0·1523	(5400) 4990	Light brown porous earthy wax; few tiny bladed iridescent crystals 7·6 % S
13	Top 6 in	0·1257	(5900) 5204	Brownish yellow tar with yellow orthorhombic crystals of sulfur 11·8 % S
Av. bitumens p.p.m. of sample			6575	—

(b)

TABLE 23 (*cont.*)

No.	Position in section	Weight of fraction	Description of fraction	p.p.m. sample	Bitu-mens (%)	Non-polar bit. (%)	Hy-dro-carb. (%)	Sul-fur	Remarks
			(a) Heptane chromatographic fractions (corrected for sulfur content)						
5	Top 12 in	0·0224	Colorless oil, frosty white substance, trace greenish white fluorescence; lathlike crystals	852	12	45·8	82·3	Yes	Heptane cut dark, elution clear
7	Top 4 in	0·0038	Mainly bladed and globular sulfur crystals; colorless oil, frosty white substance	210	3·4	26·3	65·5	Yes	Pink fluorescence in filter aid; none in alumina; a brown front developed in upper half of column
10	General	0·0015	Mainly lathlike sulfur crystals, colorless bladed crystals, frosty white substance	53	1·0	16·3	45·4	Yes	Heptane cut colored, eluant clear
13	Top 6 in	0·0024	Colorless oil, frosty white substance	114	2·2	16	96	Yes	Pink fluorescence in filter aid; developed brown zone in alumina
			(b) Benzene chromatographic fractions (corrected for sulfur content)						
5	Top 12 in	0·0048	Crystal laths of sulfur, colorless oil, frosty white substance	182	2·6	9·8	17·6	Yes	Benzene cut dark, eluant yellowish fluorescent
7	Top 4 in	0·0020	Crystal laths of sulfur; colorless oil with weak greenish yellow fluorescence, colorless crystals, frosty white substance	110	1·8	13·9	34·5	Yes	No fluorescence in benzene in column
10	General	0·0018	Mainly lathlike crystals and globules of sulfur	64	1·3	19·5	54·5	Yes	Cut colored, eluant clear, no fluorescence
13	Top 6 in	0·001	Crystals of sulfur, trace frosty white substance	4·7	0·09	19·5	4·0	Yes	
			(c) Pyridine+methanol chromatographic fractions (corrected for sulfur content)						
5	Top 12 in	0·0217	Yellowish brown wax	825	11·7	44·4	77	Yes	Pyridine cut very dark, elution also dark, purplish red color disseminated in alumina slightly fluorescent
7	Top 4 in	0·0086	Yellowish brown wax, frosty white substance	475	7·7	59·7	87	Yes	Pyridine front had red-orange fluorescence
10	General	0·0059	Yellow-tan to brown wax, frosty white substance	209	4·2	64	93	Yes	Pyridine cut light colored, elution dark, good purplish red band in alumina, slightly fluorescent
13			Yellowish green wax, frosty white substance	594	11·2	83·3	86		

4.2 HIGH-ASPHALT BITUMENS OF LAKE AND BOG SEDIMENTS OF UNGLACIATED REGIONS

Very little study has been made of hydrocarbons and related substances in non-glacial lakes and bogs. The following are cited as examples: Pyramid Lake, Nevada (Swain & Meader, 1958); Lake Nicaragua (Swain, 1966a);

TABLE 24 *Chromatographic analyses of lipoid extracts of Lake Nicaragua and other lake deposits in Minnesota and Nevada (Swain, 1966a, Journal of Sedimentary Petrology,* **36**, *pp. 522–40)*

	L. Nica. Sta. 20*	L. Nica. Sta. 18**	Eutrophic Lake, Minn.	Oligotr. Lake Minn.	Peat bog, Minn.	Pyramid Lake, Nev.	Lake Superior Minn.
Total extract % of sample	0·090	0·243	0·835	0·380	6·52	0·612	0·029
Saturated hydrocarbons % of sample	0·008	0·006	0·038	0·058	0·47	0·021	0·005
Aromatic hydrocarbons % of sample	0·011	0·007	0·051	0·042	0·50	0·011	0·007
Asphaltenes % of sample	0·024	0·023	0·025	0·016	1·60	0·047	0·004
Polar compounds % of sample	0·053	0·208	0·536	0·13	4·29	0·593	0·12
Saturated hydrocarbons % of extract	8·8	2·5	4·8	15·2	6·5	3·4	17·2
Aromatic hydrocarbons % of extract	11·4	2·8	5·1	12·5	4·8	1·8	24·1
Asphaltenes % of extract	25·0	8·9	25·5	42·5	25·5	7·7	15·5
Polar compounds % of extract	54·6	85·6	64·6	27·8	63·2	87·1	53·1

Analyses by D. A. Peterson and F. M. Swain.
* Extracted cold in ultrasonic tank, with benzene 80%, methanol 20%.
** Extracted with Sohxlet apparatus with benzene 80%, methanol 20%, at boiling temperature of solvent.

Mississippi Delta, Louisiana (Smith, 1954); and Mud Lake, Florida (Bradley, 1966).

The generalized distribution of benzene–methanol extractable materials in Pyramid Lake, Nevada sediments was studied by Swain & Meader (1958). Some limnologic characteristics of the lake were described above (chapter 3). The results of extraction of Pyramid Lake sediments with benzene and methanol (80:20) and column chromatography of the extracts is shown in tables 23a and 23b. The values for total lipoids (0·65%) are intermediate between those for peat deposits (6·5%) and those for oligotrophic lakes (0·30%) and are of the same order of magnitude as those of typical eutrophic lakes (0·83%) of glaciated regions. Sulfur was present in all the Pyramid Lale extracts. It probably originated from anaerobic bacterial reduction of sulfates in the interstitial sediment waters.

The chromatographic fractions eluted from the columns of activated alumina with *n*-heptane are a colorless oil and form about 72% of the hydrocarbon fractions; the fractions represent mainly saturated hydrocarbons. The fractions eluted with benzene, taken to represent the aromatic hydrocarbons, also are colorless oil and represent 28% of the hydrocarbons. White and colorless crystals occur in the dried extracts; the white cystals seem to represent anthracene, judged from ultraviolet absorption spectra

(λ_{max} 254 nm) and color of fluorescence of the extracts. The non-hydro-carbon asphaltic fractions eluted with pyridine and methanol are yellowish brown wax, containing red, orange, and purple pigmenting substances that include chlorinoid and carotenoid pigments, as shown by their absorption spectra; furthermore, the pink and red fluorescence of the solutions observed during chromatography is suggestive of chlorophyll degradation

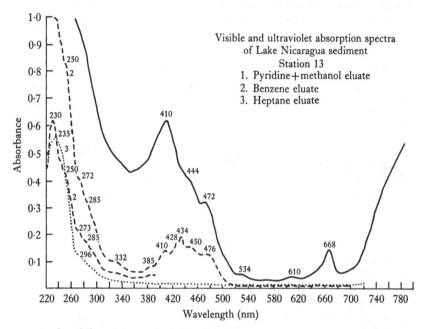

FIGURE 60   Ultraviolet and visible absorption spectra of chromatographic fractions of benzene + methanol extracts of Lake Nicaragua sediments (Swain, 1966a, *Journal of Sedimentary Petrology*, **36**, pp. 522–40, fig. 10).

products. The polar compounds remaining adsorbed on the alumina columns comprising some resins, pigments and mineral substances form a generally higher percentage of the Pyramid Lake lipoids than those of several other types of lakes (Swain, 1956).

The lipoid substances of the bottom sediments at two stations in Lake Nicaragua, western Nicaragua, central America, were extracted and chromatographed as in the preceding case (Swain, 1966a). The total extract free of sulfur, as shown in table 24 is relatively small and compares more closely with an oligotrophic lake (Swain, 1956) than with other types studied.

The heptane eluates, taken to represent saturated hydrocarbons are small in quantity and are similar in amount to the benzene eluates which represent mainly aromatic hydrocarbons. The proportions of hydrocarbons are

TABLE 25 *Ultraviolet spectral data on organic substances extracted from Lake Nicaragua sediments.*

(*a*) Benzene and methanol (80:20), followed by separation on activated alumina, data presented on benzene eluates redissolved in *n*-heptane; (*b*) acetone, followed by drying and redissolving in hexane; prominent absorption maxima in italics. The spectra suggest presence of aromatic hydrocarbons, phenols and pyridine, but further characterization was not made (Swain, 1966 *a*, *Journal of Sedimentary Petrology*, **36**, pp. 522–40).

Station and type of sample	Solvent	Absorption 200–300 nm	Absorption 300–400 nm
13, Dredge	*a*	230, 252, 272, 285, 296	332, 385
20, Dredge	*a*	231, 252, 275, 286	302, 328, 333
35, Top of core	*b*	234, 256, 288	320, 336, 360
35, Middle of core	*b*	256, *288*	320, 336, 384
35, Bottom of core	*b*	234	
36, Top of core	*b*	236	300, 310
36, Middle of core	*b*	236	310
36, Bottom of core	*b*	236	310
41, Top of core	*b*	236, 272	300
45, Top of core	*b*	236	—
45, Middle of core	*b*	*250*, 290, 294	300
45, Bottom of core	*b*	222, 236, 272, 296	300
47, Top of core	*b*	238	300
47, Middle of core	*b*	236, 286	—
47, Bottom of core	*b*	236	—
48, Top of core	*b*	236, *272*	300
48, Middle of core	*b*	236	300
48, Bottom of core	*b*	234, 244	310
51, Top of core	*b*	*229*, 234, 263, 284	—
51, Middle of core	*b*	*229, 234*, 244, 248, 251, 292	306, 330
51, Bottom of core	—	234, 252	—
52, Davis core	*b*	282	318, 330
53, Top of core	*b*	235	—
53, Middle of core	*b*	235	—
53, Bottom of core	—	238	300

thus quite different from those of Pyramid Lake in which the saturated hydrocarbons greatly exceeded the aromatics. The total hydrocarbon values of the Lake Nicaragua sediments are much lower in proportion to total bitumens than in oligotrophic lakes and are more like those of Pyramid Lake.

The asphaltic factions, eluted with pyridine and methanol, and the polar compounds left on the alumina columns also are relatively higher than in

TABLE 26   (*a*) *Chromatographic analyses of miscellaneous Recent sediment samples.* (*b*) *Spectro-chemical analyses of ash in organic matter extracted from Recent sediments* (*Smith*, 1954, *American Association of Petroleum Geologists Bulletin*, **38**, no. 3, pp. 377–40, *tables* 14 *and* 15)

(*a*)

Source	Organic matter extracted, p.p.m. of dried sediment	Chromatographic analysis of extract Paraffin-naphthene (%)	Aromatic (%)	Total hydro-carbons p.p.m. of dried sediment
Mississippi Delta samples				
BC 236–52	2,550	1·6	2·7	110
PL 242–52	326	17·6	10·5	92
GM 311–52	457	4·8	2·2	32
MP 321–52	362	1·3*	6·1	27
Louisiana Coast samples				
Marsh sample	490	0·8	1·1	9
Mud-flat deposit	1,040	2·7	2·3	52
California Coast sample				
Estuary pond	7,600	6·6	3·9	800
Non-marine samples				
Stony Lake, N.J.	11,200	1·4	0·6	224
Lake Wapalanne, N.J.	2,230	1·6	3·1	105
Mirror Lake, N.J.	2,300	1·1	1·1	51
Marsh sample, N.J.	6,810	0·9	0·8	116
Rancocas Creek, N.J.	1,130	4·3	3·7	90
Hemlock Forest, N.J.	3,870	2·2	1·3	135
Soil sample, Indiana	450	2·8	4·0	31
Peat sample, Minnesota	10,500	3·8	5·6	986
Great Salt Lake, Utah				
Salt-water sediments	5,090	1·3	1·5	144
Freshwater sediments	2,740	0·8	0·5	37

* Most of sample lost by spilling.

(*b*)

Sediment source	Laguna Madre, Texas	Pelican Island Core, La.	Stony Lake, N.J.	Hemlock woods Humus, N.J.	Freshwater marsh, La.
Major	Fe, Mg	Mg, Ca	Fe	Fe	Mg
Minor	—	Si, Fe	Na	—	Fe, Zn
Trace	Si, Cu	B, Zn	B, Cu, Al, Si, Mg	B, Mg, Si, V, Al, Ca	Si, P, B, Cu, Ni, Mn
Present	Sn, Pb, B, Mo, Zn, Ni, Ca, Mn	Mn, Pb, Cu, Ti	V, Ca, Pb, Mo, Sn, Zn, Ni, Mn, Cr, Ag, P	Ni, Zn, P, Pb, Mn, Cr, Cu, Na	Ca, Pb, Sn, V, Ag, Co, Al, Cr

TABLE 27   *Ultimate analyses of two samples of algal ooze from Mud Lake, Florida compared with the ultimate analysis of the organic matter from Green River oil shale (Bradley, 1966)*

	Mud Lake		Green River Oil Shale
	MW-0	MW-3	
C	57·50	58·23	80·50
H	7·77	7·70	10·30
O	29·06	26·25	5·75
N	3·75*	4·46	2·39
S	1·92	3·36	1·04
	Thermal values		
cal/g	6600 *ca.*	6600 *ca.*	9530

Subsequent analysis by W. H. Bradley gave 4·6 per cent.

oligotrophic lakes and are comparable to Pyramid Lake. The data for Lake Nicaragua appear to reflect the high content of non-hydrocarbon pigmenting material (fig. 60), perhaps mostly from diatoms and other algae, compared to hydrocarbons.

A wide variety of other substances of aromatic or heteroaromatic nature was detected in ultraviolet and visible absorption spectra of benzene-methanol and acetone extracts of the sediments of Lake Nicaragua. These are shown in table 25. The number of such substances in Lake Nicaragua is considerably greater than in temperate-climate lakes studied.

Samples of freshwater sediments from the Mississippi Delta, Louisiana were studied for hydrocarbons by Smith (1954). The analytical data are shown in table 26. The first sample from H_2S-rich mud yielded much more hydrocarbons than the other samples of more clastic nature and from slightly deeper water. Aromatic hydrocarbons exceeded paraffin-naphthene hydrocarbons in the first sample, perhaps owing to the anaerobic environment (Meader, 1956) which in Kandiyohi Lake, Minnesota seems to favor the preservation of aromatic hydrocarbons.

The bottom sediments of shallow subtropical Mud Lake, Florida consist (Bradley, 1966) of copropelic algal ooze about one m thick underlain by peaty sediment. The ultimate analyses of two samples from Mud Lake compared to a sample of Green River Oil Shale are shown in table 27. Organic chemical analyses of Mud Lake copropel show that it contains 0·03–0·05% of higher fatty acids (C_{12}–C_{34}) in which palmitic acid (C_{16}) is predominant; unsaturated fatty acids, palmitoleic and stearic also occur; 0·3 % of n-alkanes in which C_{29} is dominant also occur and branched cyclic alkanes, stearanes and triterpanes are present.

TABLE 28   *Measurements of pH and Eh in Burntside Lake, Minnesota cores (Swain, 1956, American Association of Petroleum Geologists Bulletin, 40, no. 4, pp. 600–53)*

Station	pH	Eh, mv	Type of sediment
1, Bottom water	5·5	+437	——————
1, 0–3 in	6·2	+359	Silty clay, few pollen grains
1, 16–19 in	6·5	+299	Silty clay, abundant pollen
2, Bottom water	5·8	+503	——————
2, 0–3 in	6·2	+419	Silty clay, few pollen grains
2, 13–16 in	6·3	+353	Silty clay, abundant pollen
3, Bottom water	6·05	+527	——————
3, 0–3 in	6·1	+419	Silty clay, few pollen grains
3, 9½–12½ in	6·5	+302	Silty clay, abundant pollen grains
4, 0–3 in	6·5	+410	Silty clay, abundant pollen, diatoms, cladocerans
4, 3–5 in	6·3	+413	Silty clay, abundant pollen, diatoms, cladocerans
5, Bottom water	7·0	+410	——————
5, 0–3 in	6·5	+455	Silty, diatomaceous clay, pollen
5, 4–7 in	6·3	+365	Silty, diatomaceous clay, pollen
6, NR	—	—	Sand and pebbles
8, 0–3 in	6·4	+491	Silty, hackly clay, pollen
7, 11–14 in	6·9	+323	Silty, hackly clay, pollen
8, Bottom water	6·7	+449	——————
8, 0–3 in	6·5	+299	Silty clay, abundant pollen
8, 18–21 in	6·5	+281	Darker gray silty clay, pollen
9, Bottom water	6·5	+459	——————
9, 0–3 in	6·4	+459	Sandy silty clay, abundant pollen
9, 18–21 in	6·7	+254	Silty clay, fish bones, etc.
10, 0–3 in	6·6	+476	Sandy silty clay, pollen, etc.
10, 27–30 in	7·2	+293	Sandy clay and sand, diatoms

4.3   LOW-ASPHALT BITUMENS OF OLIGOTROPHIC LAKE DEPOSITS OF GLACIATED REGIONS

The lipoid substances of the sediments of three oligotrophic lakes in Minnesota and Wisconsin are discussed here: Burntside Lake, St Louis County, Minnesota (Swain, 1956); Trout Lake, Wisconsin (Judson & Murray, 1956); and Silver Bay, Lake Superior, Minnesota (Swain & Prokopovich, 1957).

Burntside Lake lies in early Precambrian igneous and metamorphic rocks. Its bottom sediments (fig. 61) are gray and tan silty clay. Some properties of the sediments, the lipoid extracts, and the chromatographic analysis of the extracts are shown in tables 28 and 29. The extracts are of significance in that although the total lipoids are low as compared to those

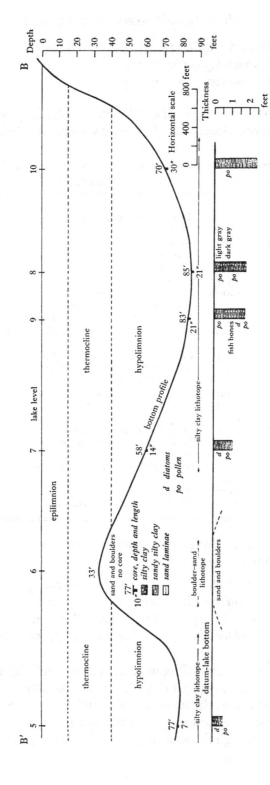

FIGURE 61   Cross-section, Burntside Lake, Minnesota (Swain, 1956, *American Association of Petroleum Geologists Bulletin*, 40, no. 4, pp. 600–53, fig. 14).

TABLE 29   (*a*) *Lipoid extracts of Burntside Lake core samples.* (*b*) *chromato-graphic analyses of extracts from core no.* 10, *Burntside Lake,* (*Swain,* 1956, *loc. cit.*)

(*a*)

Station	Bottom sediment	Weight of extract (g)	Sample (%)	Remarks
10, Top 6 in	Sandy, silty clay, pollen	0·0040	0·65	Greenish yellow wax
10, Bottom 6 in	Silty clay, sand laminae	0·0164	0·12	Greenish gray brown 'frosty' wax

(*b*)

	Top 6 in	Bottom 6 in	Description
Total lipoids, % of sample	0·65	0·12	Top: yellow green wax
			Bottom: green-brown wax with long crystals
Saturated hydrocarbons, % of sample	0·04	0·027	Colorless oil
Saturated hydrocarbons, % of lipoids	6·2	24·3	
Saturated hydrocarbons, % of non-polar lipoids	10·6	29·5	
Saturated hydrocarbons, % of hydrocarbons	37·3	62·5	
Aromatic hydrocarbons, % of sample	0·068	0·016	Top:greenish yellow oil,
Aromatic hydrocarbons, % of lipoids	10·4	14·6	yellow green fluorescence
Aromatic hydrocarbons, % of non-polar lipoids	18·0	17·6	Bottom: almost colorless oil,
Aromatic hydrocarbons, % of hydrocarbons	62·7	37·5	weak brown fluorescence, few needle crystals, granules
Asphaltenes, % of sample	0·27	0·05	Top: greenish yellow and
Asphaltenes, % of lipoids	41·3	43·8	yellowish brown oily wax,
Asphaltenes, % of non-polar lipoids	71·3	53·0	weak brown fluorescence
			Bottom: light green and brown oily wax, weak brown fluorescence, small crystals

of the sediments of more productive lakes, the saturated and aromatic hydrocarbons are relatively high in proportion to total lipoids than in the more productive lakes. The saturated hydrocarbon fraction consisted of colorless non-fluorescent oil. The aromatic fractions from the uppermost sediments were more highly colored and had stronger fluorescence than in the deeper sediments. The organic material in Burntside Lake that presumably contributed to the hydrocarbons consists of evergreen pollen, cladoceran exo-skeletons, diatoms, and testate protozoans, together with other microorganisms.

The clayey sediments of Silver Bay, Lake Superior described above contain the lipoid extracts and hydrocarbon residues shown in table 30.

TABLE 30   *Chromatographic analyses of lipoid extracts from Silver Bay cores, Lake Superior (Swain & Prokopovich, 1957)*

Lipoid substances	Station 13 (p.p.m.)	Station 22 (p.p.m.)
Total lipoids	290	360
n-Heptane eluant	50	60
Benzene eluant	70	45
Pyridine + methanol eluant	45	45
Polar compounds	125	210

A difference from the Burntside lipoids is that fluorescence was mainly lacking from the Silver Bay aromatic fractions. The fluorescent substances are probably anthracene or a similar aromatic hydrocarbon.

The bottom sediments of Trout Lake, northern Wisconsin (fig. 62) consists in 32 m of water of olive green gyttja (copropel), and the lake itself is oligotrophic and has 6·5–7 ml fixed CO_2/l (Judson & Murray, 1956). The hydrocarbons were reflux-extracted for 8 hours with a mixture of benzene, methanol and acetone, followed by extraction with carbon di-sulfide, methanol and acetone. After drying and removal of the benzene-insoluble portion the extract was chromatographed on alumina using *n*-heptane and benzene as eluants. Little or no free sulfur was found in the Trout Lake samples. The results of the analyses are given in table 37, together with the carbon-14 content of one sample. All samples analyzed contained paraffin-naphthene and aromatic hydrocarbons, but these were not reported separately nor were the types and molecular weights obtained.

The ^{14}C activity shown by the Trout Lake sample indicates that the hydrocarbons are not 'dead' migratory contamination but were generated in the lake sediments. The hydrocarbon content of the Trout Lake sediments was about the same as that of eutrophic Lake Mendota of central Wisconsin whereas the total organic matter of the two lakes was 45 and 11% respectively. Two possible suggestions were offered (Judson & Murray, 1956) for these differences: (1) the organic source matter of eutrophic Lake Mendota is mainly heavy green algal blooms while that of Trout Lake is largely transported forest humic material; (2) the bottom waters of the deep eutrophic lake are oxygen-deficient part of the year thus favoring the preservation of oxidizable hydrocarbons, while oligotrophic Trout Lake retains some oxygen throughout the year which would favor partial destruction of hydrocarbon by oxidation. The present writer, as discussed under Burntside Lake above, favors a different source as the principal explanation of the hydrocarbon differences in lakes of this type. A high silt content of the eutrophic Wisconsin Lake caused by land cultivation is suggested to account for its lower organic content.

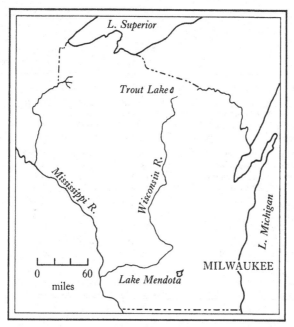

FIGURE 62   Map of Wisconsin to show location of Lake Mendota and Trout Lake (Judson & Murray, 1956, *American Association of Petroleum Geologists Bulletin*, 40, no. 4, pp. 600–53).

4.4 BITUMENS OF HIGHLY CALCAREOUS LAKE SEDIMENTS IN GLACIATED REGIONS

Very little study has been made of the hydrocarbon content of freshwater marl deposits. The bottom sediments of a small lake in Cedar Creek Bog, Anoka and Isanti Counties, Minnesota are diatomaceous copropelic marl (Swain, 1956) (fig. 55). The lipoid extracts of this material are shown in table 31.

These extracts and fractions are probably not representative of typical alkalitrophic lakes because of the high organic content of Cedar Creek Lake. The extracts are greenish brown wax. The saturated and aromatic hydrocarbon fractions are comparable to those of eutrophic lakes in amount. The origin of the lipoid extracts is believed to be similar to that in eutrophic lakes mainly settled phytoplankton and pondweed debris that have been worked over by benthic organisms.

TABLE 31   *Average values of lipoid substances and their fractions from Minnesota lakes (Swain, 1956, American Association of Petroleum Geologists Bulletin, **40**, no. 4, pp. 600–53)*

	Lake Minne-tonka	Cedar Lake	Burntside Lake	Cedar Creek bog organic sediment	Cedar Creek bog-marl
Lipoids, % of sample	0·835	2·15	0·38	6·52	1·56
Saturated hydrocarbons, % of sample	0·038	0·19	0·058	0·47	0·04
Saturated hydrocarbons, % of lipoids	4·77	9·32	15·2	6·5	5·3
Saturated hydrocarbons, % of non-polar lipoids	11·6	15·26	20·0	13·6	8·7
Saturated hydrocarbons, % of hydrocarbons	45·38	48·20	49·9	47·8	45·2
Aromatic hydrocarbons, % of sample	0·051	0·17	0·042	0·50	0·10
Aromatic hydrocarbons, % of lipoids	5·14	8·5	12·5	4·8	6·9
Aromatic hydrocarbons, % of non-polar lipoids	14·02	15·83	17·8	8·5	11·8
Aromatic hydrocarbons, % of hydrocarbons	54·61	51·79	50·1	52·2	56·6
Asphaltenes, % of sample	0·205	0·64	0·16	1·60	0·43
Asphaltenes, % of lipoids	25·47	30·13	42·5	25·5	27·1
Asphaltenes, % of non-polar lipoids	73·3	68·26	62·1	44·4	44·9
Combined hydrocarbons, % of sample	0·089	0·36	0·10	0·68	0·19
Combined hydrocarbons, % of lipoids	9·91	17·82	27·7	11·3	12·2
Polar compounds, % of sample	0·536	1·15	0·13	4·29	0·95
Polar compounds, % of lipoids	64·62	52·05	21·1	63·2	60·7

TABLE 32   *pH and Eh values of samples from Lake Minnetonka, Minnesota (Swain, 1956, loc. cit.)*

Station	Position of sample	Depth of water	pH	Eh	Ox. (+) or Red. (−)	Type of sediment
1	Bottom water	30	7·23	+425	+	—
1	11 in	—	7·30	+389	+	Calcareous copropel
2	Bottom water	45	7·60	+413	+	—
2	26 in	—	7·10	+269	+	Molluscan sand
3	Bottom water	35	7·40	+467	+	—
3	23 in	—	6·73	+215	−	Diatomaceous copropel
4	Bottom water	22	7·45	+485	+	—
4	19 in	—	7·03	+215	−	Copropelic, calcareous sand
13*b*	Bottom water	4	7·23	+431	+	—
13*b*	15 in	—	7·02	+173	−	Peat
13*b*	24 in	—	7·61	+179	−	Peat
13*b*	36 in	—	6·91	+351	+	Peat
13*b*	48 in	—	6·60	+359	+	Copropel
13*b*	60 in	—	6·70	+305	+	Copropel

TABLE 33   *Partial chemical analyses of interstitial waters from cores,
Lake Minnetonka (Swain, 1956, loc. cit.)*

Sta. no. and position in core (in)	Total alkalinity (p.p.m.)	SO$_4^-$ (p.p.m.)	Total P (p.p.m.)	Total Kjeldahl N p.p.m.	Generalized type of sediment
1 Top 6	n.d.	n.d.	0·54	0·63	Copropelic sand
2 Top 6	319·5	n.d.	0·12	4·30	Sandy copropel
2 Middle 6	125·0	n.d.	0·24	n.d.	Sandy copropel
2 Bottom 6	222·1	100·0	1·42	n.d.	Sand
3 Top 6	249·3	29·1	0·42	n.d.	Copropel
3 Middle 6	127·8	n.d.	0·59	9·4	Copropel
3 Bottom 6	257·7	68·6	2·04	n.d.	Copropel
4 Top 6	264·5	15·1	0·23	n.d.	Calcareous sandy copropel
4 Bottom 6	255·0	n.d.	0·29	n.d.	Calcareous sandy copropel

TABLE 34   *Measurements of some properties of Cedar Lake core
samples, Minnesota (Swain, 1956, loc. cit.)*

Station	Coarse fraction > 0·074 mm (%)	Moisture < 105 °C	pH	Eh, mv
1 Top	42·5	85·5	7·3	353
1 Middle	76·0	92·0	7·4	203
1 Bottom	56·0	91·0	7·1	233
2 Top	47·7	92·3	—	—
2 Middle	83·5	92·5	—	—
2 Bottom	32·0	92·2	—	—
3 Top	27·7	95·5	—	—
3 Bottom	30·0	92·0	—	—

**4.5   ASPHALTIC AND POLAR BITUMENS OF EUTROPHIC LAKE
SEDIMENTS IN GLACIATED REGIONS**

The lipoid components of eutrophic lake sediments are characterized by
generally greater proportions of asphaltic and other polar residues, than in
oligotrophic lakes.

The sediment properties and lipoid contents of other lakes in Minnesota
are given in tables 32–35 (Swain, 1956). The aromatic and saturated
hydrocarbon fractions are about equal in amount. The hydrocarbons of the
deep copropelic and sapropelic diatomaceous hypolimnetic sediments of
Cedar Lake, Wright County, Minnesota form a greater portion of the

TABLE 35   *Chromatographic analyses of lipoid substances,*
Lake Kandiyohi, Minnesota (Meader, 1956)

Fraction	Top 6 in Silty Copropel	Bottom 6 in Copropel	Description
	Core no. 3; 31 in long; 14·5 ft deep		
Total lipoids			
Weight extract (gm)	0·1951	0·17	*Top:* black tarry wax. *Bottom:*
total sample (%)	0·81	0·53	black to dark brown tarry wax. Creosote odor
Heptane fraction			
Weight of fraction	0·0013	0·0074	*Top:* colorless to slightly yellow
Sample (%)	0·0054	0·023	clear wax. *Bottom:* clear solid wax
Lipoids (%)	0·67	4·35	containing cloudy granular
Non-polar lipoids (%)	2·02	14·42	particles. Weak white
Hydrocarbons (%)	17·33	36·27	fluorescence
Benzene fraction			
Weight of fraction	0·0062	0·013	*Top:* milky, discoid to subspherical
Sample (%)	0·0259	0·041	S globules in clear wax. Weak
Lipoids (%)	3·18	7·65	yellowish fluorescence. *Bottom:*
Non-polar lipoids (%)	9·64	25·34	greenish brown to black wax
Hydrocarbons (%)	82·67	63·73	
Pyridine–methanol fraction			
Weight of fraction	0·0568	0·0309	*Top:* black tarry wax.
Sample (%)	0·237	0·0968	*Bottom:* black tarry wax
Lipoids (%)	29·11	18·18	
Non-polar lipoids (%)	88·34	60·23	
	Core no. 9; 18·5 in long; 15·5 ft deep		
Total lipoids			
Weight of extract (gm)	0·265	0·2468	*Top:* black, tarry wax creosote odor.
Total sample (%)	0·90	1·13	*Bottom:* brownish black tar
Heptane fraction			
Weight of fraction	0·0015	0·005	*Top:* milky S globules in clear oil.
Sample (%)	0·0051	0·0023	*Bottom:* clear wax. Weak
Lipoids (%)	0·57	0·20	light-green fluorescence
Non-polar lipoids (%)	1·68	0·74	
Hydrocarbons (%)	8·67	5·38	
Benzene fraction			
Weight of fraction	0·0158	0·0088	*Top:* yellow-brown wax containing
Sample (%)	0·0537	0·0402	few milky S globules. Dark yellow
Lipoids (%)	5·96	3·57	green fluorescence. *Bottom:*
Non-polar lipoids (%)	17·73	13·01	clear waxy solid mat with yellow
Hydrocarbons (%)	91·33	94·62	waxy segregations
Pyridine–methanol fraction			
Weight of fraction	0·0718	0·0583	*Top:* black to greenish brown tarry
Sample (%)	0·2437	0·266	wax. *Bottom:* black tarry wax
Lipoids (%)	27·09	23·62	
Non-polar lipoids (%)	80·58	86·24	

TABLE 36 *Hydrocarbon contents of Lake Mendota, Wisconsin,*
samples (Judson & Murray, 1956)

Sample no.	Depth in inches below surface of lake bottom	Type of sediment
M-56-1	0–10	Black sludge
M-56-2	10–16	Buff marl
M-56-3	20–26	Buff marl

* The interface between M-56-1 and M-56-2 is knife sharp.

	Hydrocarbon p.p.m.	Content weight (g)	Organic matter (%)
Lake Mendota			
M-56-1	120	0·06	10
M-56-2	350	0·1	10
M-56-3	225	0·04	12

lipoid extracts (table 31) than the copropelic sediments of Lake Minnetonka, Hennepin County, Minnesota. This difference is believed to be
due to the greater extent of bacterial and bottom scavenger activity on the
bottom sediments of Cedar Lake, and perhaps to their greater content of
oil-bearing diatoms.

The lipoids of Big Kandiyohi Lake, Kandiyohi County, Minnesota
(Meader, 1956) differ from the other eutrophic examples in their higher
content of aromatic hydrocarbons (table 35). This is a shallow epilimnetic,
sulfate-hardness, eutrophic lake lying in glacial till in which has been
incorporated a good deal of Cretaceous clay. There is no explanation at
present for the preponderance of aromatic hydrocarbons.

The hydrocarbon contents of Lake Mendota sediments (Judson &
Murray, 1956) are shown in table 36. The lake is a hardwater eutrophic
lake with hardness of 35–40 ml of fixed CO_2/l, and is subject to contamination from motor boats, farmland and municipal street drainage. The
bottom sediments are green copropel to depths of 0·8 m. ^{14}C measurements
of hydrocarbon residues showed ages of $5,400 \pm 2,000$ years, indicating that the hydrocarbons are not 'dead' contamination but are being
generated in the lake. A discussion of the genesis of the hydrocarbons is
given in connexion with Trout Lake under oligotrophic lakes above.

The lipoids of the sediments of three eutrophic lakes in New Jersey
were studied by Smith (1954) as shown in table 26. The saturated and

TABLE 37   *Hydrocarbon contents of Trout Lake, Wisconsin samples (Judson & Murray, 1956)*

Sample no.	Depth in inches below surface of lake bottom	Type of sediment
T-20-1	0–9	Olive green gyttja
T-20-2	9–18	Olive green gyttja
T-20-3	18–27	Olive green gyttja

No recognizable horizons existed although a uniform increase in plasticity from top to bottom was noted.

Comment.—The reported hydrocarbon concentration is essentially free of elemental sulphur.

	Hydrocarbon p.p.m.	Content weight (g)	Organic matter (%)
Trout Lake			
T-20-1	275	0·03	43
T-20-2	285	0·03	46
T-20-3	270	0·05	50

aromatic hydrocarbons in quantity are variable with respect to each other. Lack of information as to the bottom sediment types of the New Jersey lakes prevents discussion as to possible reasons for these variations.

4.6   ASPHALTIC AND HUMIC BITUMENS OF DYSTROPHIC DEPOSITS IN GLACIATED REGIONS

The peat deposits of two dystrophic bogs in Minnesota have been examined for their lipoid contents: Cedar Creek Bog, Anoka and Isanti Counties (Swian & Prokopovich, 1954); and Rossburg Bog, Aitkin County, (Swain, 1967).

The lipoid contents of the peat sediments of Cedar Creek Bog are given in table 49 (Swain & Prokopovich, 1954). The sediments of this eutrophic and succeeding dystrophic sequence were described above. The paraffin-naphthene fractions (fig. 25) varied from 0 to 93% by weight of the hydrocarbons and averaged 49%. The residues were colorless oils with a slight petroleum odor and some have pale-blue fluorescence. The aromatic hydrocarbons averaged 52% of the hydrocarbons. The aromatic fractions consist of odorous pale-yellow heavy oil or wax, many exhibiting yellow-green to gold fluorescence.

The asphaltic portions of the Cedar Creek lipoids greatly exceeded the hydrocarbons in amount, averaging 26% and attaining 44% of the total lipoids. The fractions consist of yellowish brown heavy oil and wax with a tarry odor and with only weak golden fluorescence or none at all. A comparison with the composition of the lipoids from Cedar Creek Bog and those of hydrocarbons from the Gulf of Mexico region (fig. 63) shows that the

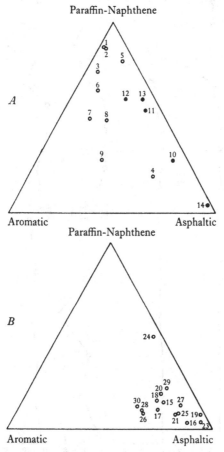

FIGURE 63   Comparison of chemical composition of various crude oils and hydrocarbons from the Gulf of Mexico with that of the hydrocarbons and asphalts from Cedar Creek Bog (Swain & Prokopovich, 1954).

Cedar Creek extracts are similar to those of asphalt from Bermudez Lake, Venezuela.

The lipoid contents of the peats of Rossburg Bog, Aitkin County, Minnesota were studied by Swain (1967b). The bog accumulation comprises basal clays, silty pyritic sand about 0·3 m thick, overlain successively by up to 5 ft of pale gray calcareous clay having a 6 in peat layer (first eutrophic

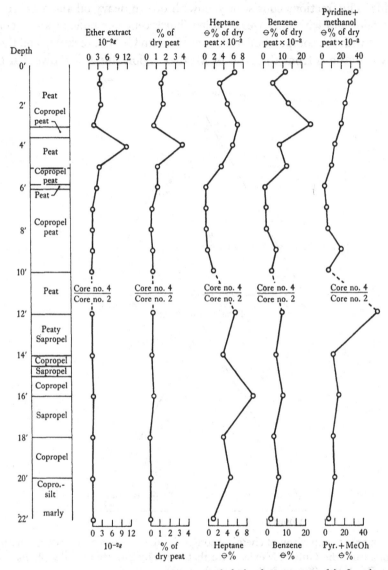

FIGURE 64 Distribution of bitumens and their chromatographic fractions
in Rossburg peat (Swain, 1967*b*).

episode) in upper third, about 0·7 m of copropelic sandy-silty clay (early
second eutrophic episode), up to 2·5 m of diatomaceous copropel and
sapropel, up to 15 ft of brown moss peat interbedded with copropel and
sapropel. Other general properties of the peat are shown in fig. 24. The
total ether extractable material and its column chromatographic fractions

are shown in fig. 64. The heptane-eluted fractions representing mainly
saturated hydrocarbons, mostly colorless, odorless oil, had petroleum odors
in the copropelic peat 3·5–5·5 m. Infrared spectra of the fractions showed
no carbonyl absorption. High values of a saturated hydrocarbons in the
middle part of the moss peat occur in a zone of highly resinous peat-

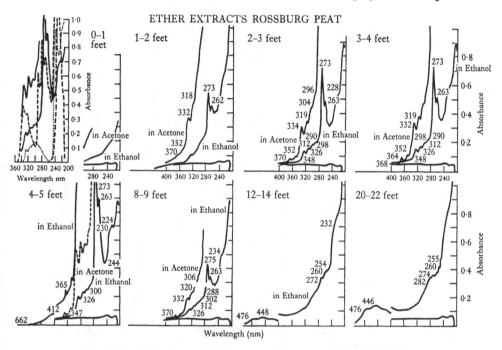

FIGURE 65   Ultraviolet absorption spectra of benzene eluates of column chromatographic fractions of ether extracts of Rossburg peat; spectra run in *n*-heptane
(Swain, 1967*b*). Insert shows spectra of 1-naphthol (dashed line) and 2-naphthol
(dotted line) compared to the peat substance (solid line) and suggests that both
compounds are present in the peat substance.

matrix. Saturated hydrocarbons are higher in amount in the copropelic
peat than in the overlying moss peat probably representing original differences. The benzene-eluted aromatic hydrocarbons are distributed in
more or less parallel, but somewhat higher quantities than the saturated
hydrocarbons. These show aromatic structure in infrared spectra, and are
colored oils and waxes having odors that include crude oil, mercaptans and
allyl disulfide (garlic). Ultraviolet absorption spectra of the benzene eluates,
dried and taken up in ethanol suggest the presence of a naphthol-like
compound, possibly in large part 2-naphthol together with smaller amounts
of 1-naphthol (fig. 65). The naphthols while present only in p.p.m. quan-

tities are at a maximum in the upper third of the moss peat and are absent or nearly so in the underlying copropel. Three possibilities were suggested to account for the presence and distribution of the naphthols: (1) formation

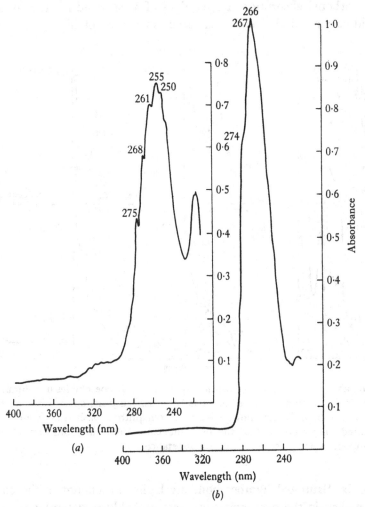

(a)

(b)

FIGURE 66(a) Ultraviolet absorption spectrum of Rossburg peat 0–1 ft, station 9, distilled at 125 °C, principal organic product separated from distillate in gas chromatographic column and scanned in *n*-heptane; operating temperature of column, 110 °C; carrier gas, nitrogen, 100 ml/min. (b) Same at 14–17 ft (Swain, 1967 b).

from a protein such as naphthylamine by the Bucherer reaction; (2) through activity of plant-growth accelerators or auxins such as naphthylacetic acid; or (3) by the decomposition of β-carotene. Refer to Chapter 1 for these reactions.

In the first case and in the third case a rather general distribution in the peat might have been expected rather than concentration in the middle of the moss peat. The second possibility might have resulted from microbial production of auxins that were not used because of the scarcity of rooted plants in the bog.

Samples of the Rossburg moss peat and the copropelic peat were heated at 125 °C and the resulting distillation products were separated isothermally on a gas chromatograph at 110 °C, using a thermal conductivity detector. In each case the effluent of the principal product was collected by bubbling it into *n*-heptane and it was scanned in the ultraviolet range (fig. 66). The principal distillation product of the copropel has an absorption spectrum like toluene; this may have been produced by thermal decomposition of β-carotene which occurs in the copropel as discussed above. The principal distillation product of the moss peat has a u.v. spectrum like that of hydrated phenol; the latter may have formed artificially from sphagnol or other such peat organic acid.

4.7   BITUMENS OF SALINE LAKE SEDIMENTS

The lipoid contents of two samples from Great Salt Lake, Utah were reported by Smith (1954); one was from salt-water sediments and the other from freshwater sediments (table 26). In both instances the paraffin-naphthene hydrocarbon fractions and the aromatic fractions were nearly equal to each other and were comparable to the amounts obtained from eutrophic lakes. The salt-water sediment lipoids were noticeably higher in amount near those of the underlying freshwater sediments.

4.8   DISCUSSION AND SUMMARY OF LACUSTRINE BITUMENS

In contrast to general levels of organic productivity and concentration in the sediments of some other organic residues, the lipoid residues of non-glacial lakes are not greatly different from those of glacial lakes.

The total lipoids, hydrocarbons and asphaltic residues of apatotrophic Pyramid Lake, Nevada are more or less similar to those of eutrophic lakes of temperate glaciated regions. The origin of the hydrocarbons of Pyramid Lake sediments is plausibly accumulated fatty acids, oils of diatoms and other algae and perhaps of invertebrates, such as sponges that live in the lake and of bacteria. Carotenes (Breger, 1963*a*; Swain, 1967*b*) which were noted in chromatographic separations of the sediments may have contributed to the aromatic hydrocarbons, and chlorophyll degradation products may have provided some of the hydrocarbons (Eglinton *et al.* 1966*c*).

Bitumens of non-marine sediments

The hydrocarbons and related substances in Lake Nicaragua sediments are small in total amount considering the tropical setting of the lake. Aromatic and saturated hydrocarbons are about equal but both are low compared to asphaltic and polar residues. Diatoms and boghead type algae (*Botryococcus* = *Elaeophyton*) are apparently the source of much of the lipoid material, but the environment is apparently not conducive to good preservation of the lipoids, despite the low-positive to negative Eh values in the sediments. The wide variety of u.v.-visible-absorbing lipoid compounds in trace amounts differs from that of most of the other lakes studied, especially the cool-temperate types. In general it appears that the lower latitudes favor a greater variety of aromatic and heteroaromatic residues in freshwater sediments.

The lipoid extracts of the cool-temperate oligotrophic lake sediments that have been studied are apparently characterized by low total extracts but relatively high hydrocarbon fractions, and correspondingly low asphaltic and polar fractions. In some oligotrophic sediments, evergeen pollen, arthropod exoskeletons, diatoms and thecamoebae are important sources of the lipoids. In other oligotrophic lakes green algae blooms are believed to represent an important lipoid source. Although total lipoid content is less, the distribution of hydrocarbons and asphaltic residue in alkalitrophic lake marls is similar to that of eutrophic lakes.

The lipoids of eutrophic low-sulfate lake sediments are high in asphaltic and polar fractions the source of which is in the green plants that grow in the lake, both benthonic and planktonic. Saturated and aromatic hydrocarbons are similar in amount in this type of eutrophic lake. In a high-sulfate eutrophic lake the aromatic fractions were relatively high.

In dystrophic lake and bog sediments of glaciated regions the saturated and aromatic hydrocarbons are more or less equal in amount; the asphaltic and polar compounds are high as in eutrophic lakes.

Aside from an increase in total lipoid content over those of freshwater sediments, saline lake sediments yield similar distribution of lipoid fractions to those of eutrophic lakes.

4.9 BITUMENS OF FLUVIAL DEPOSITS

Relatively little study has been made of the bitumens of fluvial sediments. Smith (1954) showed that sediments of Rancocas Creek, New Jersey contained small but nearly equal amounts of paraffin-naphthene and aromatic hydrocarbons (table 26), both of which comprised only about 2% of the benzene–methanol extractable material. Mississippi Delta sediments (table 26) yielded a greater variety in amount of hydrocarbons, probably owing to the variable source materials, but in general were among the richer samples studied by Smith.

The bituminous materials of sediments and associated organisms of the Delta–Mendota Irrigation Canal, California (Swain and Prokopovich, 1969) were extracted and separated chromatographically. There is a

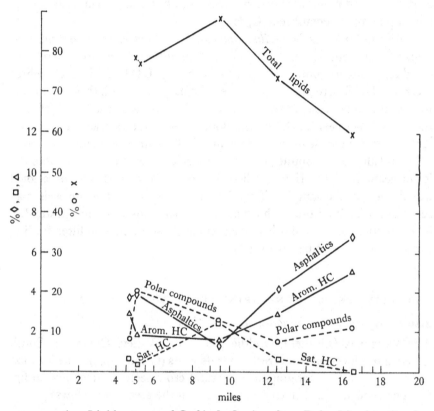

FIGURE 67   Lipid extracts of *Corbicula fluminea* from Delta–Mendota Canal, California (Swain, 1967a).

general increase in total lipids, aromatic hydrocarbons and polar lipids in *Corbicula fluminea* in the upper part of the Canal (4·7–16·43 miles) (fig. 67). This possibly reflects storage of food in *Corbicula* because it is becoming scarcer down the canal. Farther down the canal the amount of aromatic hydrocarbons and asphaltic fractions levels off or decreases perhaps because of changing food supply. Gas chromatographic analyses of some of the extracts gave the following results: (1) *Corbicula fluminea* (asiatic clam) showed presence of small amounts of C_{16}–C_{28} alkanes; (2) extracts of the sponge, *Spongilla fragilis* contained a wide variety of saturated hydrocarbons ranging from C_{16} to C_{33}. The bryozoans *Plumatella* sp. yielded only small amounts of saturated hydrocarbons possibly including C_{16}–C_{18} types. The silty, sandy sediments associated with the large number of *Corbicula* yielded relatively little hydrocarbon material; the saturated fractions included C_{21}–C_{28} compounds. Neither the organisms or the sediments in the canal contained much aromatic hydrocarbon, but a bryozoan extract contained possible phenanthrene $C_{14}H_{10}$.

Freshwater mussels *Lampsilis siliquoidea* and *Anodonta grandis* from Red Cedar River, south-central Michigan, at six different localities were analyzed for pesticide content (Bedford *et al.* 1968). DDT and its metabolites TDE and DDE were present in all the animals analyzed. All these chemicals were found to increase downstream and to increase with the length of time the mussel had been in the water. Both types of values reached a plateau level along that course of the river sampled. Two other pesticides methoxychlor and aldrin were found; the former was in mussels introduced into the lower sections of the river studied, while the latter had only occasional distribution in the samples. Thus it was demonstrated that mussels may serve as useful monitors of chlorinated hydrocarbons and other pesticides. The possible build-up of other biogeochemical substances in filter-feeding invertebrates should be investigated.

4.10 BITUMENS OF NON-MARINE SEDIMENTARY ROCKS

4.10.1 *Hydrocarbons in non-marine Tertiary rocks*

The Velasquez Oil Field, Middle Magdalena Valley, Columbia, South America produces oil from freshwater deposits (Olson, 1954). The location of the field and a stratigraphic–structural cross-section are shown in fig. 68. The sequence of Tertiary Formations in the area is as follows:

Diamante Formation, Quaternary to Pliocene age: lacustrine massive sheet sandstones and claystones, 1,500 ft thick containing silicified wood.

Zorro Formation, Miocene age: lacustrine thin sandstones and varicolored claystones 3,000–10,000 ft thick; containing carbonaceous material lignitic streaks and leaf impressions; contains fresh water.

Tuné Formation, Oligocene age: fluvial lenticular sandstones, siltstones and claystones, 1,600–2,000 ft; contains brackish water; produces some oil in Velasquez Field.

Avechucos Formation, Eocene age: fluvial sandstone, with minor red and green claystone and siltstone, 2,000–10,000 ft +; highly saline water content; contains most of oil producing horizons of field.

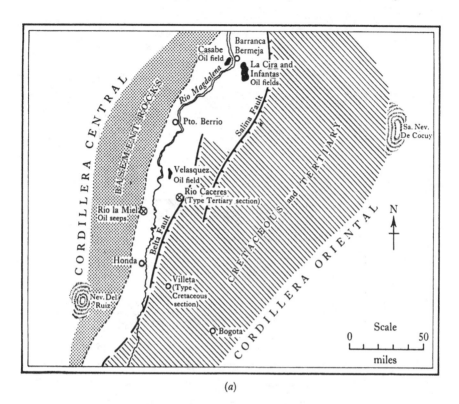

(a)

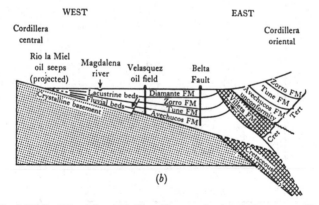

(b)

FIGURE 68   Middle Magdalena Valley, Venezuela. (*a*) Index map. (*b*) Diagrammatic east–west cross-section (not to scale) in Middle Magdalena Valley (Olson, 1954, *American Association of Petroleum Geologists Bulletin*, **38**, no. 8, pp. 1645–52, figs. 1 and 2).

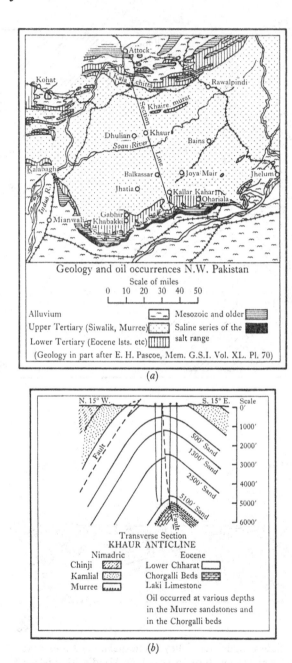

FIGURE 69 (a) Geologic map of north-western Pakistan, showing position of oil fields and oil seepages. (b) Cross-section of Khaur oil field, showing oil occurrences in Nimadric (Murree) Beds.

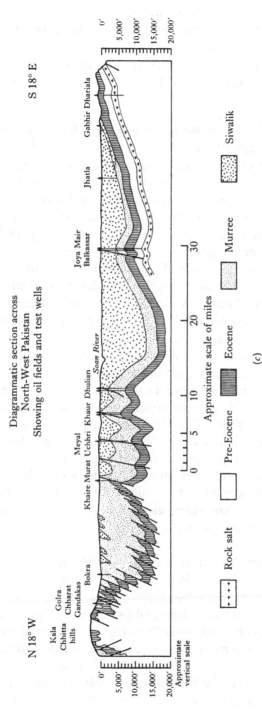

FIGURE 69 (c) Diagrammatic transverse section across Soan basin, north-western Pakistan (Pinfold, 1954, *American Association of Petroleum Geologists Bulletin*, **38**, no. 8, pp. 1635–60, figs. 2, 3 and 4).

Unconformity and thrust fault contact.

Villeta Formation, Cretaceous: dark marine shale, minor limestone and sandstone, 30,000 + ft thick.

The oil field water in this area has the composition of marine brines (p.p.m.): Na+K, 10,916; Ca, 6,416; Mg 302; SO_4, 32; Cl, 29,006; CO_3, 36; HCO_3, 18.

The author (Olson) believes that the oil and formation water of the Velasquez Field migrated into the freshwater reservoirs from the Villeta Formation because of the absence of apparent source material in the Tertiary freshwater beds, the presence of abundant organic matter in the Villeta Formation, and the marine character of the formation fluids in the Velasquez Field. The present writer feels that the evidence for the marine origin and subsequent migration into non-marine beds is probably correct but the oil of the non-marine Green River Formation, Utah, discussed below also occurs in a similar geologic situation to that in the Velasquez Field and the Green River oil is generally considered non-marine in origin. Further study should be made of the composition of, and variations in, the Velasquez oil and the residual oil in the surrounding formations before a final decision can be made as to its origin.

Petroleum has been produced in relatively small amounts from Neogene non-marine deposits in north-western Pakistan (Pinfold, 1954). The stratigraphic and structural conditions in the area are shown in fig. 69. The Neogene rocks range in age from Miocene to Pleistocene and are referred to as the Murree–Siwalik or Nimadric Formation. The underlying Eocene rocks are marine limestones and shales, and the Nimadric beds are sandstones, shales and conglomerates with vertebrate fossils and exceeding 30,000 ft in thickness. Oil has been produced from these beds since 1915 (fig. 69). The oil is believed to have migrated into the Nimadric from the Eocene foraminiferal limestones, along bedding planes. High reservoir pressures up to 7,000 p.s.i. at depth of 5,400 ft probably aided considerably in the migration.

Hydrocarbons are known in several localities in lacustrine Tertiary deposits in Utah, Colorado, Wyoming and Nevada, in some places in commercial quantities (Felts, 1954). Natural gas has been produced from vesicular basalts with which arkosic coal-bearing continental sediments are associated in the Columbia Plateau, Washington. The distribution of these occurrences and the Tertiary stratigraphic sequences are shown in figs. 70 and 71.

Gas has been found in the Yakima Basalt of Miocene age of south central Washington. The basalt overlies thick Eocene continental arkosic sediments of the Manastash and Roslyn Formations that contain coal-beds and the underlying Eocene Series that has coaly streaks and in which shows of oil and gas were found in a well south-west of Wenatchee, Washington (Glover, 1947, 1953). The source of the Yakima Basalt gas may be the

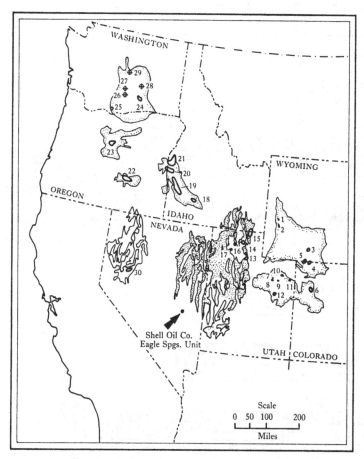

FIGURE 70 Production and principal shows of Tertiary oil and gas, north-western and west-central United States (Felts, 1954. *American Association of Petroleum Geologists Bulletin*, **38**, no. 8, pp. 1661 and 1670, figs. 2 and 3).

coal-bearing Eocene sequence, or in the peaty and diatomaceous deposits that are intercalated with the basalt. There are examples of geodes filled with tar and vesicles filled with viscous tar in flows above some of the peat deposits.

The Pliocene Idaho Formation of west-central Idaho is a non-marine sequence of more than 18,000 ft of partly lacustrine beds including coal,

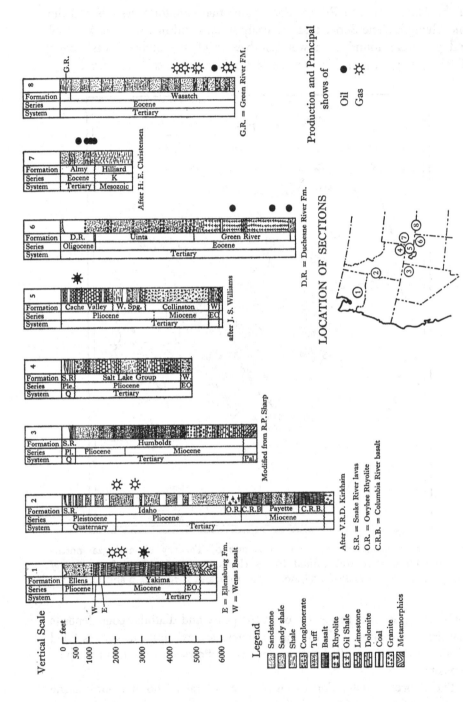

FIGURE 71 Typical stratigraphic sections in Tertiary non-marine petroliferous deposits of western United States (Felts, 1954, *loc cit.*).

154

lignite and basalt flows. Gas flows are known from the Idaho Formation at several localities but none in commercial quantities.

The Humboldt Formation of north-eastern Nevada has surface seeps of oil at several localities but no commercial oil production. It consists of

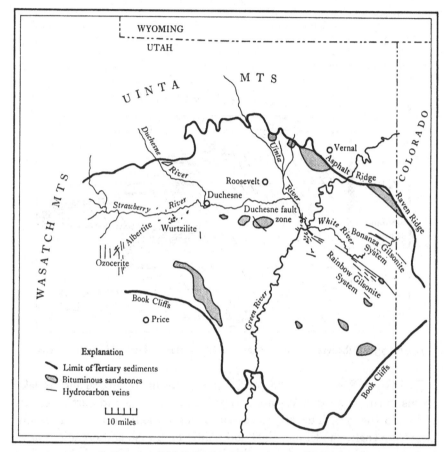

FIGURE 72   Index map, Uinta basin, Utah, showing hydrocarbon deposits
(Hunt *et al.* 1954).

lacustrine and fluvial shales, limestones and sandstones more than 5,800 ft thick. The Upper Humboldt of Plio-Pleistocene age rests with angular unconformity on the Lower Humboldt of Oligocene? and Miocene age around the margins of some of the mountain uplifts. Near the base of the Humboldt Formation is the lacustrine Elko Oil Shale, near Elko, Nevada. It produces a waxy, highly paraffinic crude oil on distillation in retorts, and was once mined for a short time. The oil is similar to that distilled from the Green River Oil Shales of Utah and Colorado.

Bitumens of non-marine sediments

The Salt Lake Group of Pliocene age, south-eastern Idaho and north-central Utah contains a good deal of lacustrine marly limestone in which tar seeps and sub-commerical flows of gas are known. The Salt Lake Group is about 4,000 ft thick and includes both fluvial and lacustrine beds.

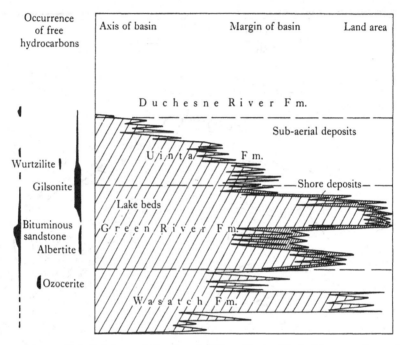

FIGURE 73   Subdivisions of Tertiary in Uinta Basin, Utah (Hunt *et al.* 1954).

The Green River Formation of Eocene age occurs in several ancient lake basins in south-western Wyoming, north-eastern Utah and north-western Colorado (fig. 72). The sequence comprises 5,000 ft or more of mostly lacustrine shale, sandstone, algal and oölitic limestone and organic, shaly-textured, thinly laminated dolomitic limestones known as 'oil shales' and 'marlstones' (fig. 73). Liquid petroleum has been discovered in commercial amounts in the lower Green River oölitic limestone and sandstones. The oil is highly paraffinic and in some wells yields so much paraffin wax that heating of the production casing and storage tanks is necessary to keep the oil from setting up. The upper part of the Green River Formation contains oil shales from which shale oil may be obtained by distillation in retorts or by hydrogenation.

A comprehensive study of the distribution of hydrocarbons in the Green River Formation was made by Hunt *et al.* (1954). In addition to oil shale and oil producing strata, the Green River Formation the underlying

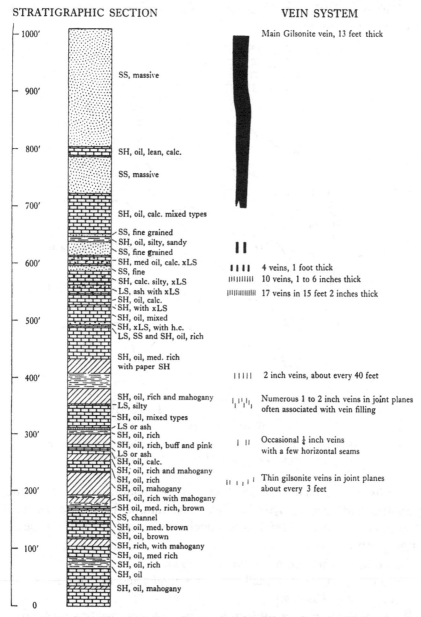

STRATIGRAPHIC SECTION

VEIN SYSTEM

Main Gilsonite vein, 13 feet thick

1000'

900'

800'   SS, massive

        SH, oil, lean, calc.

        SS, massive

700'   SH, oil, calc. mixed types

        SS, fine grained
        SH, oil, silty, sandy
        SS, fine grained
600'   SH, med oil, calc. xLS   I I I I   4 veins, 1 foot thick
        SS, fine   IIIIIIIIII   10 veins, 1 to 6 inches thick
        SH, calc. silty, xLS   IIIIIIIIIIIII 17 veins in 15 feet 2 inches thick
        LS, ash with xLS
        SH, oil, calc.
        SH, with xLS
500'   SH, oil, mixed
        SH, xLS, with h.c.
        LS, SS and SH, oil, rich

        SH, oil, med. rich
        with paper SH
400'   I I I I I   2 inch veins, about every 40 feet

        SH, oil, rich and mahogany   I, I¹, I,   Numerous 1 to 2 inch veins in joint planes
        LS, silty   often associated with vein filling
        SH, oil, mixed types
        LS or ash
300'   SH, oil, rich
        SH, oil, rich, buff and pink   I II   Occasional ¼ inch veins
        LS or ash   with a few horizontal seams
        SH, oil, calc.
        SH, oil, rich and mahogany
        SH, oil, rich   II I I I I Thin gilsonite veins in joint planes
200'   SH, oil, mahogany   about every 3 feet
        SH, oil, rich with mahogany
        SH oil, med. rich, brown
        SS, channel
        SH, oil, med. brown
        SH, oil, brown
100'   SH, rich, with mahogany
        SH, oil, med rich
        SH, oil, rich
        SH, oil
        SH, oil, mahogany
0

FIGURE 74   Generalized section, Bonanza vein, Uinta County, Utah, showing termination of vein in Green River oil shales, Sec. 32, T. 9 S., R. 25 E (Hunt *et al.* 1954).

Wasatch Red Beds and overlying Uinta Formation in the Uinta Basin, north-eastern Utah, are notable for the contents of solid or highly viscous hydrocarbons (fig. 74).

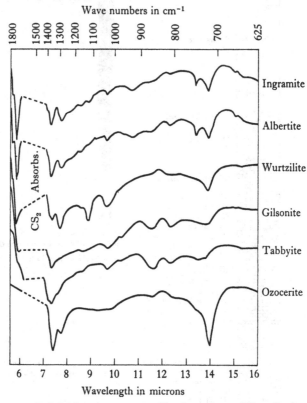

FIGURE 75 Infrared spectrograms of hydrocarbons, Uinta Basin (in CS₂ solution) Hunt *et al.* 1954.

The hydrocarbons of this type found in the Uinta Basin are: ozocerite, gilsonite, wurtzilite, tabbyite, argulite, ingramite and albertite. Infrared absorption spectra are useful in distinguishing several of these substances (figs. 75, 76):

Ozocerite: absorption maximum at 13·9 μm.

Gilsonite and tabbyite: small abs λ at 8·6, 9·65 and 10·3 μm, strong abs λ at 11·5 and 12·3 μm.

Tabbyite: less general absorption than gilsonite in the range from 8 to 10 μm.

Wurtzilite: strong abs λ at 8·9, 9·65 and 13·9 μm, the 8·9 and 9·65 μm λ being stronger than the other hydrocarbons of this area.

Albertite and ingramite: abs λ at 5·9, 10·8 and 13·4 μm.

Chromatographic separation of ozocerite and gilsonite fractions shows that the ozocerite is much higher in content of paraffinic hydrocarbons than the gilsonite.

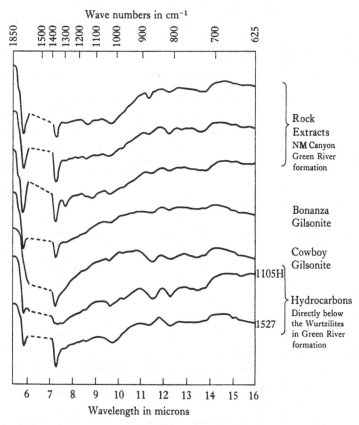

FIGURE 76   Infrared spectrograms of hydrocarbons and rock extracts from Green River Formation (Hunt *et al.* 1954).

The ozocerite of this area is believed to have originated in lacustrine shales of the basal Green River or upper Wasatch Formations. This conclusion is based on similarity of properties of ozocerite and extracts of basal Green River–Upper Wasatch sediments. The grahamite and albertite veins occur 300–600 ft above the base of the Green River Formation in the Uinta Basin. Infrared spectra of extracts of the lower Green River shales, starting 50 ft above the base of the formation, indicate that the albertite and ingramite probably originated therein.

The gilsonite of the Uinta Basin, based on infrared spectra of the associated rock extracts evidently originated in the middle and upper Green

River shales beginning about 1,000 ft above the base of the formation and continuing upward for about another 1,000 ft. Wurtzilite comes from that part of the Green River and overlying Uinta Formations that succeed the gilsonite-bearing beds. These conclusions are based on infrared studies of the shales with which the wurtzilite beds are associated. Thus there is a

Age	Controlling environment	Results of environment	
Group	Increasing relative salinity and H₂S conc.	Dominant components of sediments	
		Minerals	Hydrocarbons
Uinta		Siliceous Dolomites Pyrite	WURTZILITE Naphthenes, sulphur and nitrogen compounds
Green River		Calcareous Dolomites Pyrite	GILSONITE Aromatics and nitrogen compounds
		Dolomite	ALBERTITE Condensed aromatic rings
Wasatch		Calcite	OZOCERITE Paraffin chains

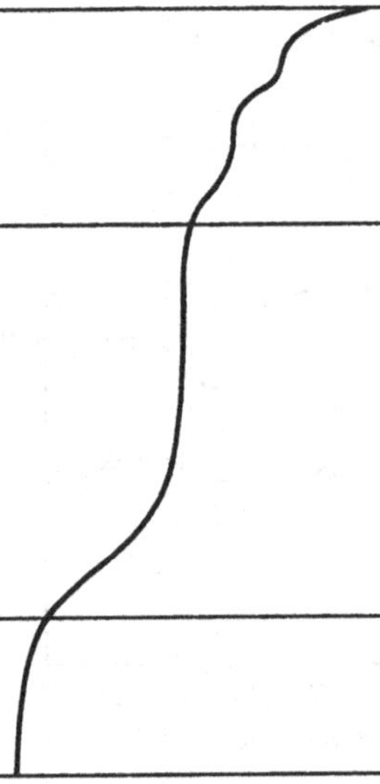

FIGURE 77 Change in composition of hydrocarbons and minerals with environment in Green River and associated formations, Uinta basin, Utah (Hunt *et al.* 1954).

definite stratigraphic relationship of these hydrocarbons in the Green River Formation (fig. 77).

There is also an environmental control on the distribution of hydrocarbons in the stratigraphic sequence. With the progress of time the Green River lacustrine environment became progressively more saline (alkaline) and the H₂S contents of the waters increased. During early Green River and Late Wasatch time Uinta Lake supported a large population of ostracodes (Swain, 1964) (not mentioned by Hunt *et al.* 1954) as well as mollusks (*Goniobasis, Elliptio*, etc.) that form coquina layers at certain horizons.

These forms of life may have contributed importantly to the ozocerite as well as to the high paraffin crudes of the lower Green River. As the lake became saline the ostracodes and mollusks were eliminated, planktonic life became predominant and the hydrocarbons became more aromatic and less paraffinic. A relative increase in the organic pigment content of the accumulating organic matter in the saline lake stages may have been responsible for the change in hydrocarbons (Swain, 1956).

Gas chromatographic and mass-spectrometric studies of the extractable hydrocarbons of Green River oil shale have shown (Eglinton *et al.* 1966a) the presence of phytane and pristane (fig. 26). These isoprenoid hydrocarbons originate very probably from phytyl alcohol attached to the chlorin nucleus of chlorophyll, and are a further clue to the pigment source of some of the Green River hydrocarbons.

Oil and gas shows and some indigenous production occur in southwestern Wyoming in lacustrine and fluvial beds of the Wasatch and Almy Formations (fig. 70) (Felts, 1954).

In recent years oil was discovered in potentially large quantities in Tertiary freshwater deposits in Cook Inlet, Alaska (Sweet, 1964). The discovery well in the Middle Ground Shoal pool had what is believed to be more than 550 ft of oil sand in the 'Hemlock Sandstone' of the Upper Eocene Kenai Formation. This formation (Dall & Harris, 1892; Barnes & Cobb, 1959) consists of lacustrine and paludal sands, conglomerates, clays and coal beds 3,000 ft thick or more.

Other occurrences of petroleum in Tertiary non-marine sediments are listed in table 38. In some of these the hydrocarbons are believed to be indigenous, in others the petroleum may have migrated from a marine facies into the present reservoir beds.

The organic contents of the red beds of the Cheleken Formation, U.S.S.R., were studied from drill cores ranging from 300 to 2778 m in depth (Khanov, 1959). The bitumens of these rocks ranged from 0 to 0·06%, organic carbon 0 to 0·129%, and humic acids 0 to 0·08%.

In an examination of the bitumens of Tertiary rocks of the central and northern Caucasus region Gimpelwich (1959) noted the presence of the aromatic hydrocarbon perylene together with a high humus content in the marine(?) Upper Maikop, Karagan and Chokrak Formations. The perylene and humic nature of the rocks can be traced for long distances in the region. The humic acid contents of the Aleksandrovsk and Sertunaisk Formations (Neogene) of Sakhalin Island, U.S.S.R. were found to decrease with increasing depth in the formations (Bazanova, 1960).

The humic substances of marine(?) Tertiary Formations of the Caspian–Kuban Territory, U.S.S.R. ranged from 0·01 to 0·05% and showed a

TABLE 38 *Additional occurrences of hydrocarbons in non-marine sedimentary deposits*

Locality	Description of rocks	Age	Type of hydrocarbons	Origin and reference
Capitan, South Mountain, Shiells Canyon, Bardsdale Pools, Ventura region, California	Continental red bed facies of Sespe Formation, 7,500 ft thick arkosic sandstones, vari-colored shales and mudstones grading westward into a marine facies	Eocene to Lower Miocene	Oil 13–38° API, partly high paraffin	Marine Eocene shales mainly (Bailey, 1947)
Quirequire Field, eastern Venezuela	Quirequire Formation, 1,600 ft, clays, unsorted sands, lignite, asphalt seeps	Plio-Pleistocene	Oil 10–16° API	Indigenous to marine and brackish facies of Quirequire Formation in part, as well as from pre-Pliocene truncated marine beds (Borger, 1952)
Meotian, Roumania Pools	Meotian Formation, sandstones and limestones of freshwater origin	Pliocene	Oil	Origin? Baturin (1937)
Baku Area Apsheron Peninsula U.S.S.R.	'Productive Series', lenticular sands, clays and silts with freshwater fossils, 9,500 ft thick	Miocene–Pliocene	Oil	Origin? Kugler (1939)
Turkeman Republic, U.S.S.R.	'Red-bed Series' similar to Productive Series at Baku	Miocene–Pliocene	Oil	Origin? Kugler (1939)
South Kahetia Georgia, U.S.S.R.	Clays, sands and conglomerates	Pliocene–Miocene	Oil showings	Origin? Goguetidze (1937)
Near Yumen, Kansu Province and Wuser, Sinkiang Province	Continental sandstones	Tertiary	Oil	Origin? Huang (1947)
Assam	Tipam sands and clays	Miocene	Oil seeps	Derived from underlying marine Eocene-Oligocene, Lepper (1938)
Szechuan, China	Tzuliuching Limestone of lacustrine origin	Lower Cretaceous	Oil seeps	Origin? Huang (1947)
North Shensi, China, Yenang Pool	Shensi Series, fluvial and lacustrine beds 6,600 ft thick	Lower Jurassic–Triassic	Oil seeps and tar springs	Origin? Wade (1929)
West side of Madagascar	Karroo Formation, continental beds	Carboniferous–Lower Jurassic	Oil seeps and tar springs	Origin? Wade (1929)
West side of Madagascar	Ankavandra Formation, sands, conglomerates, shales, 2,400 ft thick	Lower Jurassic–Triassic	Oil seep and tar springs	Origin? Wade (1929)

TABLE 38 (*cont.*)

Locality	Description of rocks	Age	Type of hydrocarbons	Origin and reference
San Pedro Field, Argentina	Tupambi Formation glacial sands, silts, tillites	Basal Permo–Carboniferous	Oil	Migrated from underlying Devonian marine shales Reed (1946)
Schuler Sand Fields of Southern Arkansas	Schuler Formation, sands, and varicolored clays, up to 3,000 ft	Upper Jurassic	Oil and gas distillate	Alluvial and deltaic; migration from marine offshore equivalents is possible (Swain, 1944)
Bahia, Brazil	Bahia Formation, lacustrine limestone and clay more than 6,500 ft	Upper Jurassic–Lower-Cretaceous	Oil	Lake beds (Hedberg, 1945; de Oliviera & Leonardos, 1943)

relationship to mineral composition, color, carbonate content and geologic age of the country rock (Merktiev & Aliev, 1963). The shales contained the largest amount of bitumens, organic carbon, humic acids and nitrogen. These amounts decrease with increasing amount of sand-size particles. Darker colored rocks of the sequence were generally richer in all the organic constituents than the lighter colored rocks except for nitrogen which was higher in the latter type. The organic components decreased with increase in amount of carbonates and increased with increasing age of the formations.

4.10.2   *Hydrocarbons in non-marine pre-Tertiary rocks*

The cyclically deposited Coal Measures (Pennsylvanian System) of the Lancashire and East Midland coalfields of England have numerous oil showings (Kent, 1954). The geographic and stratigraphic distribution of the important occurrences are shown in figs. 78, 79. Some of the occurrences can be traced to migration from marine source beds in underlying Millstone grit. An example of this is the Eakring Field, Nottinghamshire which produces both from the Millstone grit and the overlying coal Measures. In another case at Harlequin, Nottinghamshire, oil was found in a jointed dolerite that intrudes the Coal Measures. Most of the occurrences are high in paraffin which is reminiscent of the lower Green River Eocene oil in Utah. Commonly the Coal Measures oil is lacking in low-boiling fractions and is a greenish brown, waxy or greasy crude, high in lubricating-oil fractions, up to and including the pure paraffin pentacontane ($C_{50}H_{102}$). The latter may indicate a source in cannel coal.

Bitumens of non-marine sediments

The scarcity of marine horizons in the East Midland coalfield and the unusually high molecular weight character of the oil in the Coal Measures from that area strongly suggests a non-marine origin of these occurrences.

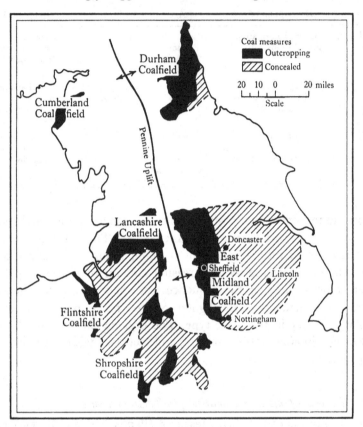

FIGURE 78   Relation of coalfields of northern England (Kent, 1954, *American Association of Petroleum Geologists Bulletin*, **38**, no. 8, pp. 1699–1713, fig. 1).

Some other pre-Tertiary occurrences of hydrocarbons in non-marine rocks are given in table 38. Some of these examples are of instances where the oil has migrated into non-marine deposits from a marine source.

Hydrocarbon materials were extracted from alluvial deposits of the Jurassic Twist Gulch Formation of Utah (Swain, 1963a) and were separated chromatographically (table 39). These extracts showed the presence of saturated and aromatic hydrocarbon fractions and asphaltenes but are noteworthy for their lack of polar substances such as pigments, carbohydrates, etc. The Twist Gulch deposits are gypsiferous (selenite crystals) and represent arid conditions that apparently favored oxidation of the higher molecular weight organic matter.

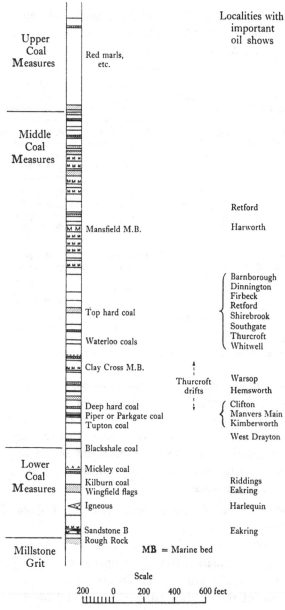

FIGURE 79   Vertical succession in East Midland coalfield (Kent, 1954, *loc. cit.*, fig. 4).

Recent work by Hedberg (1968) has shown a relationship of freshwater and brackish-water sedimentary environments to high-wax content of certain crude oils. In general, high wax crudes are restricted to certain types of stratigraphic environments: (1) shales and sandstones, (2) non-marine

TABLE 39 *Chromatographic analyses of benzene–methanol extracts of Jurassic sediments from Utah (Swain, 1963a, American Association of Petroleum Geologists Bulletin, **47**, pp. 777–803, table 5)*

Formation, well and depth	Petra and total extract (p.p.m.)	Heptane fract. (p.p.m.) and description	Benzene fract. (p.p.m.) and description	Pyridine fract. (p.p.m.) and description	Polar compounds and sulfur
Twist Gulch beds, 65 ft above redbeds, 3 miles W. of Mayfield, Utah	Alluvial marly shale: 22	18; colorless oil, crystalline particles	8; yellowish wax, rod-like crystals; trace fluorescence	28; orange wax, crystal particles, weak fluorescence, petroleum odor	60 S; no polar compounds
Same locality, 950 ft above redbeds	Alluvial marly shale; 104	42; colorless oil, crystalline particles; blue fluorescence, petroleum odor	22; colorless oil; blue fluorescence, petroleum odor	381; orange grease, yellow crystal particles, fluorescence, petroleum odor	64 S; no polar compounds
Same locality, 1,285 ft above redbeds	Alluvial marly shale; 42	18; colorless oil; crystal rods, petroleum odor	16; colorless oil; crystal particles, blue fluorescence	12; orange grease, crystalline particles, trace blue fluorescence	44 S; no polar compounds
Same locality, 1,600 ft above redbeds	Alluvial marly shale; 26	12; colorless oil, clear crystals	6; clear rod-like crystals	18; yellow-orange wax	54 S; no polar compounds

or sub-normal salinity waters, (3) common association with coal or other highly carbonaceous strata, (4) stratigraphic ranges from Devonian to Pliocene inclusive, (5) continental, paralic, or nearshore marine environment, (6) low-sulfur content. Hedberg believes that the high wax crudes reflect the contribution of important amounts of terrigenous organic matter. He considers the following to be typical examples of nearshore to non-marine high wax crudes:

(1) Cretaceous of the Wyoming basin (Lakota, Dakota, Muddy and Wall Creek sandstones);

(2) Jurassic of the Wyoming basin (Sundance and Morrison sandstones);

(3) Lower Pennsylvanian of the Mid-Continent region (Morrowan and Desmoinesian Stages of Kansas, Oklahoma and Texas);

(4) Cretaceous of the Denver basin (Dakota and Muddy 'D' and 'J' sandstones);

(5) Cretaceous–Eocene of north-west Colorado;

(6) Eocene of the Louisiana–Texas Gulf Coast (Wilcox, Cockfield and Queen City sandstones);

(7) Oligocene of the Louisiana–Texas Gulf Coast (Frio sandstones);

(8) Lower Upper Cretaceous of east Texas (Woodbine sandstone);

(9) Lower Upper Cretaceous of Mississippi (lower Tuscaloosa sand-stones);
(10) Lower Tertiary of the Uinta basin (Wasatch and Green River Formations);
(11) Cretaceous of Alberta (Viking, Cardium and Belly River sandstones);
(12) Mississippian of New Brunswick (Albert Formation);
(13) Oligocene–Miocene of the eastern Venezuela basin (Merecure, Oficina, Chaguarramas and La Pica Formations);
(14) Tertiary of the Barco region of Colombia;
(15) Cretaceous of eastern (Oriente) Peru;
(16) Cretaceous (Chubutiano) of Comodoro Rivadavia, Argentina;
(17) Triassic of Mendoza, Argentina;
(18) Upper Cretaceous of southern Chile;
(19) Lower Cretaceous of the Reconcavo basin, Brazil;
(20) Carboniferous 'Coal Measures' of Great Britain;
(21) Tertiary of Pechelbronn, France;
(22) Tertiary of Poland;
(23) Pliocene of Rumania;
(24) Tertiary of the Baku–Grosny region of U.S.S.R.;
(25) Paleogene of Ferghana, U.S.S.R.;
(26) Miocene of the Suez graben, Egypt;
(27) Cretaceous–Tertiary of the southern Sirte embayment, Libya;
(28) Devonian of western Libya (maximum pourpoints only moderately high; i.e. 20–45 °F)
(29) Miocene of Nigeria;
(30) Lower Cretaceous of Gabon and adjacent parts of West Africa;
(31) Karroo of South Africa;
(32) Oligo–Miocene of north-central Iran;
(33) Tertiary of the Gujarat embayment, India;
(34) Tertiary of Assam, India;
(35) Tertiary of Burma;
(36) Miocene of Indonesia;
(37) Miocene of northern Borneo;
(38) Tertiary of Taiwan;
(39) Mesozoic of both eastern and western Australia; and
(40) Miocene of New Zealand.

The remains of spheroidal objects believed to be unicellular, algae-like organisms were described as *Archaeosphaeorides barbertonensis* (Schopf & Barghoorn, 1967) from the early Precambrian Fig Tree series of South Africa, more than $3 \cdot 1 \times 10^9$ years old. A possible bacterium *Eobacterium isolatum* also occurs in the Fig Tree cherts (Barghoorn & Schopf, 1967).

Bitumens of non-marine sediments

It is possible that these and similar Precambrian deposits were of fresh-water origin (McLaughlin, 1955) or that no salinity distinction existed between marine and non-marine environments in those early times (Rubey, 1951). Organic geochemical studies of the Fig Tree chert and shale yielded alkane hydrocarbons ranging from 17 to 35 carbon atoms (Meinschein, 1967; Hoering, 1965). The isoprenoid hydrocarbons pristane and phytane occurred in the Fig Tree suggesting origin in chlorophyll and thus implying the existence of photosynthesis in Fig Tree time. These hydrocarbons are also known from other Precambrian rocks (Oló et al. 1965; Meinschein et al. 1964). The carbon isotopic composition of the Fig Tree organic matter lies in the range of many crude oils and other known organic materials of photosynthetic origin.

4.10.3 Geochemistry of fossil resins

An extensive review of fossil resins or amber has recently been presented by Langenheim (1969). Natural resins are complex mixtures of mono-, sesqui-, di- and triterpenoids. The isoprene (C_5H_8) precursor is mevalonic acid and the active isoprene is isopentyl pyrophosphate. Geranyl pyrophosphate, formed by the linking of isopentenyl pyrophosphate and dimethylalyl pyrophosphate, is the initial compound in most of the plant terpenes. Variation in the subsequent mode of condensation accounts for the multiplicity of natural terpenoids. The volatile fractions of resins are mainly mono-, sesqui- and some di-terpene hydrocarbons; the non-volatiles are largely carboxylic diterpene acids and some triterpene acids; other components, may be alcohols, aldehydes, esters and neutral unsaponifiable substances (resenes). Although there is a tendency for the volatile fractions to evaporate rapidly and be lost to fossilization, small amounts may be trapped and preserved. About 10 per cent of 280 plant families, and 25 per cent of 338 genera studied, synthesize resins. Two-thirds of the abundant resin producers are tropical. The coniferous families, which are mainly temperate, all produce resins, but only the Pinaceae, and Araucariaceae yield large quantities. The angiospermous tropical members of the Leguminosae and Dipterocarpaceae are prolific producers, but other angiosperms are also important.

Fossilization of resins to amber involves progressive oxidation and polymerization probably by a free-radical mechanism. Ambers in lignite have infrared spectra similar to those of modern resins, but those of amber in medium-volatile bituminous coal are much different. Amber is known from Carboniferous through Pleistocene deposits but is most common in Cretaceous and Tertiary rocks. Two broad categories of amber are established on the presence of succinic acid (succinite) as in Baltic amber, or absence of it (retinite). Succinic acid, however, is non-terpenoid and is not helpful in characterizing fossil resins. In recent years X-ray diffraction and infrared absorption spectra have been helpful in elucidating the structure of amber. Both modern resins and amber have similar spectra in the $2 \cdot 5$–8 μm ($4,000$–$1,250$ cm^{-1}) region. Included are the following absorption bands: $2 \cdot 9$ μm ($3,500$ cm^{-1}) due to stretching of hydrogen–oxygen bonds; $6 \cdot 1$ μm ($1,650$ cm^{-1}), bending motion of hydrogen–oxygen bonds; $3 \cdot 4$ μm ($2,950$ cm^{-1}) due to stretching of carbon–hydrogen bonds; $6 \cdot 8$ μm (1470 cm^{-1}) and $7 \cdot 25$ μm ($1,380$ cm^{-1}) bending motion of carbon–hydrogen bonds; and $5 \cdot 8$ μm ($1,700$ cm^{-1}) due to stretching of carbon–oxygen double bonds. Absorption bands between 8 and 10 μm ($1,250$–625 cm^{-1}) are variable between different resins and different ambers; these are due to carbon–oxygen single bonds but are difficult to assign to specific structural features. A sharp band near $11 \cdot 3$ μm (885 cm^{-1}) can be assigned to out-of-plane bending of the two hydrogen atoms of a terminal methylene group. Several resin acids, such as agathenedicarboxylic and copalic acid have this band. These have been separated from recent resins and the band has also been found in fossil resins.

X-ray diffraction of crystalline components of resins and amber may serve to differentiate between different ages and types of samples.

Gymnospermatous plants are the earliest types to have produced resin. Coniferales are the only kinds known to have synthesized resin but the extinct Cordaitales also contain dark resinous material in the secondary wood, pitch, and parenchyma cells that may be related to true resins. Resin receptacles first appear in *Araucarioxylon arizonicum* of the Chinle Formation (Triassic) of Arizona but no amber is known from these deposits. Early and late Cretaceous deposits of the Atlantic Coastal Plain, United States contain amber from a variety of sources, some of which can be traced to its gymnospermatous source types by infrared spectra. Some Cretaceous amber from Manitoba that has been transported into marine sediments, is similar chemically to Cretaceous walchowite from Europe which is gymnospermatous. Amber from the Tertiary of New Zealand and Australia is thought to have been derived from *Agathis*, a living genus that produces copious resin. The geochemistry of Baltic amber suggests that araucarian plants were the source of it even though no plants of this type have been found in the Eo–Oligocene deposits in which the amber occurs. The large quantities of Baltic amber suggest that some disease of the plants may have been responsible for the amber production (i.e. 'succinosis' from succinic acid content of the amber). This concept, however, overlooked the large quantities of resin being produced by certain types of modern tropical plants, both gymnosperms and angiosperms.

Amber of possible angiospermous origin is recorded from the late Cretaceous of New Jersey and Montana. This has i.r. spectra like that of resin of modern *Liquidambar*. The similarity lies in i.r. bands at 13·3 and 14·3 μm (750 and 700 cm^{-1}) which is evidence for monosubstituted benzenoid rings. The cinnamic acid and styrene of *Liquidambar* contain this structure. Siegburgite of Tertiary lignite of the Rhine Valley also contains this structure owing to associated *Liquidambar*.

X-ray diffraction analyses of Tertiary amber from London Clay (copalite), the Baltic Eo–Oligocene (glessite), and from Equador (guayaouillite) all show similarities due to the presence of a crystalline triterpenoid alcohol, α-amyrin. Several modern Burseraceae and Rutaceae contain this alcohol. A correlation between resins from the tropical tree *Hymenea* and amber from the Tertiary of Mexico, Columbia and Brazil has been noted. The infrared spectra are similar except that a broad band at 14·2–14·4 μm (695–705 cm^{-1}) due to unassigned skeletal vibrations in *Hymenaea* is absent in the fossil amber probably because of polymerization. In addition, a 11·3 μm (885 cm^{-1}) band assigned to $=OH_2$ out-of-plane deformation of terminal carbon–carbon bonds decreases in the fossil amber owing to oxidation of the $=CH_2$.

The causes and functions of modern resins and their fossil counterparts is not well understood. Possible factors are: physiological functions of the resins in metabolic processes, ecological adaptation in protection against injury, and effects in repelling attacks of, or attracting, certain insects. Chemical studies of the resins are being used to shed light on some of these problems.

4.11 CARBON ISOTOPIC COMPOSITION OF NON-MARINE ORGANIC MATTER AND SEDIMENTS

The study of the relative abundances of the stable carbon isotopes of non-marine organic matter has shown significant differences from those of marine organic matter (Silverman & Epstein, 1958; Ecklemann *et al.* 1962). These studies have proved useful in recognition of non-marine *v.* marine sources of petroleum and petroleum source-beds.

$^{13}C/^{12}C$ ratios may vary as much as 5 % in natural carbonaceous substances and are relatively high (isotopically heavy) in carbonate rocks and relatively low (isotopically light) in organic carbon. In modern natural inorganic carbon systems such as carbonate–carbon dioxide in equilibrium at room temperature and atmospheric pressure and in the methane–carbon dioxide system, CO_2 and CH_4 respectively tend to become enriched in ^{12}C. According to Park & Epstein (1961) fractionation of stable carbon isotopes takes place in tomato plants during photosynthesis; the process occurs in two steps, the first, of kinetic nature during uptake of atmospheric CO_2 by plant tissue, the second during conversion of CO_2 to 3-phosphoglyceric acid through activity of the carboxydismutase enzyme. In marine

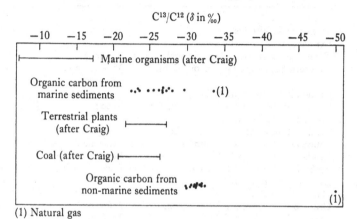

(1) Natural gas

FIGURE 80 $^{13}C/^{12}C$ δ-values of various organic materials (Silverman & Epstein, 1958). *American Association of Petroleum Geologists Bulletin*, **42**, no. 5, pp. 998–1012, fig. 1).

plants it is suggested that the first fractionation stage is omitted, and only the second stage of ^{12}C enrichment is represented. The relatively isotopically heavier carbon compounds of marine as compared to terrestrial plants can be thus explained. However, the characteristically isotopically lighter freshwater aquatic organic material as compared to marine organic matter (Eckelmann *et al.* 1962) cannot be explained on this basis.

$^{13}C/^{12}C$ analyses are obtained by converting the geologic sample to carbon dioxide by combustion over copper oxide in an oxygen atmosphere at 800–900 °C (Silverman & Epstein, 1958); the carbon dioxide is analyzed in a Nier 60° sector-type mass spectrometer; mass spectrometric analyses are reported as deviations in per mil. (δ-values from the $^{13}C/^{12}C$ ratio of a Maestrichtion (Upper Cretaceous) belemnite, *Belemnitella americana*, Peedee Formation, South Carolina. The delta values are obtained in the following way:

$$\delta\ (\text{‰}) = \frac{^{13}C/^{12}C\ \text{sample} - {}^{13}C/^{12}C\ \text{standard}}{^{13}C/^{12}C\ \text{standard}} \times 1000. \qquad (99)$$

Some δ values of marine and non-marine materials are shown in fig. 80, and those of a variety of samples are listed in tables 40, 41. Available data

TABLE 40 *Carbon isotopic composition of organic matter of marine origin (Silverman & Epstein, 1958, American Association of Petroleum Geologists Bulletin, 42, no. 5, pp. 998–1012, table 1)*

No.	Sample	Age	$^{13}C/^{12}C$ $\delta(\text{‰})$
—	Marine plants	Modern	$-12 \cdot 0$*
—	Marine invertebrates	Modern	$-13 \cdot 3$†
9654	Extract—organic shale, Calif.	Pliocene	$-22 \cdot 2$
9652	Total organic matter—organic shale, Calif.	Miocene	$-22 \cdot 8$
9493	Southern Segregated crude oil, Calif.	Miocene and Pliocene	$-23 \cdot 0$
11580	Crude oil—Oxnard field, Calif.	Oligocene	$-23 \cdot 1$
2622-C	Extract—recent offshore sediment, Santa Barbara basin, Clif.	Recent	$-23 \cdot 3$
11705	Crude oil–Bay Marchand field, La.	Miocene	$-24 \cdot 6$
11061	Crude oil—Minas field, Sumatra	Miocene	$-25 \cdot 3$
12929	Crude oil—Carraño Valley field, Venez.	Eocene	$-25 \cdot 3$
9188	Seep oil—Whiskey Creek, Utah	Eocene	$-26 \cdot 0$
9960-41	Extract—recent offshore sediment, Gulf Coast, La.	Recent	$-26 \cdot 4$
10012	Crude oil—Lagunillas field, Venez.	Miocene	$-26 \cdot 4$
11730	Crude oil—Barataria field, La.	Miocene	$-26 \cdot 5$
13327	Crude oil—Bents Fort field. Colo.	L. Pennsylvanian	$-27 \cdot 0$
9368	Crude oil—Rangley field, Colo.	U. Cretaceous	$-27 \cdot 3$
13125	Crude oil—Gas City field, Mont.	Silurian	$-27 \cdot 3$
9351	Crude oil—Ashley Valley field, Utah	Pennsylvanian	$-27 \cdot 8$
11325	Crude oil—Cabin Creek field, W.Va.	Mississippian	$-29 \cdot 2$
12915	Crude oil—Poplar field, Mont.	Mississippian	$-29 \cdot 4$
200-SP	Gas—south Coles Levee field, Calif.	Miocene	$-33 \cdot 4$

* Average value; range is $-7 \cdot 6$ to $-16 \cdot 5$‰ (after Craig, 1953).
† Average value; range is $-10 \cdot 2$ to $-17 \cdot 4$‰ (after Craig, 1953).
‡ Southern Segregated crude oil is a mixture containing major amounts of West Coyote, Seal Beach, Signal Hill and Santa Fe Springs productions and minor amounts of Rosecrans and Alondra productions. All of these fields are in Los Angeles County, California.

suggest that residual organic matter and petroleum derived from marine sediments have δ values that range from $-22 \cdot 2$ to $-29 \cdot 4$‰, while those from the Green River Oil Shale, a lacustrine deposit are in the $-29 \cdot 9$ to $32 \cdot 5$‰ range (Silverman & Epstein, 1958). Eckelmann *et al.* (1962), however, found that the δ-values of 128 crude oils of Devonian and younger age were concentrated in the -27 to -29‰ range (fig. 80). These data indicated to them a major fresh or brackish origin of the organic matter of these oils. Alternative possibilities are (1) an increase of $^{13}C/^{12}C$ ratio of oceanic bicarbonate during the past few million years; (2) isotopically heavy gases are released in significant amounts during diagenesis of the organic matter; (3) there has been a change in the fractionation factor for

TABLE 41   *Carbon isotopic composition of organic matter of non-marine origin* (*Silverman & Epstein*, 1958 *loc. cit.* table 2)

No.	Sample	Age	$^{13}C/^{12}C$ $\delta(\text{‰})$
—	Freshwater plant (*Rhizoclonium* sp.)	Modern	−22·2*
—	Coal	Various	−23·7†
—	Land plants (wood)	Modern	−24·9‡
—	Land plants (leaves and plants)	Modern	−25·8§
8878	Gilsonite, Utah	Eocene	−29·9
10886	Crude oil—Duchesne field, Utah	Eocene	−30·0
5429	Extract—Green River shale, Utah	Eocene	−30·5
11521	Crude oil—Brennan Bottom field, Utah	Eocene	−30·6
—	Retorted oil—Green River shale, Utah	Eocene	−31·0
11066	Crude oil—Red Wash field, Utah	Eocene	−31·0
10885	Crude oil—Duchesne field, Utah	Eocene	−31·2
10424	Crude oil—Red Wash field, Utah	Eocene	−31·3
10881	Crude oil—Duchesne field, Utah	Paleocene	−31.3
10169	Crude oil—Flat Mesa field, Utah	Eocene	−31·7
—	Total organic matter–Green River shale	Eocene	−31·9
8764	Ozocerite, Utah	Paleocene	−32·0
6337	Ozocerite, Utah	Paleocene	−32·5
11342	Gas—Red Wash field, Utah	Eocene	−49·3

* After Craig, 1953.
† Average value, range is −21·1 to −26·7 (after Craig, 1953).
‡ Average value, range is −22·0 to 27·4 (after Craig, 1953).
§ Average value, range is −23·1 to −28·6 (after Craig, 1953).

the overall process of photosynthesis. There is no concise evidence to support any of these hypotheses.

The Witwatersrand system of middle Precambrian age ($\sim 2\cdot15 \times 10^9$ years), South Africa contains, in association with its well-known gold deposits, carbonaceous matter or thucolite in sandstones and conglomerates (Hoefs & Schidlowski, 1967). The material seems to represent hydrocarbons that migrated into the clastic rocks and were afterwards polymerized to a solid state by α- and γ-radiation from the uranium minerals that occur with the thucolite. $^{13}C/^{12}C$ ratios obtained for the thucolites show ^{13}C values ranging from −22·4 to −32·8% (PDB standard). These values are somewhat closer to those for crude oils than for coals. The carbon isotope data together with the recognition of cellular bodies in the carbonaceous material (Schlidowski, 1965) are indications of biogenic nature. Whether the Witwatersrand beds are non-marine or marine is not definite, but the coarsely clastic character of much of that sequence suggests that it may be non-marine.

5 Protein amino acids of non-marine deposits

5.1 AMINO ACIDS IN AQUATIC SOURCE ORGANISMS

Under ordinary circumstances all twenty-two of the protein α-amino acids should be expected to occur in aquatic plants and animals as well as in the terrestrial species that contribute to organic matter in sediments. It is not typical, however, for all twenty-two protein amino acids to be found in geochemical preparations because of unequal stability of the various amino acids and losses of some of them in laboratory preparations and analysis.

The following are generally not present in geochemical preparations: hydroxyproline (occasionally present), tryptophan (destroyed by acid hydrolysis), thyroxine, and iodogorgoic acid. The latter two are evidently quite rare in organisms that supply geochemical residues. Cystine and cysteine are generally found as cysteic acid in laboratory hydrolysates. Methionine is also uncommon in laboratory preparations, apparently because of the ease with which it alters to alanine, glycine and aminobutyric acid (Vallentyne, 1964). Non-protein amino acids such as aminobutyric acids, hydroxylysine and others are sometimes found in geochemical samples where they may have formed diagenetically by bacterial action on other amino acids (Swain, 1967*b*).

The amino sugars glucosamine and galactosamine together with several amines are present in some laboratory preparations and apparently formed by treatment of residues of chitin, mucopolysaccharides and mucoproteins. These substances are not typically found in older geochemical samples, although chitin which consists of units of *n*-acetyl-D-glucosamine, joined by β-1 $\rightarrow$ 4 glycosidic bonds, has been reported from a Cambrian mollusk *Hyolithellus* (Carlisle, 1964). The chitin was identified by means of a chitinase enzyme preparation.

Examples of amino acid analyses of freshwater aquatic plants are given in fig. 81. These show the variability in amino acid suites from one species to another as well as the seasonal variation that occurs within a species (Swain, Venteris & Ting, 1964).

5.2 AMINO ACIDS OF NON-GLACIAL LAKE AND BOG SEDIMENTS

The acid hydrolyzable amino acids of several lake-and-bog sediment sequences have been studied in both glaciated and non-glaciated areas (Blumentals & Swain, 1956; Züllig, 1956; Swain, 1961*a*; Meader, 1956;

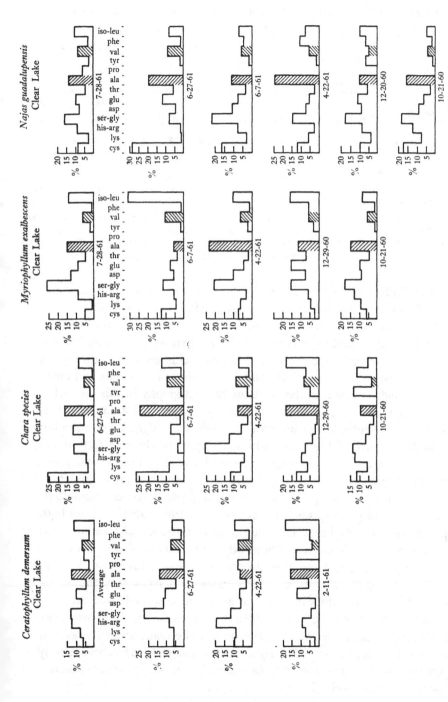

FIGURE 81 Histograms of amino-acid distribution in acid hydrolyzates of aquatic plants from Blue Lake, Minnesota (Swain, Venteris & Ting, 1964), arranged in order of their R_F values on paper chromatograms developed in butanol: acetic acid : water (4:1:5). Dates on which collections were made are given under each histogram. Alanine and valine are shown by inclined ruling.

TABLE 42   *Vertical distribution of amino acids (in parts per 10,000) in acid hydrolyzates of Dismal Swamp peat, Virginia (Swain et al. 1959)*

Depth ft	Type of sample	Bog sample zones	Wt of wet* sample (g)	Glycine	Lysine	Glu- amic acid	Threo- nine	Alanine	Valine	Leucine	Histi- dine	Σ
1–2	Reddish brown woody peat	II$_0$	28	0·32	0·08	0·16	0·08	2·49	0·99	0·59	—	4·71
2–3	Reddish brown woody peat	II$_f$	27	—	—	—	—	0·005	0·005	0·005	—	0·015
3–4	Reddish brown woody peat	β_f	40	1·46	0·37	7·16	3·54	11·81	2·16	1·51	0·50	28·52
4–5	Dark brown copropelic fibrous peat	β_c	60	0·44	0·11	0·54	0·27	2·88	0·59	0·95	0·82	6·60
5–6	Dark brown copropelic fibrous peat	β_c	45	0·53	0·13	0·91	0·45	4·91	0·86	1·74	0·18	9·71
6–7	Sandy, dark fibrous copropel	β_c	45	0·14	0·04	0·37	0·18	1·43	0·37	0·66	0·11	3·30

* Moisture content approximately 90 %.

Jones & Vallentyne, 1960; Swain, Venteris & Ting, 1964; Swain, 1965; Swain & Meader, 1958; Swain *et al.* 1959; Kleerekoper, 1957).

The amino acid contents of samples of peat from Dismal Swamp, Virginia–North Carolina were studied to determine: (1) whether the apparent downward concentration of amino acids noted in other peats occurs in Dismal Swamp as well, and (2) whether the amounts and kinds of amino acids showed similar relations to stratigraphy in this and other peats (Swain *et al.* 1959). The stratigraphy of Dismal Swamp peat was described in Chapter 3. The amino acids attain their highest concentration in the lower part of the reddish-brown upper woody peat (4 ft thick) (table 42). No amino acids were detected by acid hydrolysis and paper chromatography in the 0·7–1 m sample. Because of the rapid growth of vegetation in this swamp, the surface layers probably contain undecomposed protein or polypeptides plus humus. The amino acids occur in both types of compounds here. The humic acid and associated amino acids appear to undergo downward concentration to the base of the woody peat where they are precipitated. The absence of amino acids in the 0·7–1 m layer is believed to be a result of the downward movement of the water-soluble humic acids (to which amino acids are adsorbed) that are continually leached from the upper layers of peat as fast as they form. The possible effect on amino acid distribution of the many swamp fires that have swept the area is not definitely known; perhaps there has been merely a general loss of organic constituents rather than of particular constituents. It does not seem likely that the reduction of amino acids in the 0·9–1 m

TABLE 43   *Amino-acid content of Pyramid Lake core samples (parts per 10,000) of wet sample (Swain & Meader, 1958)*

Amino acid	Core no. 17	Core no. 18
Aspartic acid + glycine	1·62	2·13
Glutamic acid + threonine	0·05	0·46
Alanine	0·26	0·12
Valine	0·36	0·43
Leucine	0·01	0·06
Histidine	0·12	0·01
Total	2·42	3·19

layer resulted from their destruction by heat, because the peat of this layer is not observably different from that above or below.

The amino acids of the profundal sediments of Pyramid Lake, Nevada (Swain & Meader, 1958) are shown in table 43. The quantities are similar to those of a late oligotrophic or early eutrophic lake. The glacial-drift and glaciated-bedrock lakes of Minnesota yielded sedimentary amino acids that range from $0·1 \times 10^{-4}$ g/g in early eutrophic marl to 50×10^{-4} g/g in late eutrophic and dystrophic peat. The relatively low amino acid content of this apatotrophic lake is believed to be due in part to the rapid inorganic sedimentation in Pyramid Lake as compared to that in lakes of humid regions.

Deadman Lake, north-western New Mexico (fig. 82) is a mountain plateau lake in which about 8 m of silt and sand have accumulated (Megard, 1964). The amino acid content of the lake sediments (Swain, 1965) increases with depth, despite the relatively uniform sedimentary sequence and lack of evidence from pollen of an earlier major change in vegetation in that area. It is possible that downward post-depositional concentration of amino acids occurred through groundwater activity as discussed elsewhere in this chapter for Dismal Swamp, Virginia and Cedar Creek Bog, Minnesota.

The protein amino acid contents of the copropelic diatomaceous sediments of Lake Nicaragua (Swain, 1966a) are shown in table 44 and fig. 83. Amino acid analyses of several other lake sediment-types are also listed. The results of the Lake Nicaragua amino acid analyses correspond best to those of eutrophic cool temperate lake sediments. In comparison to the other lakes the basic amino acids, lysine, histidine and arginine are higher than seem to be typical of lake sediments that are low in carbonate content (Swain, 1961a).

Protein amino acids of non-marine deposits

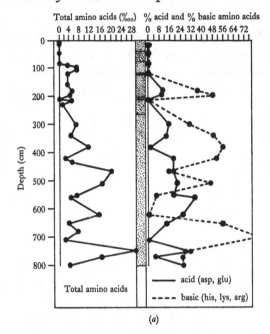

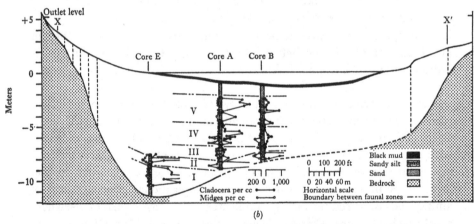

FIGURE 82(a) Vertical distribution of amino acids in sediments of Deadman Lake, Chuska Mountains, New Mexico (Swain, 1965). Shown are depths, sediment types (small dots, silt; large dots, sand), total amino acids in part per 10,000 of wet sediment (left), and percentage of acidic and basic amino acids to total amino acids (right). (b) Cladocera in Deadman Lake sediments. (Megard, 1964, *Ecology*, **45**, 529–46.)

Catahoula Lake, Louisiana (Swain, 1961 a) represents an acidic environment of high positive redox potentials unfavorable to the preservation of organic matter. The total amino-acid content of Catahoula Lake sediments is low and neutral types predominate statistically (table 45).

LAKE NICARAGUA

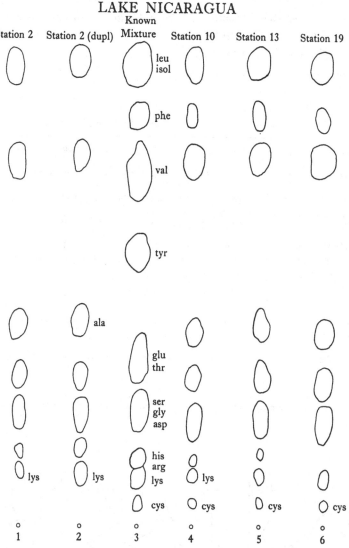

FIGURE 83 Paper chromatograms of amino acids in Lake Nicaragua sediments (Swain, 1966 *a*, *Journal of Sedimentary Petrology*, **36**, pp. 522–40, fig. 12).

The peat of a swampy area bordering Lake Ponchartrain contains relatively low amounts of amino acids (table 48) and suggest that humification was somewhat more complete in these peats than in others studied. Neutral and acidic amino acids predominate in the Lake Ponchartrain sample, whereas in other acidic peats, such as those in Dismal Swamp (fig. 84), basic amino acids are also present.

TABLE 44 *Amino-acid content of Lake Nicaragua sediments (in parts per 1,000 of wet sample) compared to other lakes in Minnesota, Montana, Nevada & Virginia (Swain, 1966a, loc. cit.)*

Lake	Wet weight (g)	Moisture	Cys	Lys	Hist. + arg.	Ser. + gly. + asp.	Glut. + thr	Ala. pro.	Tyr	Val.	Phe.	Leu. + isol.	Sum
Lake Nicaragua no. 2	8·4	82·0	1·12	1·08	2·66	1·89	1·15	0·38	0·06	0·71	0·09	0·52	9·66
Lake Nicaragua no. 10	8·7	—	9·48	1·66	3·30	2·22	1·53	0·79	0·22	0·76	—	0·47	13·43
Lake Nicaragua no. 13	9·2	—	0·86	1·47	3·12	2·59	1·62	0·78	0·12	0·13	—	0·89	11·58
Lake Nicaragua no. 19	10·2	86·0	1·51	1·33	0·41	2·41	1·76	0·68 0·22	0·09	1·21	0·08	1·03	10·74
Green Lake, Minn.	22·0	94·8	0·58	—	5·19	0·54	1·78	2·23	0·87	0·30	1·23	1·61	14·33
Lake of the Woods, Minn.	167·0	76·7	—	—	—	0·20	1·23	0·11	—	0·14	—	0·11	1·79
Rainy Lake, Minn.	70·0	80·2	—	—	—	0·04	0·02	0·02	0·02	—	0·08	0·12	0·30
Kabekona Lake, Minn.	92·0	74·6	—	—	—	0·01	tr.	0·01	—	—	—	—	0·02
Flathead Lake, Mont.	50·0	76·0	0·08	—	—	0·06	0·12	0·05	0·01	0·01	—	0·19	0·52
Pyramid Lake, Nevada	62·0	65·0	—	—	0·01	2·13	0·46	0·12	—	0·43	—	0·06	3·21
Dismal Swamp, Va.	40·0	90·0	—	0·37	0·50	1·46	7·16 (G) 3·54 (t)	11·81	—	2·16	—	0·15	28·52

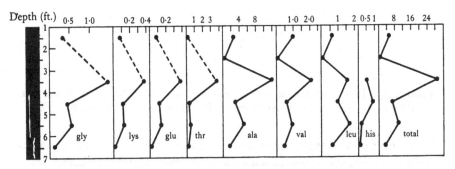

FIGURE 84 Vertical distribution of amino acids in sediments from Dismal Swamp, Virginia–North Carolina (Swain, 1965).

TABLE 45 *Amino acids obtained by acid hydrolysis of sediments from lakes in Minnesota and elsewhere (Swain, 1961a)*

Lake	Type of sediment	Wet wt (g)	Moisture (%)	Cys.	Gly.	Asp. acid	Glut. acid	Thr.	Ala	Tyr.	Val.	Isol.	Leu.	Total
Green Lake no. 5	Diatomaceous silt	22	94·8	0·58 / 5·19a	—	0·54†	1·78*	—	2·23	0·87	0·30	1·23p	1·61	14·33
Green Lake no. 6	Diatomaceous silt	17	90·5	0·010	4·94	6·85	10·59	—	3·37	—	14·12	0·42	1·30	41·59
Fannie Lake no. 5	Copropelic sand	1¾ (dry)	44·3	5·02 / 3·301a / 5·34a	—	0·252†	7·62	—	—	1·07	3·62	1·07p	2·91	30·78
Blue Lake no. 5	Silty copropelic marl and sapropel	45	88	0·48 / 0·21h	—	0·05	0·11	—	0·47	—	0·16	—	0·14	1·62
Spectacle Lake	Sandy peaty copropel	7½ (dry)	94·2	3·48 / 4·01a	—	1·78†	—	1·97	0·763	0·360	3·05	1·89p	0·57	17·87
Rush Lake no. 8	Diatomaceous marl overlying sapropel	29	91	—	0·04	0·80	0·06	0·23	1·05	0·11	0·63	0·10	0·59	3·61
Eagle Lake no. 2	Diatomaceous copropelic marl	70	88·5	0·01 / 0·231 / 4·69a	0·23†	—	0·38*	—	0·32	0·07‡	—	0·75	0·37	7·05
Stanchfield Lake no. 5	Diatomaceous copropelic marl	80	91·4	—	0·12	0·12	0·07	0·06	0·23	0·02	0·09	0·01	0·09	0·81
Clear Lake no. 7	Silty copropelic marl	43	88	0·30l / 0·44a	—	0·16†	0·47*	—	—	—	0·25	0·04p	0·13	1·79
Lake of the Woods no. 3 A	Diatomaceous clay and silt (Desalted)	167	76·7	—	0·20†	—	0·89	0·34	0·11	—	0·14	0·04	0·07	1·79
Rainy Lake no. 1	Pale-red clay	30	78·6	0·11 / 0·13l	—	0·48†	0·58*	—	0·78	—	0·18	0·03p	0·22	2·51
Rainy Lake no. 2	Pale-red clay	70	80·2	—	0·03	0·01	—	0·02	0·02	0·02	0·08	0·05	0·07	0·30
Rainy Lake no. 8	Gray and pale-red clay and silt	80	75·4	—	0·01	0·03	0·02	0·07	0·33	0·03	0·20	0·03	0·18	0·90
Kabetogama Lake no. 1	Copropelic sand	80	20·2	—	0·19	0·15	0·04	—	0·45	0·13	0·08	—	0·12	1·16

* Includes threonine. † Includes aspartic and/or serine. ‡ May include valine.

a – arginine+histidine; h – histidine; l – lysine; p – phenylalanine; s – serine.

TABLE 45 (cont.)

Lake	Type of sediment	Wet wt (g)	Moisture (%)	Cys.	Gly.	Asp. acid	Glut. acid	Thr.	Ala.	Tyr.	Val.	Iso.	Leu.	Total
							Amino acids, parts per 10,000							
Kabetogama Lake no. 3	Diatomaceous copropelic clay	80	90·5	—	0·61	1·62	0·07	0·27	0·76	0·62	0·07	—	0·92	4·94
Pelican Lake no. 1	Diatomaceous shelly copropel	64	93·2	0·09^l	0·10 / 1·05^a	0·18	1·04	—	1·07	0·06	0·05	0·07^p	0·11	3·82
Kabekona Lake no. 5	Marl	92	74·6	—	0·01†	—	Trace	Trace	0·01	Trace	—	—	—	0·02*
Reno Lake no. 12	Sandy copropel	58	n.d.	0·10	0·07†	—	0·10*	—	0·10	Trace	—	—	0·03	0·40
Flathead Lake, Montana no. 6	Gray clay	50	76	0·08^l	0·06^a	—	0·12*	—	0·05	0·01	0·01	—	0·19	0·52
Flathead Lake, Montana no. 9	Peaty sandy clay	44	64	0·37^l	0·49^a	—	0·20*	—	0·14	0·03	0·03	0·12^p	0·17	1·55
Catahoula Lake, Louisiana Indian Bluff	Red clay	52	n.d.	—	0·01	—	—	0·02	0·03	—	0·04	0·01	0·06	0·17
Catahoula Lake, Louisiana, 2 ft below surface	Gray clay	74	n.d.	—	Trace	—	—	0·02	0·01	—	0·02	Trace	0·03	0·08
Lake Ponchartrain, Louisiana	Peat	20	n.d.	—	0·05	—	0·45	0·12	0·18	0·03	0·10	—	0·20	1·13
Lake Minnetonka Big Island no. 2, 1½ ft below surface	Peat	40	92·4	—	0·57	—	0·71	0·31	0·34	0·53	0·38	0·07	0·28	3·19
Prior Lake, Minn.	Sapropel	58	90·4	—	0·08	0·30	2·54	0·84	0·74	—	0·40	0·06	0·09	5·05
Prior Lake, Minn.	Copropel	50	84·7	Trace	0·41	1·73	0·29	0·30 / 0·26^s	0·68	—	0·57	0·14	0·27	4·65
Pyramid Lake, Nevada	Sapropelic silt	62	65	0·01^a	2·13†	—	0·46†	—	0·12	—	0·43	—	0·06	3·21
Cedar Creek Bog, Minn.	Sedge peat and copropel	17	89	—	3·53	1·95	5·68	2·83	21·92	—	2·83	—	9·33	48·90
Dismal Swamp, Virginia	Peat	40	90	0·37^l / 0·50^a	1·46	—	7·16	3·54	11·81	—	2·16	—	1·51	28·52

* Includes threonine.   † Includes aspartic and/or serine.   ‡ May include valine.

a — arginine + histidine;   h — histidine;   l — lysine;   p — phenylalanine;   s — serine.

5.3 AMINO ACIDS OF GLACIAL LAKE AND BOG SEDIMENTS

The nitrogen content of the peats of glaciated regions reaches 4% or more, but probably averages 1% or less (Waksman, 1938). The nitrogen content of Cedar Creek Bog peat, Minnesota (Swain *et al.* 1959) is 2·3% at 1·3–1·7 m, 2·8% at 2·3–2·7 m, 3·4% at 3–3·3 m and less than 1% at greater depths where the marl content of the peat is higher. The nitrogen of glacial region peats has been considered to occur in the form of protein-aceous compounds that have originated as cell constituents of micro-organisms (Waksman, 1938). The latter are thought to have grown in the bog from the decomposition products of the accumulating higher plants. The downward increase in total nitrogen, typically found in low-moor peats (Waksman, 1938) is cited in support of this idea. The information gained from more recent studies (Swain *et al.* 1959; Swain, 1961 *a*) suggests, on the other hand, that the nitrogenous materials in low-moor bogs, both in glacial and non-glacial regions, are associated mainly with humic acids rather than with proteins in the strict sense. The presence of some peptide-linked material in the peat sample cannot be ruled out, however.

The vertical distribution of amino acids in the upper 2–4 m of lake sediments in southern Ontario were studied by Kleerekoper (1957). He found that the amino acids showed little change with depth in the sediments of some lakes, an increase in depth in others and a decrease in still others. The changes apparently are related to variations in the sand clay and perhaps marl content of the lake sediments, rather than to migrations within the sediments as suggested herein for Cedar Creek peats and Dismal Swamp peats. Kleerekoper was able to isolate a rather large number of amino acids by sulfuric acid hydrolysis from the lake sediments, including alanine, arginine, aspartic acid, cystine, glutamic acid, glycine, hydroxy-proline, leucine, isoleucine, lysine, methionine, proline, serine, taurine, threonine and valine. An excess of nitrogen not represented by the total amino acids was believed to occur in lignin–nitrogen complexes, such as humic acid.

The amino acids in peats of Cedar Creek Bog, Anoka and Isanti Counties, Minnesota were studied by Swain *et al.* (1959). The bog zones in Cedar Creek bog are shown in table 46. In the upper 1·7 m of the bog (table 46, fig. 85) there is a downward increase in all the amino acids from as little as two-fold in aspartic acid to as much as ten-fold in alanine. The upper 1·3–1·7 m of the sediments of Cedar Creek bog are predominantly the remains of forest and Sphagnum peat, below which lies sedge peat copropel and marl. Part of the downward increase in amino acids is believed to be due to the originally greater protein content in the sedge peat and copropel

TABLE 46 *Vertical distribution of amino acids in hydrolyzates of Cedar Creek peat, Minnesota (Swain et al. 1959)*

Depth ft	Type of sample	Bog zones	Wt of wet sample (g)	Moisture (%)	Glycine	Asp. acid	Glut. acid	Thr	Ala	Val	Leu	Σ	C %	N %
2–3	Copropelic peat	II$_f$	17	89	1·7	0·94	0·60	0·29	2·94	1·12	2·06	9·65	47·23†	2·31
5–6	Peaty copropel	β$_p$	21	90	3·53	1·95	5·68	2·83	21·92	2·83	9·33	48·07	47·59	2·77
8–9	Sapropelic copropel	β$_c$	20	87	0·90	0·60	3·55	1·77	11·25	5·82	7·82	31·71	48·01	3·36
11–12	Sapropelic copropel	β$_c$	21·5	68	1·74	1·16	3·49	1·74	19·77	3·02	7·37	38·29	9·78	0·90
14–15	Copropelic marl	M$_c$	23	61	0·43	0·28	1·58	0·80	0·49	1·67	1·48	6·73	15·13	0·75
17–18	Copropelic marl	M$_c$	29	60	0·42	0·28	1·37	0·67	—	0·15	0·15	3·04	—	—
20–21	Copropelic marl	M$_c$	26	60	0·01	0·07	0·03	0·02	—	—	—	0·13	—	—
23–24	Copropelic marl	M$_c$	23	60	0·05	0·03	0·48	0·24	—	0·14	0·20	1·14	—	—
26–27	Dark copropel	β$_c$	12·5	79	0·28	0·19	—	—	0·27	0·17	0·40	1·31	—	—
29–30	Copropelic marl	M$_c$	30	60	0·02	0·01	0·25	0·14	—	0·25	0·45	1·12	—	—
32–33	Copropelic marl	M	37	60	0·04	0·03	0·02	0·10	0·03	0·02	0·01	0·25	—	—
35–36	Sideritic marl	M$_{sid}$	24	70	0·07	0·05	0·80	0·39	0·63	0·19	—	2·13	8·47	0·9
37–38	Sideritic sand	Γ$_s$	42·9	60	0·007	0·005	0·24	0·12	1·66	0·21	0·31	2·55	—	—
41–42	Sand	Γ$_s$	28	35	—	—	—	—	—	—	—	—	—	—
Bison rib	Dark, humus impregnated	—	10 (dry)	—	0·236* 0·224h	2·996 ly	0·254	0·030	0·098a 0·864pa	0·198	0·288	4·410	—	—

* Probably includes serine and aspartic acid.

† C and N determinations were made previously on cores collected near 'b' on Figure 1. Depths are, respectively: 4–5, 7–8, 9–10, 11–12, 12–14 and 35–36 ft.

ly = lysine; *a* = alanine; *pa* = phenylalanine; *h* = histidine.

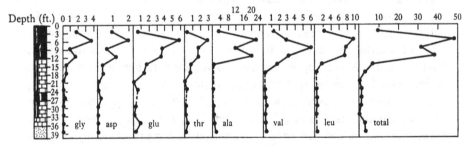

FIGURE 85 Vertical distribution of amino acids in hydrochloric acid hydrolyzates of peat and marl from Cedar Creek Bog, Anoka County, Minnesota (Swain, Blumentals & Millers, 1959; Swain, 1965). Values are in parts per 10,000 of wet sediments. Solid black peat; broken black copropel and sapropel; blocks marl, circles sand.

than in the forest peat. Additional downward concentration of amino acids may be accomplished by means of aqueous solutions of humic acid to which the amino acids are attached. Because the copropelic material is relatively impervious there is probably little bog water circulation from above, but later circulation may have occurred. The amino acids brought into the upper part of the copropel by the moving bog waters are likely to be concentrated and held there.

Water humus is the soluble or gel-suspension brownish material of bogs and dystrophic lakes (Waksman, 1938, p. 290). According to Shapiro (1957) these humolimnic acids are of enolic nature and have a general empirical formula of $C_{50}H_{62}O_{27}N$. He thought that the nitrogen probably is an impurity. Swain *et al.* (1959, p. 121) suggest that the amino acids may be attached to these humic materials by means of an amine bridge as was proposed by Thiele & Kettner (1953). These authors suggest that humic acids of this type formed from quinones, such as *p*-benzoquinone, probably by bacterial decomposition:

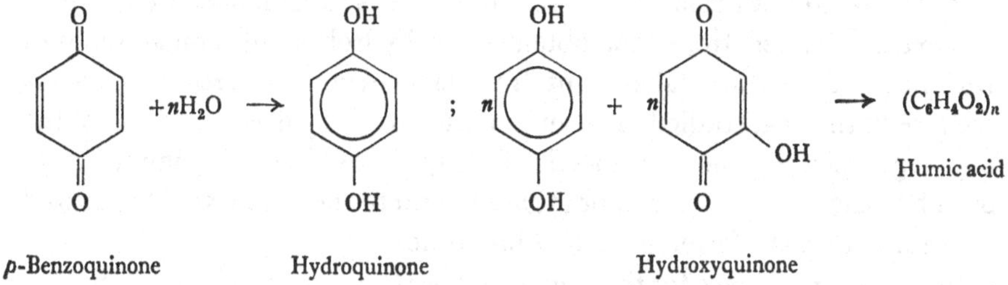

| *p*-Benzoquinone | Hydroquinone | Hydroxyquinone |

$$(100)$$

The humic acid benzenoid system may form by polymerizing the hydroxyquinone, as follows:

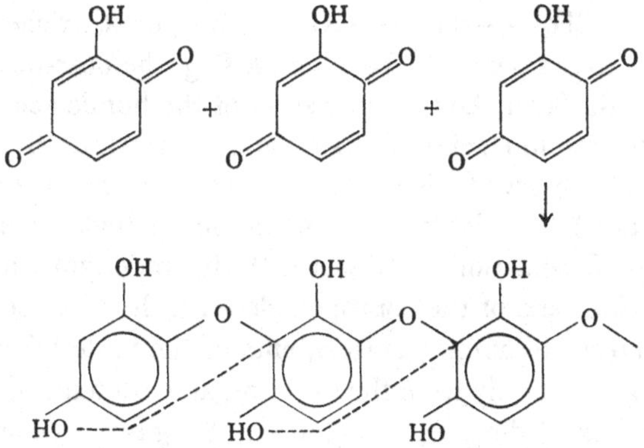

$$(101)$$

Thiele & Kettner (1953) visualized that nitrogen might occur in the humic acid molecule as a bridge substance of the nature of oxazine:

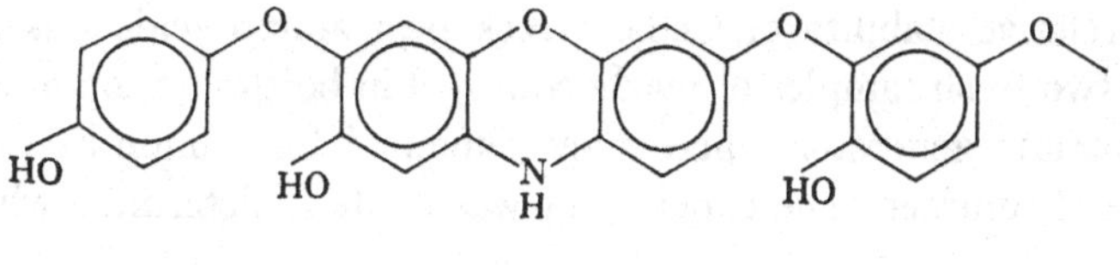

Protein amino acids of non-marine deposits

This mechanism may also provide a means of incorporation of amino acids in the humic acid micelle (Swain *et al.* 1959). The amino acids would thus be subject to transportation vertically or laterally within a bog, attached to the water-soluble humic acids.

In the bog waters studied (Swain *et al.* 1959) water-soluble proteins apparently are absent and free amino acids are lacking or nearly so. Samples of surface bog water and of water squeezed from the peat samples were concentrated at room temperature *in vacuo* and were tested for proteins and free amino acids by paper chromatography with negative results. However, when the water samples were extracted with dilute hydrochloric acid protein amino acids were obtained. When both peat and water samples were extracted with cold dilute HCl, differences in amino acid content of the extracts varied from that obtained by hydrolysis of similar samples with hot acid only in total amount. The data suggest that free amino acids are rare in the peat studied. The amino acids freed by decomposition of the bog vegetation and not completely destroyed are either: (1) immediately used by other bog organisms, or (2) are bound to the humic acid benzenoid structures that are forming at the same time.

Supporting evidence for this conclusion is the presence of amino acids in, or associated with, humic acid preparations (Okuda & Hori, 1954, 1956; Hayashi, 1956; Bremner, 1955*b*; Blumentals & Swain, 1956). It is uncertain whether the amino acids are adsorbed by the humic acid micelles, or are linked to it by —CH_2—, —$\overset{H}{N}$—, —CO—, —S—, or other chemical groups.

With regard to the peats of Cedar Creek Bog, the near-neutrality of the bog waters would favor the transportation of the humic acid micelles and their associated amino acids laterally and downward, thus helping to account for the observed distribution in the upper 3·3–4 m (copropel, sapropel and peat). The downward-moving humic acids would be precipitated in part by flocculation in contact with the underlying marl, as well as by the imperviousness of the copropel. Beneath the copropelic peat there is a sharp decrease in amino acid contents of the underlying marl that is believed to result not only from the lower organic contents of the marl, but also from the lack of downward migration of high-molecular weight substances through the copropel. The increase of several amino acids in the dark copropel layer at 8·7–9 m may be the result both of original high protein composition and of downward concentration.

The relative stability of Cedar Creek peat amino acids was tested by placing two fresh samples of marly copropel in bottles on top of a Van der Graff nuclear generator where they remained for 6 months during the spring and summer. The experiment was made to determine whether the

X-rays and gamma rays from the generator, together with bacterial action inside the glass bottle would affect the amino acid content. Extracts of these samples were co-chromatogrammed with a sample that had been stored in their own water appeared to have changed little during the 6-month period.

The amino acid contents of the surface sediments in several lakes of the Anoka Sand Plain a deglaciated region in east-central Minnesota which includes the Cedar Creek Bog area are discussed above. The Anoka Sand Plain is an area of sandy Mankato (Late Wisconsin) outwash covering more than 1,300 square miles. During wastage of the Grantsburg sublobe of the Des Moines lobe of Mankato ice, meltwater flowing generally eastward formed a pitted sandy outwash plain. Several elevated areas comprising red Superior lobe till and gravel and of gray Grantsburg till project above the plain. Lakes that occupy depressions on the sand plain, on the hills of red and gray drift or on the St Croix moraine lying east of the sand plain were selected for this study. Acid analyses are shown in tables 47 and 48.

The amino acid contents of copropelic sediments of Green Lake, Chisago County were up to 41 parts per 10,000 wet weight, the highest of the lakes studied in this area. Blue-green algae and diatoms which bloom plentifully in the lake probably supplied most of the amino acids. Variations in valine, glutamic acid, alanine, and arginine + histidine are shown in table 47. Diagenetic changes are probably responsible for these variations because the cores analyzed are in similar bottom sediments. The marl sediments of Clear Lake, Sherburne County are low in total amino acids, and alanine was absent or rare in analyzed samples perhaps because of its relative instability during laboratory processing rather than to instability in the natural condition.

Fannie Lake, Isanti County, has copropelic sediments which are higher in basic amino acids, cystine and other amino acids that are less common in deposits which are more completely humified than those of Fannie Lake. Blue Lake, Isanti County, on the other hand has marly copropel that is highly humified and contains relatively little amino acids. The sapropel of deep, thermally stratified Spectacle Lake, Isanti County, is marked by rather high amino acid contents compared to other Sand Plain lakes, probably owing to anaerobic preservation of the settled organic matter.

Among other Anoka Sand Plain Lakes studied, Rush Lake, Chisago County, with diatomaceous marly copropel sediment, was low in total amino acids but with relatively high aspartic acid; Eagle Lake, Sherburne County, having diatomaceous copropelic marl has moderate amounts of total amino acids with rather abundant arginine and histidine; and South Stanchfield Lake, Isanti County, diatomaceous copropelic marl, has low

TABLE 47 *Description of lake samples, total amino acids, and other properties of Minnesota lakes and bogs (Swain, 1961)*

Lake and surrounding terrain	Sample no.	Description of sample	Wet wt (g)	Moisture (%)	Total amino acids in wet sediments (‰)	Other properties bottom sediments
		Lakes of Anoka Sand Plain, Minnesota				
Green Lake, Chisago County, in gray till of Grantsburg lobe	5	Medium-dark-gray, slightly peaty, sapropelic, silty to very finely sandy, very diatomaceous copropel; fragilaroid diatoms, cladocerans, testate Protozoa, *Candona* sp., *Cypria?* sp.	22	94·8	14·33	88·5% < 0·074 mm; pH of bottom water 7·5, of sediment 6·9–7; Eh of water +293 mv, of sediment +143 to +257 mv
Green lake, Chisago County, in gray Grantsburg till	6	Medium-gray-brown to brownish gray, slightly peaty, silty copropel to copropelic silt, very finely sandy; abundant cladocerans, including ephippia, fragilaroid, and coscinodiscoid diatoms	17	90·5	41·59	63·6% < 0·074 mm
Fannie Lake, Isanti County, in Anoka sand plain	5	Fine- to coarse-grained tan and white peaty sand; seeds, charophyte oögonia, gastropods, worm-tube aggregates, *Cypridopsis vidua*	1·5 (dry) 83·5	44·3	30·78	91·3% sand-sized particles
Blue Lake, Isanti, County, in red gravel and till of Superior lobe	5	Light-gray, very diatomaceous, sapropelic, copropelic marl or calcareous copropel; fragilaroid, naviculoid and campylodiscoid diatoms, chlorophytic algae, testate protozoans, cladocerans including ephippia, *Candona* spp., *Cypria* cf. *lacustris*, *Cypridopsis vidua*	45	89–90	1·62	pH 6·6; Eh–96; pH of water 7·8–8·4
Spectacle Lake, Isanti County, in Anika sand plain	5	Medium-dark-gray, finely sandy, silty, diatomaceous copropelic sapropel; fragilaroid and cocconeoid diatoms	7·5 (dry) 43·3	94·23 94·23	17·87 17·87	65·32% sand-sized particles; pH of water 7·6–8·4
Rush Lake, Chisago County, in gray Grantsburg till	8	Light-gray copropelic, silty, diatomaceous marl, finer texture than in other samples from this lake; fragilaroid and melosiroid diatoms, cyprinotid ostracodes	29	91·15	3·61	81% < 0·074 mm
Eagle Lake Sherburne County, in Anoka sand plain	2	Light-gray, silty, very diatomaceous marl; melosiroid and fragilaroid diatoms	70	88·5	7·05	74·9% < 0·074 mm; pH of water 7·3–7·9
S. Stanchfield Lake, Isanti County, in Anoka sand plain	5	Light-gray silty, finely sandy, copropelic, diatomaceous marl and marly silt; campylodiscoid diatoms, cladocerans, *Cypria* sp., *Cyclocypris* sp.	80	91·4	0·81	89% < 0·074 mm; pH 7·5, Eh +371 m Eh of core +407 mv pH of water 7·5

TABLE 47 (*cont.*)

Lake and surrounding terrain	Sample no.	Description of sample	Wet wt (g)	Moisture (%)	Total amino acids in wet sediments (‰)	Other properties of bottom sediments
Clear Lake, Sherburne County, in Mississipi Valley train	7	Medium to light-gray, fine texture, copropelic marl; thin-shelled ostracodes, testate protozoans, cladocerans, abundant mayfly? wings	43	88·1	1·79	—

Other Lakes and Bogs in Minnesota

Lake and surrounding terrain	Sample no.	Description of sample	Wet wt (g)	Moisture (%)	Total amino acids in wet sediments (‰)	Other properties of bottom sediments
Lake of the Woods, Lake of Woods County, in Precambrian granite and schist	3A	Light- to medium-grayish tan, slightly copropelic clayey silt; melosiroid diatoms and others, testate protozoans, cladocerans including ephippia, *Cypria* sp., arthropod? egg cases, pondweed fragments	167	76·7	1·79	98% < 0·074 mm; general pH of sediments 7–7·4; Eh $^+$395 to $^+$422
Rainy Lake, St Louis and Kootchiching counties, in Precambrian granite and schist	1	Pale-grayish tan slightly micaceous silty clay, contains scattered diatoms, pollen grains, pondweed fragments, cladocerans, dark-brown shiny chironomid? egg cases	30	78·6	(de-salted) 2·51	97·8% < 0·074 mm; water: pH 6·6, Eh $^+$467 mv; sediment: pH 6·6, Eh $^+$491 mv
Rainy Lake	2	Pale-tannish gray, finely sandy diatomaceous clay; melosiroid diatoms, cladocerans, many small reddish brown pellets (coprolites or concretions?) in sand-sized fraction	70	61·5	0·30	97·6% < 0·074 mm; pH of sediment 6·6, Eh $^+$419 mV
Rainy Lake	8	Pale-grayish tan very silty, siliceous, glistening clay: sand-sized fraction contains abundant shiny brown chironomid? egg cases	80	53·7	0·90	97·4% 0·074 mm; pH of sediment 6·95, Eh $^+$429 mV
Kabetogama Lake, St Louis and Kootchiching Counties, in Precambrian granite and schist	1	Fine- to medium-grained, angular to subrounded peaty sand	80	20·2	1·16	95·5% is sand-sized; water: pH 7·4, Eh $^+$467 mV; sediment: pH 7·05, Eh $^+$481 mV
Kabetogama Lake	3	Medium-gray, very diatomaceous, copropelic silty clay; melosiroid diatoms, cladocerans including *Bosmina?* sp.	80	90·6	4·94	89·4% 0·074 mm; water: pH 7·3, Eh $^+$473 mV; sediment: pH 7·3, Eh $^+$485 mV
Pelican Lake, St Louis County, in Precambrian granite and schist	1	Dark-gray diatomaceous copropel; naviculoid and other diatoms, testate protozoans, small gastropods	64	93·3	3·82	87·4% 0·074 mm; water: pH 7·1, Eh $^+$413; sediment: pH 6·4, Eh $^+$285 mV

TABLE 47 (*cont.*)

Lake and surrounding terrain	Sample no.	Description of sample	Wet wt (g)	Moisture (%)	Total amino acids in wet sediments (o/ooo)	Other properties of bottom sediments
Kabekona Lake, Hubbard County, in gray till of Wadena? lobe	5	Very pale-gray microcrystalline marl; melosiroid, cymbellaceoid, and other diatoms, cladocerans, *Candona* cf. *candida, Candona* cf. *caudata, Ilyocypris* sp., *Cypria* cf. *lacustris*, egg cases	92	74·6	0·02	75·3% < 0·074 mm
Reno Lake, Pope County, in Wadena lobe glacial drift	12	Medium-gray copropel; abundant gastropods	58	n.d.	0·40	Water: pH 5·35, Eh $^+$569 mV to $^+$683 mV; sediment: pH 5·5–5·8, Eh $^+$569 mV to $^+$719 mV or] higher
Cedar Creek Bog, Anoka County, Minnesota, in Anoka sand plain	5–6 ft	Medium-gray-brown, coarsely fibrous copropel-peat; matrix of dark-brown, resinous coprogenic? pellets and irregular aggregates; few cladoceans and other chitinous exoskeletons	17	90	48·07	Summer pH 7·3, Eh $^+$405 mV; bitumens 6·1%
Cedar Creek Bog	17–18 ft	Light-gray-brown microcrystalline copropel-marl pondweed fragments	29	60	3·04	Winter pH 7·5, Eh $^+$224 mV; bitumens 1·5%
Cedar Creek Bog	35–36 ft	Light-rust-brown (when dried), microgranular, very sideritic marl; fresh samples dark gray to black; few ostracodes including *Candona* sp., *Cyclocypris* sp.	24	70	2·13	pH 7·15, Eh $^+$125 mV; C 8·47%, N 0·9%, bitumens 1%, CaCO$_3$ 38·6%, MgCO$_3$ 1·13%, S 0·43%, Fe 15·63%, P$_2$O$_5$ 0·255%
Lake Minnetonka, Hennepin County (Big Island Marsh), in gray drift	1·5–2·5 ft	Dark-brown, coarse-textured, sandy, copropelic and sapropelic peat and light-gray copropelic diatomaceous silt; sedges, fragilaroid diatoms, cladocerans	40	92·4	3·19	59% < 0·074 mm; water: pH 7·23, Eh $^+$431 mV; sediment: pH 7·02, Eh $^+$173 mV
Prior Lake Scott County, Minnesota, in gray drift	1	Light-gray, copropelic, slightly calcareous sandy silt or silty copropel; coscinodiscoid, naviculoid, fragilaroid diatoms, testate, Protozoans, cladocerans	46	81·7	1·50	Water: pH 7·5, Eh $^+$400 mV; sediment: pH 6·9, Eh $^+$335 mV
Prior Lake	5	Medium-dark-gray, slightly calcareous (magnesian) sapropelic silt and silty sapropel; a few copropelic peat laminae; in part very diatomaceous with naviculoid and campylodiscoid diatoms	58	90·4	5·05	Water: pH 7·1, Eh $^+$411 mV; sediment: pH 7·2, Eh $^+$273 mV

TABLE 47 (*cont.*)

Lake and surrounding terrain	Sample no.	Description of sample	Wet wt (g)	Moisture (%)	Total amino acids in wet sediments (o/ooo)	Other properties of bottom sediments
		Lakes and Bogs outside Minnesota				
Pyramid Lake Nevada, in Tertiary volcanic rocks	18	Pale-gray, silty, and finely sandy clay; *Candona* sp., *Limnocythere* sp., melosiroid and campylodiscoid d'atoms; black in wet state	62	70	3·21	Water: pH 9·1–9·2, Eh $^+$275 to 300 mV; H_2S odor in sediment
Catahoula Lake Louisiana, in Miocene sands	Surface layer	Light-gray and reddish-brown finely sandy, silty clay; plant fragments, seeds, cladocerans including ephippia, insect parts, few naviculoid diatoms	52	—	0.17	Water: pH 4.0, Eh $^+$324 mV; red surface sediment: pH 4·0, Eh $^+$420 mV
Catahoula Lake	2 ft below surface	Gray silty clay	74	—	0·08	pH 6·8, Eh $^+$129 mV
Lake Pontchartrain, Louisiana, peat bog, in Quaternary alluvium	1–2 ft	Sedge peat in bog near shore of Lake Pontchartrain	20	—	1·13	—
Flathead Lake, Montana, in Precambrian argillites	Surface layer	Gray and reddish brown clay	—	—	0·53	pH 7·1, Eh $^+$185 mV

total amino acids in which alanine is the most abundant. Thus, although the more copropelic and less humified lake deposits of the Anoka Sand Plain have the greater amount and variety of amino acid residues, no obvious relationship exists between the known source organisms and individual amino acids.

The bottom sediments of several of the larger lakes of Minnesota and Ontario were examined for their amino acid contents (Swain, 1961 *a*). The sediment types and amino acid contents of Lake of the Woods, Rainy Lake, Kabetogama Lake, and Pelican Lake are given in tables 47 and 48.

The protein amino acids of diatomaceous mesotrophic copropelic clays of Lake of the Woods, Minnesota–Ontario are low in amount and glutamic acid is more abundant than in most of the other glacial lakes studied. Oligotrophic reddish-brown and gray clays of Rainy Lake, Minnesota–Ontario, contain only small amounts of amino acids in which glutamic

TABLE 48   *Amino acids from sediments of north-eastern Minnesota lakes*

(10^{-4} g/g of sample, dry wt)

	Cys	Lys	Hist+arg	Ser+gly+asp	Glu+thy	Ala	Tyr	Meth val	Phe	Leuc isol.	Σ
Gunflint Lake no. 3											
Top 5·5 g	2·27	10·0	11·48	19·66	9·09	9·34	6·49	8·72	1·95	3·76	82·76
Middle 9·4 g	4·65	13·83	21·38	18·32	10·63	8·05	5·31	7·23	2·66	7·70	99·76
Bottom 9·8 g	3·83	8·16	12·34	9·27	7·32	4·66	2·91	2·04	1·82	3·51	55·86
Snowbank Lake no. 2											
Top 2·5 g	12·00	10·00	44·00	52·68	33·16	18·84	8·56	13·32	11·42	8·56	212·54
Middle 2·3 g	8·69	16·30	34·78	45·34	20·52	19·95	21·73	17·65	15·52	18·61	219·09
Bottom	14·24	16·36	57·14	54·07	32·42	18·61	30·47	27·71	23·81	23·81	298·64
Lake Kabetogama no. 5											
Top 15·1 g	12·11	3·31	9·53	5·66	8·74	5·65	0·76	7·48	0·55	4·65	58·44
Middle 14·7 g	1·13	1·77	1·13	1·21	1·09	1·54	0·25	1·36	0·35	1·13	10·96
Bottom 13·1 g	0·30	0·19	0·57	0·83	0·95	0·07	0·13	0·11	0·15	0·52	3·82
Lake Vermillion no. 2											
Top 8·7 g	9·19	21·6	25·59	51·94	38·67	19·04	4·59	15·04	5·27	17·35	208·28
Bottom 10·25 g	14·7	14·7	15·14	33·5	28·23	14·06	3·68	14·7	5·21	15·00	158·92

acid, alanine, valine and leucine are predominant. Kabetogama Lake, St Louis and Kootchiching Counties, Minnesota is transitional to eutrophic conditions and its somewhat copropelic sediments contain moderate amounts of humic acids in which tyrosine is represented in more abundance than usual. Pelican Lake, St Louis County is also transitional to an eutrophic stage but its sediments are more marly than nearby Kabetogama Lake; its amino acid contents are also lower perhaps because of the marly nature of the sediments in which the amino acid content is generally low.

An example of highly calcareous lake sediments is that in Kabekona Lake, Hubbard County, Minnesota. In this lake (table 47) the amino acid contents of the sediments are among the lowest of any of the lakes studied and glycine and alanine are the principal constituents.

The amino acid contents of the deep-water silty and clayey sediments of oligotrophic Flathead Lake, Montana, supplied by meltwater from Glacier National Park is quite low, but is noteworthy for the variety of acids present. The Eh values of the hypolimnetic part of Flathead Lake are negative, which indicates that the accumulating organic matter though small in amount is held under reducing conditions and protected from decay.

The amino acid contents of several additional oligotrophic and early eutrophic lakes in north-eastern Minnesota are shown in table 48. These show higher concentration beneath the surface of the sediment as in the other examples discussed.

The distribution of acid-hydrolyzable amino acids in peats of Rossburg Bog, Aitkin County, Minnesota is shown in fig. 86.

As has been shown in previous studies (Swain *et al.* 1959; Swain, 1961 *a*) the greatest concentration of sedimentary amino acids typically is not in the uppermost layers of the peat, but lies several feet beneath the surface. Because this is believed to be the result of downward migration of water-

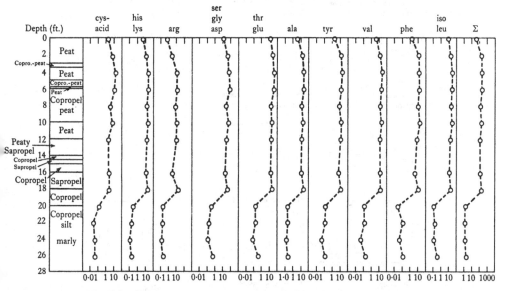

FIGURE 86   Distribution of protein amino acid in Rossburg peat (Swain, 1967*b*). Values shown are in g × 10⁻⁴/g of dry sediment, station 2.

soluble humic acid micelles, to which amino acids or peptides may be adsorbed, there is here an indication that water levels have fluctuated to some extent in the deposit. The quantities of basic amino acids (lysine, histidine, and arginine) that have been preserved in the deposit suggests that acidic conditions have generally prevailed in the bog throughout its history. Two amino acids of uncertain identity which chromatographed near arginine and near tyrosine in paper chromatograms may represent aminobutyric acid of non-protein origin, and methionine, a sulfur-bearing amino acid.

Individual amino-acid distribution in Rossburg peat varies considerably. The relative stability of common pondweed amino acids as well as sedimentary amino acids (Swain, Venteris & Ting, 1964), suggests that valine, alanine, cysteic acid, and glutamic acid among those tabulated here, are the most stable, while phenylalanine, histidine, tyrosine, and serine are fairly

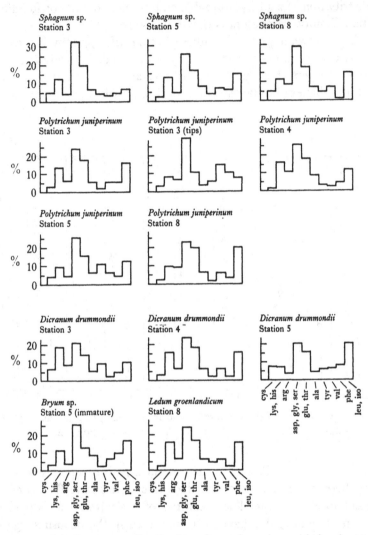

FIGURE 87 Percentage distribution of protein amino acids in plant species from Rossburg peat, Minnesota (Swain, 1967*b*).

unstable. Others, such as leucine, glycine, and arginine are intermediate in stability.

The ratios of the amounts of several amino acids in Rossburg Bog, compared with alanine show interesting relationships to stratigraphy. The ratio val/ala is fairly constant and near unity in the *Sphagnum* peat but is noticeably higher in the underlying copropelic peat, and is noticeably lower in the silty layers below the peat. If it is assumed that these two amino acids

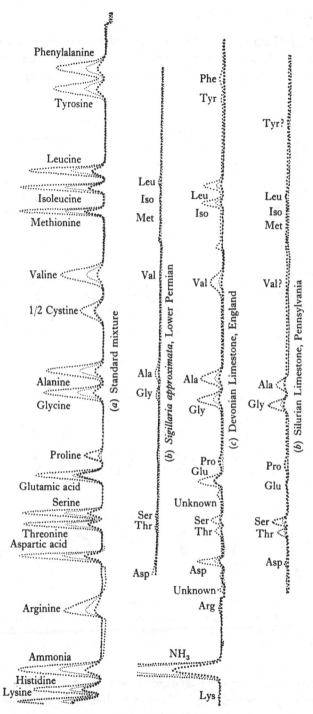

FIGURE 88   Amino acids of fossil plant and rock samples;
automatic amino-acid analyzer chromatogram.

have about equal stability in the enclosing sediments, one may conclude that the relative differences are very likely the result of differences in source organic matter. The tyr/ala ratio is slightly higher in the copropelic peat,

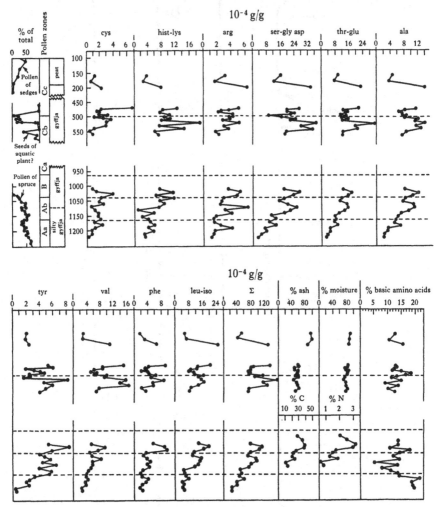

FIGURE 89 Vertical distribution of protein amino acids in sediments of core K-2, Kirchner Marsh, Dakota County, Minnesota (Swain, 1965). Shown are depths, sediment types, pollen zones, selected pollen and seed curves (after Wright *et al.* 1963, and Watts & Winter, 1965), and amino acids (analyzed by G. Venteris).

10–18 ft, than in the overlying *Sphagnum* peat and noticeably higher than in the underlying copropelic silts. The cys-acid/ala ratio, on the other hand does not exhibit any apparent relationship to peat-type above the silts. With regard to the basic amino acids the arg/ala ratio and the arg + lys/ala

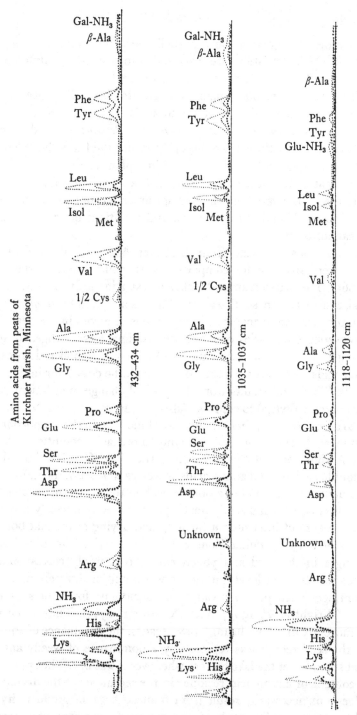

FIGURE 90   Amino acids from peats of Kirchner Marsh, Minnesota;
automatic amino-acid analyzer chromatograms.

ratio are somewhat higher in the copropelic peat and upper part of the copropelic silt. These variations are also thought to be caused by differences in source materials rather than by post-depositional changes.

Several species of plants separated from Rossburg Bog peat were studied for their contained protein amino acids (fig. 87): *Sphagnum* sp. *Polytrichum juniperinum, Dicranum drummondi, Bryum?* sp. and *Ledum groenlandicum*, the first three being bryophytes and the last a dicotyledon. The total amino-acid content of the individual plant specimens is a little higher than that of the near-surface peat as a whole. The values of each amino acid are highly variable from one specimen to another of the same bryophyte species. Whether these are natural variations or are diagenetic in origin cannot be stated.

The amino acids in peats from Kirchner Marsh, Dakota County, Minnesota, were analyzed for comparison with pollen zones in the peat (Swain, 1965). The marsh is an ice-block depression in the St Croix Moraine of late-Wisconsin age in south-eastern Minnesota (Wright *et al.* 1963). Protein amino acids were analyzed from sections of a core in the peat that cross-pollen zone boundaries or distinctive sediment types (figs. 89, 90). In the late glacial part (Zones A–a and A–b) with spruce pollen dominant, total amino acids increase upward, as shown in all the curves for individual amino acids. The trend results from the gradual change in sediment type from silty copropel (gyttja) to copropel (also see curves for C and N). Of the total amino acids however, there is an upward increase in the proportion of basic amino acids (his, arg, lys) implying more acid conditions above 1,120 cm. A short part of the core across the Zone C–b/C–c boundary shows a double peak in total amino acids as well as in most of the individual ones. The double peak matches a double peak in the curve that shows the ratio of seeds of aquatic and semi-aquatic plants to seeds of weedy annuals, a possible measure of intermittent flooding and drying of the lake bottom (Watts & Winter, 1965). Pollen Zone C–b is characterized by several such sharp changes in the seed and pollen curves that imply corresponding fluctuations in the water level during the warm and/or dry period 5,000–7,000 years ago when prairie invaded the deciduous forest of southern Minnesota (Wright *et al.* 1963; Watts & Winter, 1965). Thus when the lake intermittently dried up during a few decades, the dominance of aquatic plants in the seed record was replaced by a dominance of weedy annual plants that spread over the lake bottom whenever it was dry. The apparently lower content of amino acid residues in the sediments of the drying intervals does not necessarily result from reduced organic productivity or dilution by inorganic sediments because the curve for ash content is steady. More than likely the protein amino acids were oxidized during the drying intervals. The slightly higher proportion of basic amino acids during the drying intervals implies more acid conditions in the water during these times, favoring preservation of the basic types.

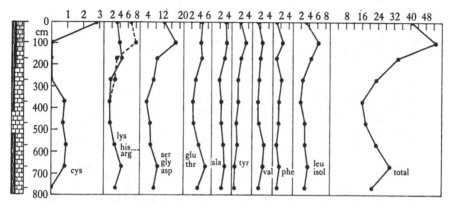

FIGURE 91 Vertical distribution of amino acids in sediments from Blue Lake, Isanti County, Minnesota (Swain, 1965). Broken black, copropel; blocks, marl; dots, silt. Values are in parts per 10,000 of wet sediment.

In the copropelic sapropelic marl sediments of Blue Lake, Isanti County, Minnesota a well-defined relationship exists between stratigraphy and total amino acids (fig. 91) (Swain, Venteris & Ting, 1964). The lower but not the basal part of the sediment shows a relative increase in concentration of amino acids as a probable result of a period of high organic productivity in the early lake history; several of the individual amino acids also increase at that depth. The absence of proline and methionine in the Blue Lake core is an indication that they are unstable under these conditions. The decrease with depth of lysine and tyrosine also reflects their relative instability.

5.4 AMINO ACIDS OF FLUVIAL DEPOSITS

Relatively little study has been made of protein amino acids in river sediments. The modern sedimentary accumulations and associated organisms of Delta–Mendota Canal, a large irrigation canal in northern California were examined for their amino acid contents (table 49) (Swain, 1967 a; Swain and Prokopovich 1969).

The amino acid cystine is low in amount or almost absent in the live *Corbicula* specimens to a distance of about 25 miles. Farther down the

TABLE 49 *Amino–acid analyses of plankton and sediment samples from Delta–Mendota Canal, California*

(a) Neutral and acidic amino acids; (b) basic amino acids, quantities are in 10^{-4} g/g of dry sample (Swain, 1967a).

	Sample	Asp	Thr	Ser	Glu	Pro	Gly	Ala	Cys
(a)	3 (plankton)	439·55	189·55	207·69	468·22	213·71	286·86	232·80	—
	11 (plankton)	19·10	9·52	10·15	16·65	8·61	15·29	11·94	—
	Sediment 95·1 miles	13·37	6·68	7·79	11·26	5·67	10·36	7·73	Tr

	Sample	Val	Meth	Isol	Leu	Tyr	Phe	Σ	Lys
(a)	3 (plankton)	209·67	27·71	172·67	261·01	353·08	254·01	3316·53	112·47
	11 (plankton)	9·36	0·93	6·54	11·17	8·27	7·71	135·24	(b) 4·71
	Sediment 95·1 miles	6·62	0·58	5·69	8·51	5·12	6·12	95·50	6·99

	Sample	His	NH₃	Arg	Glucosamine	Basic a.a.	Total a.a.	Ign loss
(b)	3 (plankton)	228·47	190·97	140·93	None	672·84	3989·37	17·14
	11 (plankton)	8·38	8·61	3·80	None	25·50	160·74	27·71
	Sediment 95·1 miles	3·10	5·30	4·24	None	19·63	115·13	6·56

Canal cystine occurs in increasing amounts. Cystine

$$(SCH_2CH(NH_2)COOH)_2$$

is a major constituent of animal keratin. Much of the cystine in acid hydrolyzates, however, is in the form of the monosulfide cysteine,

$$(HSCH_2CHNH_2COOH)$$

as a result of hydrolytic degradation. The build-up of cystine and/or cysteine in the clam samples is at present unexplained but may be due to a change in the food supply of the clams down the Canal.

In a study of the thermal stability of protein amino acids Vallentyne (1964) showed that several amino acids degrade to other amino acids and to other products, as follows:

Glutamic acid	$\longrightarrow$	γ-aminobutyric acid
Glycine	$\longrightarrow$	methylamine
Alanine	$\longrightarrow$	ethylamine
Serine	$\longrightarrow$	glycine, alanine, and ethanolamine
Threonine	$\longrightarrow$	glycine
Phenylalanine	$\longrightarrow$	phenethylamine and benzylamine
Methionine	$\longrightarrow$	glycine and alanine
Arginine-HCl	$\longrightarrow$	proline, unknown near ornithine
Aspartic acid	$\longrightarrow$	malic acid and ammonia
Proline	$\longrightarrow$	no ninhydrin-reactive products
Hydroxyproline	$\longrightarrow$	no ninhydrin-reactive products.

Vallentyne's analyses were made under the following conditions: degassed 0·01 M solutions of single amino acids were sealed in Pyrex glass tubes in the absence of oxygen; the tubes were sterilized in a boiling water bath; thermal treatment of the tubes and samples was in a sand bath at temperatures ranging from 180 to 282 °C; the residual amino acids and products were analyzed by the Moore–Stein method (i.e. see figs. 88, 90).

The thermal degradation studies of Vallentyne were made on pure compounds and are not actually relevant to natural systems. Nevertheless they are useful as a first approximation of what may happen to individual amino acids during degradation.

The ratios of these amino acids in the Canal samples of *Corbicula fluminea* and in a few of the associated sediments are shown in table 48 and figs. 92, 93. The rather irregular nature of the ratios val/gly, ser/gly, and arg/pro in *Corbicula* in the upper 30 miles of the Canal may be due to the

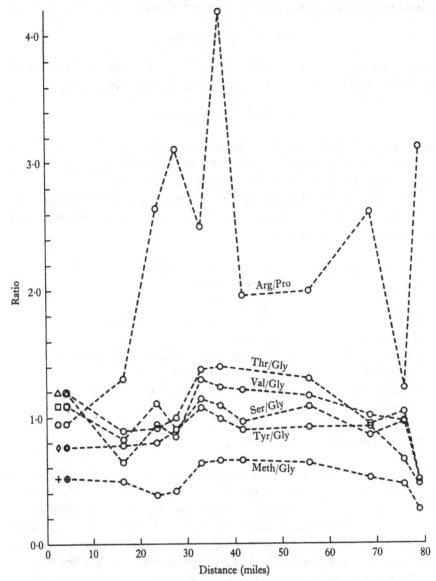

FIGURE 92 Amino-acid ratios in the body of *Corbicula fluminea* from Delta–Mendota Canal (Swain, 1967*a*; Swain and Prokopovich, 1969).

heterogeneous nature of the other organic matter of the Canal that contributes to the food supply of *Corbicula*. The relative uniformity of the ratios, other than arg/pro, between 30 and 75 miles perhaps indicates a constancy of food supply of the clams in that part of the Canal. The striking increase at 79 miles in glycine relative to threonine, valine, serine, tyrosine and methionine suggests that protein degradation may be taking place in some of the clams in the lower part of the canal. Whether this is the result of some change in the food supply or of another environmental factor at present cannot be determined.

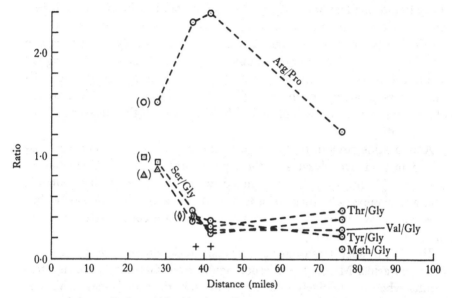

FIGURE 93 Amino-acid ratios in sediments associated with *Corbicula fluminea* from Delta–Mendota Canal (Swain, 1967a; Swain and Prokopovich, 1969).

The amino-acid contents of plankton samples in Delta–Mendota Canal is highly variable (table 49) but appear to decrease down the canal. The decrease probably is due to the gradual dying downstream of planktonic organisms that have entered the canal intake.

The free amino-acid distribution and their uptake by planktonic bacteria in York River Estuary, Virginia, were studied by Hobbie *et al.* (1968). The water contained 38 μl/l free amino acids, the most abundant of which were glycine, serine and ornithine. When examined for uptake by bacteria the greatest flux rates were shown by glycine, methionine and serine. The total flux of amino acid was 1–10 % of the daily photosynthetic carbon fixation.

Only a few studies have been made of the amino acid residues in non-marine sedimentary rocks. Jones & Vallentyne (1960) analyzed hydrochloric acid hydrolyzates of organic-rich Eocene Green River Shale from Rifle, Colorado. They found small amounts of aspartic acid, glutamic acid, alanine, leucine, non-protein γ-aminobutyric acid and an unknown compound near tyrosine on their paper chromatograms. These amounted to only 0·014% of the total organic nitrogen in the sample. The authors believe the original amino acid suite, judging from that of similar modern organic lake sediments, was probably much more diverse and through geologic time the other amino acids were complexed with the kerogen of the oil shale and lost their identity.

Amino acids amounting to 10–50 p.p.m. of alluvial sediment were obtained in acid hydrolyzates of the Upper Jurassic Schuler Formation of Louisiana (Swain, 1961b), and included cystine, arginine, glycine, glutamic acid and alanine. The rock consists of both red and green mudstones but it is not known whether the organic materials are concentrated more in one than in the other type.

Highly humified freshwater sedimentary rocks do not give good yields of amino acids. Miocene coniferous lignite from Germany gave negative results when exhaustively extracted with sulfuric acid (Jones & Vallentyne, 1960) although the lignite contained 0·26% N on a dry weight basis.

When present in high concentrations of organic matter, sedimentary amino acids apparently undergo rapid complexing reactions with the associated humus, whereas in clayey sediments the mineral particles exert a stabilizing effect that helps to preserve the amino acids (Kroplein, 1964).

In their analyses of amino-acid contents of lacustrine and marine deposits of the Ruhr region, Germany, Degens & Bajor (1962) found considerable differences between the amino acids of the two environments. The thirteen samples they analyzed were from the Upper Carboniferous and included five lacustrine samples containing *Najadites* and *Carbonicola*, four marine fossil-bearing samples containing pectinoids and goniatites, two coal samples and two other rock samples. The average contents of C, H, N, ash, total amino acids, and bitumens are given in table 50. The total amino acids are appreciably higher in the marine than in the lacustrine deposits in spite of the higher carbon contents of the latter. The individual amino-acid analyses of the samples are summarized in table 50. These analyses, however, show that except for arginine, the amino acids of the marine deposits do not greatly exceed those of the freshwater deposits in

TABLE 50   *Amino-acid contents of lacustrine and marine deposits of Ruhr region, Germany (10·5%) (Degens & Bajor, 1962)*

Sample no.	Ala	Tyr	Val	Cys	Asp	Glu	Ser	Thr
1	2·83	—*	1·42	6·34	1·17	—	3·96	Sp*
2	5·89	3·90	1·67	1·17	2·22	3·44	8·56	4·10
3	3·00	Sp	2·25	7·42	1·17	Sp	Sp	—
4	3·25	Sp	2·00	4·34	—	—	3·25	Sp
5	2·50	1·87	3·83	3·83	1·75	1·58	5·67	0·50
Av.	3·49	1·15	2·23	4·62	1·26	2·00	4·29	0·92
6	3·84	3·33	3·00	5·08	—	2·08	Sp	0·33
7	1·67	2·00	1·67	4·50	—	Sp	Sp	0·50
8	2·50	1·25	3·58	2·00	1·08	2·00	2·83	0·83
9	4·00	4·16	2·25	5·34	Sp	1·42	1·17	0·83
Av.	3·00	2·69	2·63	4·23	0·27	1·38	1·00	0·62
10	12·05	3·25	8·76	6·25	8·75	12·50	16·02	5·00
11	15·26	3·25	12·25	5·76	3·25	10·75	15·03	6·25
Av.	13·66	3·25	10·50	6·00	6·00	11·63	15·53	5·63
12	4·87	2·33	1·92	1·17	0·73	1·59	2·25	0·71
13	3·78	1·34	2·55	2·00	1·45	2·89	4·89	1·78

*Sp, present in slight amount; —, absent.

Sample no.	Lys	Arg	Pro	Gly	Lys+u.** Arg	Lys+u. Arg/Ala	Remarks
1	1·59	—	0·58	4·00	1·59	0·56	Lacustrine shales of
2	2·55	Sp	3·25	14·20	2·55	0·43	Ruhr Carboniferous
3	4·08	Sp	1·50	3·08	4·08	1·36	
4	1·83	—	1·83	2·75	1·83	0·56	
5	2·42	1·08	1·58	2·08	3·50	1·40	
Av.	2·69	0·22	1·75	5·22	2·71	0·77	
6	4·58	27·80	2·75	2·83	32·38	8·45	Marine shales of Ruhr
7	1·25	9·92	Sp	0·92	11·17	6·65	Carboniferous
8	3·50	7·32	1·33	5·75	10·82	4·32	
9	2·08	3·58	19·52				
9	8·92	10·60	2·08	3·58	19·52	4·90	
Av.	4·56	13·91	1·54	3·27	18·47	6·16	
10	10·50	4·25	7·26	33·70	14·75	1·22	Coal beds (Ruhr)
11	6·76	7·76	12·77	21·150	14·52	0·95	Coal beds (Saar)
Av.	8·63	6·00	10·02	27·60	14·64	1·07	
12	2·37	2·46	1·63	9·54	4·73	9·97	Shales (Saar)
13	5·12	32·50	—	8·88	37·62	9·60	

**u, unknown.

individual cases (tyrosine, valine and lysine) and in the majority of cases the lacustrine samples exceed the marine samples (alanine, valine, cystine, aspartic acid, glutamic acid, threonine, proline and glycine). Since arginine chromatographs on filter paper at a position which also is a troublesome one for mineral salt separation, some of the arginine reported in the Degens and Bajor study might include salts. Aside from the arginine values the amino-acid distribution is remarkably similar in the two rock types and suggests similar conditions of organic productivity and preservation.

Several amino acids have been separated from Pennsylvanian and Permian freshwater fossil plants, and from Paleozoic marine rocks (Swain & Kraemer, 1969) (fig. 88). *Calamites Suckowi* from the Upper Carboniferous of England contained: aspartic acid, serine, glycine, alanine, methionine, isoleucine, leucine, lysine and histidine totalling 0·26 μg/g, *Sigillaria approximata* from the lower Permian of Pennsylvania contained a similar suite of amino acids plus threonine, valine and proline totalling 0·38–1·19 μg/g. These may not be indigenous to the fossil plants judged from the presence of threonine and serine which Abelson & Hare (1968) believe are too unstable to have survived since the Paleozoic.

5.6 SUMMARY

The available data on amino acids in lake sediments show only little or no relationship between the lacustrine environment and the relative quantities of individual amino acids preserved in the bottom deposits. The visibly organic deposits show a slight tendency to yield large amounts of low R_F amino acids (on paper chromatograms), such as arginine, histidine and cystine.

Free amino acids were not found in the lake sediments studied or were present only in small quantities. It has been assumed therefore that the amino acids were present in proteins, other peptides, or perhaps linked or absorbed to the humic-acid micelles (Swain, Blumentals & Millers, 1959). Inasmuch as proteins soluble in water, salt solutions, and alcohol are not present or are rare in these samples, the proteins would be mainly glutelins and scleroproteins. The peptides and humic-acid-linked amino acids probably were derived from the other proteins.

The lake and bog samples studied for amino acids can be classified as slightly humified, partly humified, or well humified based on carbon–nitrogen ratios. The C:N ratio in marine plants on and in fresh marine sediments is about 9·2–1 or 9·4–1 (Trask, 1939; Sverdrup *et al.* 1942, p. 1011), that of lithified sediments is 15–1 or 16–1. In the present work, values of < 9:1, 9:1–12:1, and > 12:1 are taken roughly to indicate slightly, partly and well-humified sediments.

The total amino acids in the lake and peat samples studied show a relationship to degree of humification. The poorly humified sediments of Green Lake, Minnesota have about the same amino acid content as the peats of Cedar Creek Bog, but only 1/70 of the total carbon of the peats. The other lake samples in the partly humified state for which data are available show a gradually decreasing amino-acid content to correspond to decreasing carbon content.

The peats of Cedar Creek Bog and Dismal Swamp are characterized by a high percentage of neutral amino acids (fig. 94), 70–100% of the total. The marl underlying the peat of Cedar Creek Bog is exemplified by lower percentages of neutral amino acids and larger amounts of acidic amino acids, down to the 8·7–9 m layer of Cedar Creek Bog, wherein the marl content decreases and the organic content increases. The percentage of neutral amino acids in the 8·7–9 m layer is like that in the peats above the marl. The lower part of the peat in Dismal Swamp has increasing amounts of basic amino acids. These variations in types of amino acids in different sediment-types is believed to be at least partly caused by the dipolar (Zwitterion) characteristics of the amino acids.*

The neutral to slightly alkaline conditions in Cedar Creek Bog peat seems to favor, through processes of microbial decomposition, the persistence of neutral and acidic amino acids in proportions of about 6:1 and allow the basic amino acids to be degraded through deamination by anaerobic bacteria. Although its exact mechanism is uncertain, a possible way for the amino acids to assume the stable salt form under natural conditions: microbial decomposition of proteins under anaerobic conditions breaks the peptide linkages between the individual amino acids; the freed amino acids are in part immediately used by other micro-organisms, in part assume acid or base salt forms, depending on environmental conditions, and in part are further degraded; the stabilized salt forms enter into a ligno-humic complex, the properties of which depend on the original source material and the degree to which humification has progressed. In the copropelic marl underlying the peat of Cedar Creek Bog the alkalinity increases and there is partial elimination of the neutral amino acids through deamination, and relative enrichment of the acidic amino acids. The latter seem to be more stable biochemically under alkaline conditions because of the formation of base salts as suggested above.

* The three forms assumed by a monoaminomonocarboxylic acid are as follows:

$$Cl^-\,H_3N^+\!\!-\!\!\overset{\displaystyle R}{\underset{\displaystyle |}{C}}H\!\!-\!\!COOH \underset{HCl}{\overset{NaOH}{\rightleftarrows}} H_3N^+\!\!-\!\!\overset{\displaystyle R}{\underset{\displaystyle |}{C}}H\!\!-\!\!COO^- \underset{HCl}{\overset{NaOH}{\rightleftarrows}} H_2N\!\!-\!\!\overset{\displaystyle R}{\underset{\displaystyle |}{C}}H\!\!-\!\!COO^-Na^+$$

Acid salt form   Zwitterion   Base salt form

(102)

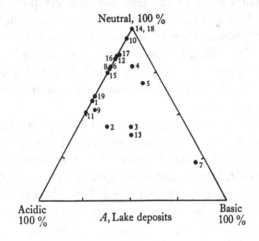

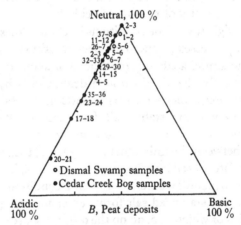

FIGURE 94   Per cent distribution of neutral acidic and basic amino acids in lake and peat deposits (Swain, 1961*a*). Lake deposits numbered as follows: 1, Green Lake, station 6; 2, Green Lake, station 5; 3, Fannie Lake, station 5; 4, Blue Lake, station 5; 5, Spectacle Lake, station 5; 6, Rush Lake, station 8; 7, Eagle Lake, station 2; 8, Stanchfield Lake, station 5; 9, Clear Lake, station 1; 10, Rainy Lake, station 1; 11, Lake of the Woods, station 3; 12, Kabetogama Lake, station 1; 13, Pelican Lake, station 1; 14, Kabekona Lake, station 5; 15, Reno Lake, station 12; 16, Prior Lake, station 1 (Swain, 1956, p. 608); 17, Pyramid Lake, station 18 (Swain & Meader, 1958, p. 287); 18, Catahoula Lake, Louisiana; 19, Lake Ponchartrain peat, Louisiana. For peat deposits: circles indicate depths in Dismal Swamp peat; discs indicate depths in Cedar Creek Bog peat (Swain *et al.* 1959, p. 120).

The peats of Dismal Swamp are more acid than those of Cedar Creek Bog, and in such acid conditions production of the acid salt forms of the amino acids are evidently favored. Therefore the acidic amino acids having several carboxyl groups would be more susceptible to bacterial degradation through decarboxylation. The basic amino acids at the same time would tend to be preserved through formation of the stable acid–salt form.

The neutral bog environment, in summary, appears to favor preservation of neutral and acidic amino acids roughly in the ratio $6n:1a$; alkaline bogs favor an increase of the proportion of acidic amino acids to provide a ratio of about $3n:1a$ neutral to acidic amino acids; acidic bogs favor a preservation of some of the basic amino acids and is detrimental to preservation of the acidic amino acids in the proportions 75–$95n:5$–$15b:0$–$10a$. Such ratios as these would be expected where the stability of the peats and marls had been reached through the formation of humus, phenolic acids, and other preservative substances. If the deposit were to be drained or undergo other changes that would disrupt its stability there would be resulting changes in the amino acid suites.

Eutrophic lake deposits differ considerably in their contents of sedimentary amino acids, depending on the source organic matter, type of lake and degree of humification of the organic matter. Among the lakes mentioned herein, the marly and copropelic deposits of Blue Lake, Rush Lake, Stanchfield Lake, Reno Lake, and Big Island Bog, Lake Minnetonka, all in Minnesota, have ratios of about $3n:1a$ amino acids (Fig. 94), and these sediments apparently were well humified at the time of sampling. On the other hand, the copropelic and sapropelic in part calcareous deposits of Spectacle Lake, Eagle Lake, and Clear Lake, Minnesota have high proportions of basic amino acids and probably contained considerable unhumified proteinaceous material at the time of sampling. As microbiological humification proceeds a sort of crude proportional stability of the neutral, acidic, and basic amino acids is reached depending on diagenetic conditions.

Alkalitrophic lakes of which Kabekona Lake, Minnesota is an example (Chapter 3), are characterized by low total amino acids, nearly all neutral types, despite the alkaline environment. This is apparently a statistical matter and results from a natural predominance of the neutral amino acids in most types of accumulating lake sediments.

Oligotrophic lakes with slightly acidic waters such as Rainy Lake, Minnesota–Ontario have low total amino acids which statistically are mainly neutral types, basic amino acids are low to absent, and acidic amino acids are also low. Rainy Lake may also be considered dystrophic since its waters are colored brown by humic substances. Introduction of

Protein amino acids of non-marine deposits

humus by surface drainage into an oligotrophic lake results in one dystrophic type; the other is that in the late stages of eutrophication in which the indigenous humic content is high but mineral substances have been depleted and productivity, as a result, declines. Conditions of preservation of the amino acids in Rainy Lake are not favorable because of the high oxidation potentials of the waters and upper sediments.

Catahoula Lake, Louisiana represents an acidic environment of high redox potentials unfavorable to preservation of organic matter, although total organic productivity is greater than that of Rainy Lake. The total amino-acid contents of Catahoula sediments is low, and neutral types statistically predominate.

The apatotrophic lake, Pyramid Lake, Nevada, which is typified by high total dissolved solids and sodium alkalinity but low organic productivity because of the restriction in biota, has amino acid ratios similar to that of the neutral peat bog ($6n : 1a$).

6 Carbohydrates of non-marine deposits

6.1 CARBOHYDRATES OF AQUATIC SOURCE ORGANISMS

6.1.1 *Carbohydrates of freshwater and marine organisms*

The four principal algal polysaccharides are agar, carageenan, alginic acid and laminaran. Galactose is the principal monosaccharide in the first two, mannuronic acid in the third type and glucose in the last type. In addition to these polysaccharides, others occur in some of the marine as well as freshwater algae (Swain, Pakalns & Bratt, 1968).

In addition to glucose and galactose, freshwater algae contain arabinose and xylose. Arabinose occurs in aqueous extractions of *Nostoc* and other blue-green algae and as part of an amorphous polysaccharide in *Anabaena* (Fogg, 1956). Xylose is believed to occur as part of a mucilaginous complex in some blue-green algae (Hough *et al.* 1952).

The various algae studied (Swain, Pakalns & Bratt, 1968) in addition to galactose and glucose have relatively large amounts of rhamnose (Chlorophyta), ribose, perhaps of microbial origin (Phaeophyta), and xylose (Chlorophyta and Phaeophyta). L-Fucose, a constituent of some marine algae, was not detected but may have chromotographed with xylose. Rhamnose was found as a component of mucilaginous algal polysaccharides (Hough *et al.* 1952).

A species of lichen (*Parmelia flaventior*) contains 44% glucose, 32% galactose, 7·8% mannose, 8·4% arabinose, 6% xylose and 5% rhamnose. A living cryptogam, *Phylloglossum*, believed to be a lycopod (White *et al.* 1967) contained sucrose, glucose and fructose as principal carbohydrates.

Several species of bryophytes contained appreciable amounts of pentoses in addition to galactose and glucose (Swain, Pakalns & Bratt, 1968): glucose 18–30%, galactose 26%, mannose 12–34%, arabinose 5–6%, xylose 7–13%, ribose 1% and rhamnose 10–13%.

The higher aquatic plants contain xylose in appreciable quantities in addition to glucose and galactose (Rogers, 1965; Swain, Pakalns & Bratt, 1968).

Marine and freshwater animals, including anemone, barnacle, freshwater ostracodes and starfish yielded principally glucose and galactose with lesser amounts of ribose (Swain, 1968a; and unpublished data).

Carbohydrates of non-marine deposits

The polysaccharides of aquatic source organisms include cellulose, starch, pectic substances, algal polysaccharides and others, but those of freshwater organisms have not been studied to any extent beyond the work of Hough *et al.* 1952, and Fogg (1956) referred to above.

6.1.2 *Chemical constitution of pollen*

The structure and chemistry of the pollen wall has recently been discussed by Heslop-Harrison (1968). Two polysaccharides are important constituents of pollen: cellulose and callose. The cellulose occurs in the inner pollen wall or intine, and in the walls of the spore protoplasts as a microfibrillar matrix. Callose, a non-microfibrillar β-1 $\rightarrow$ 3-linked glucan occurs as a wall material in the sporocytes of the pollen grain during the meiotic prophase and as a sealing and plugging agent in the spore tetrad walls. Callose is more rapidly synthesized and destroyed than cellulose in pollen. Degradation of callose is accomplished by an enzyme, callase, the early breakdown products of which include oligosaccharides, having 1 $\rightarrow$ 3 linkages such as laminaribiose and laminaritriose. In addition, the spore walls of the club-moss *Lycopodium* and of pine contain about 10 per cent of xylan and/or hemicellulose.

Three other chemical components are important in pollen: sporopollenin, lipids and carotenoids. Sporopollenin, the highly resistant main constituent of the outer pollen wall or exine, has an empirical formula ranging from $C_{90}H_{134}O_{31}$ to $C_{90}H_{150}O_{33}$. Although sporopollenin was once thought to be polyterpenoid in make-up, it is now believed to be principally of lipid and lignin-like composition. The lipid fraction contains mono- and dicarboxylic acids with up to an apparent maximum of 16 carbon atoms. The lignin fraction does not exhibit all of the microchemical properties that are shown by typical wood lignin, so presumably the lipid fraction protects the lignin part in some chemical way. Free aldehyde groups are known to occur in sporopollenin of the exine formed during the free-spore period of pollen development. This and other reactive properties decrease in the mature sporopollenin.

The pollen coat that forms on the outside of the exine contains carotenoid pigments and lipid compounds that are in part oleaginous. These seem to protect the pollen gametophytes from ultraviolet radiation, serve to attract insects, and perhaps have other functions.

Pollen constituents may contribute in a significant way to the organic geochemistry of oligotrophic lake sediments wherein the pollen may comprise the principal organic residue. The oligotrophic lake sediments of north-eastern Minnesota commonly contain spruce and pine pollen in abundance, and have been suggested as having contributed much of the hydrocarbons of the sediments (Swain, 1956). Detailed studies of the pollen organic matter in sediments have not yet been carried out.

6.2 CARBOHYDRATES OF AQUATIC SEDIMENTS AND ASSOCIATED SOURCE ORGANISMS OF EUTROPHIC LAKES

The carbohydrate contents of eutrophic lake sediments in Minnesota consisted of both free sugars and polymeric sugars that originated from both inside and outside the sediment (Rogers, 1965), but the exact sources cannot be determined for much of the carbohydrate.

The carbohydrates of lake and moss peat accumulations in Minnesota showed an increase about 1 m beneath the surface (Swain, 1967b). As in the case of the lake sediments studied by Rogers the source of the carbohydrates seems to be partly in the sediments and partly outside. Very little other study of carbohydrates in freshwater sediments has been made.

The alkalitrophic marly sediments of Clear Lake, Sherburne County, Minnesota were briefly described in Chapter 3. In midwinter 1950, the pH of the water was 7·3–7·7 and the Eh was $+108$ to $+245$ mV; the pH of the calcareous sediments was 7·0–7·3 at the surface and 6·8–7·5 at depths of 2·5 ft; Eh values of the sediments at corresponding depths were -75 to $+135$ mV and -56 to $+155$ mV at a depth of 6 ft in the sediments. The pH of medium grain sand underlying the marl was 7·0 and Eh was -45 mV. The diagenetic setting in the lake therefore is one of weakly oxidizing to weakly reducing intensities in the water and of moderately to strongly reducing intensities in the sediments (Swain, Venteris & Ting, 1964). Total carbohydrate contents of the plants and sediments were determined by the phenol–sulfuric acid method. Free sugars were extracted with ethanol and separated by paper chromatography. Polymeric sugars were extracted with dilute sulfuric acid and separated by paper chromatography. The larger aquatic plants of the lake, five species of which were analyzed for individual carbohydrates (Rogers, 1965) include: *Ceratophyllum demersum* Linn., *Myriophyllum exalbescens* Fern., *Chara* sp., *Najas guadalupensis* (Spreng) Magnus, *Potamogeton illinoiensis* Morong, *Potamogeton richardsoni* (Benn.) Rydb., and *Typha latifolia* Linn. The amino acids of these plants were also studied (Swain, Venteris & Ting, 1964). The lake is rich in diatoms representing the following: *Stephanodiscus* spp. (common), *Melosira* sp. (common), *Cyclotella* sp. (rare), *Navicula* 4 spp. (common), *Pinnularia* sp. (rare), *Cymbella* sp. (frequent), *Cocconeis* sp. (abundant), *Meridion* sp. (rare), *Rhopolodia* sp. (rare), *Epithemia* sp. (frequent), *Gomphonema* sp. (rare), *Synedra* sp. (rare), *Eunotia* ? sp. (rare),

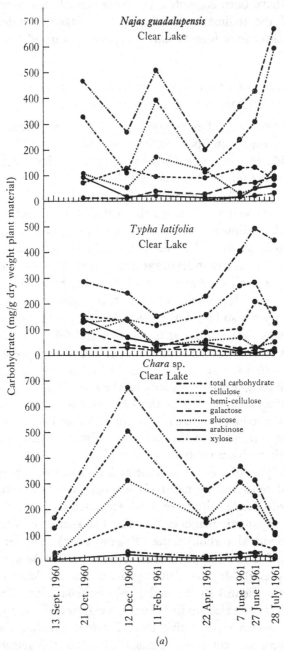

FIGURE 95 Seasonal variation in carbohydrate materials in predominant aquatic plant samples: (*a*) *Najas guadalupensis, Typha latifolia,* and *Chara* sp.

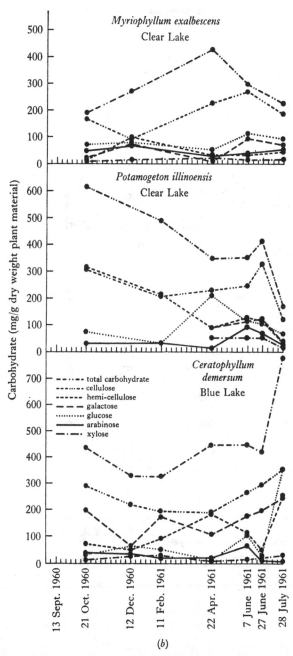

FIGURE 95 Seasonal variation in carbohydrate materials in predominant aquatic plant samples: (*b*) *Myriophyllum exalbescens* (Clear Lake), *Potamogeton illinoiensis*, and *Ceratophyllum demersum*.

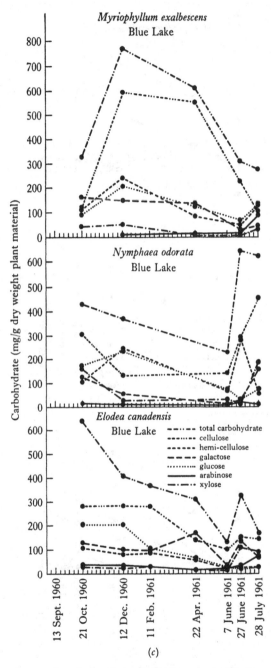

FIGURE 95 Seasonal variation in carbohydrate materials in predominant aquatic plant samples: (c) *Myriophyllum exalbescens* (Blue Lake), *Nymphaea odorata*, and *Elodea canadensis*.

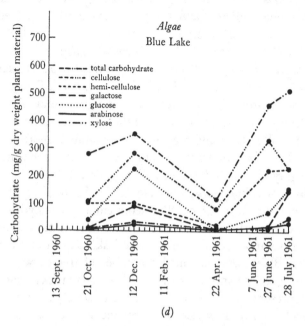

Carbohydrate (mg/g dry weight plant material)

Algae
Blue Lake

----- total carbohydrate
----- cellulose
···--- hemi-cellulose
-- -- galactose
········· glucose
——— arabinose
-·-·· xylose

(*d*)

FIGURE 95   Seasonal variation in carbohydrate materials in predominant aquatic
plant samples: (*d*) Algae. (Rogers, 1965)

Fragilaria sp. (rare). The littoral areas of Clear Lake contain abundant
scavenging gastropods and pelecypods; cladocerans, ostracodes, insects and
sponges are also present. All these organisms in addition to fish, worms and
Protista contribute to the carbohydrate contents of the sediments.

The carbohydrates of several of the aquatic plant species of Clear Lake,
from a seasonal point of view are shown in fig. 95 (Rogers, 1965). The
greatest carbohydrate contents are more related to the species than to
season of the year. Glucose was found to be the commonest monosaccharide
and the content of other simple sugars as well as of cellulose and hemicellulose shows a direct relationship to total carbohydrate content.

Blue Lake, Isanti County, Minnesota is an elongate ice-block depression
in red gravel and till of the Superior ice lobe (Swain, 1961 *a*). The lake is
in a eutrophic–alkalitrophic condition that in 1949 had the following water
properties: total alkalinity 112 p.p.m., sulfate 4·5 p.p.m., total P 0·012
p.p.m., total N 0·36 p.p.m. There is no well-developed thermocline but
the bottom waters typically are depleted of oxygen in deeper parts of the
lake in the summer. The bottom sediments are diatomaceous, sapropelic,
and copropelic marl. The pH of the water in midsummer 1958 was 7·8–8·4;
in the upper part of the sediments the pH was 7·4–7·5 and at a depth

217

of 2·5 ft the pH was 6·6–6·7 and the corresponding Eh values were −66 to +459 in the upper sediment and +173 to +431 in the lower sediment. The larger aquatic plants of Blue Lake include those noted above for Clear Lake, plus *Elodea canadensis* Michx., *Lemna trisulca* Linn.; *P. zosteriformis* Fern.; *Sagittaria latifolia* Willd. forma *gracilis, Scirpus validus* Vahl., and *Vallisneria americana* Michx. The algae *Anabaena* and *Polycystis* form waterblooms, and ostracodes, cladocerans, gastropods and diatoms are abundant in the present lake deposits. The ratio of producers to consumers is much higher in Blue Lake and organic sedimentation is more rapid than in Clear Lake. The lake sediments beneath 3–5 ft of copropelic and sapropelic marl are gray marl to a depth of 25 ft or more.

The carbohydrates obtained by acid hydrolysis of several aquatic plants in Blue Lake are shown in fig. 95. The sedimentary carbohydrates, both combined and free, are listed in tables 51 and 52. The vertical distribution of monosaccharides in a deep core from Blue Lake is given in fig. 96.

The water of Blue Lake contained a number of free sugars: glucose, mannose, sucrose, xylose and 'near raffinose'. Rogers (1965) suggests that this list may represent mainly a measure of the relatively higher solubility of these particular sugars in water. The absence of galactose, a relatively insoluble sugar in water, but widespread in the other samples from Blue Lake substantiates this belief.

The low yield of monosaccharides in the surface sediments compared to total carbohydrates suggests either that experimental recovery of monosaccharides was very incomplete or that some unidentified sugars or their derivatives were present in appreciably large amounts. Both possibilities are believed by the present writer to be effectual in these samples.

It appears likely that the xylose and mannose detected in analyses of acid extracts of the sediments were also recoverable as free sugars by ethanolic extraction.

From 1 to 10% of the amount of the sugars of the aquatic plants occurs in the associated sediments. The loss is attributable to the weakly oxidizing to weakly reducing conditions that prevail in the lake and allowed slow but incomplete breakdown of the polysaccharides. The presence of glucuronic acid in the sediments indicate that micro-organisms may have been the source in addition to aquatic plants. Vallentyne & Bidwell (1956) and Rogers (1965) viewed the problem of the preservation of free as well as hydrolyzable sugars in the lake water and sediments in terms of a steady-state equilibrium. In the case of free sugars the balance is between recruitment of free sugar molecules, from sinking plankton and hydrolysis of polysaccharides, and emigration, by the breakdown of sugars during bacterial metabolism. The presence of free sugars in the lake samples indicates

TABLE 51 *Sugars released by acid hydrolysis of lake bottom sediments*
(Rogers, 1965)

(mg sugar/g dry sediment)

Lake (sediment) type	Date	T.C.T.	% Cellulose and hemicell	% T.C.T.*	Cellulose	Hemicell	Ribose	Mannose	Rhamnose	Glucuronic acid	Galactose	Glucose	Arabinose	Xylose	Total sugar	% T.C.T.
Blue Lake (sapropel)	28 July 1961	30·1	36·5	121·3	19·5	17·0	0·3	0·5	0·5	—	1·8	1·6	2·0	1·6	8·3	27·6
Blue Lake (sapropel)	21 Oct. 1960	35·8	29·6	82·7	7·2	22·4	1·0	0·1	0·6	—	1·4	2·1	1·9	0·2	7·3	20·4
Clear Lake (marl)	28 June 1961	5·3	4·7	88·7	2·3	2·4	0·3	—	0·3	—	1·5	—	0·8	1·0	3·9	73·6
Clear Lake (copropel & marl)	28 June 1961	26·2	26·2	100·0	8·7	17·5	1·1	2·4	3·2	—	2·9	—	1·8	2·1	13·5	51·5
Clear Lake (copropel & sand)	21 Oct. 1960	21·7	23·3	107·4	9·9	13·4	0·5	1·3	—	—	0·8	—	—	2·5	5·1	23·5
Clear Lake (copropel & marl)	22 Apr. 1961	34·1	32·4	95·0	11·0	21·4	0·2	—	—	—	1·1	—	2·2	1·2	4·7	13·8
Clear Lake (copropel)	28 June 1961	44·0	49·4	112·3	24·2	25·2	—	—	—	5·3	0·6	1·3	19·1	2·3	28·6	65·0
Clear Lake (copropel)	22 April 1961	18·2	12·0	65·9	3·5	8·5	0·1	—	—	—	0·4	—	0·8	0·3	1·6	8·8
	Total	215·4	214·1	99·4	86·3	127·8	3·5	4·3	4·6	5·3	10·5	5·0	28·6	11·2	73·0	33·9
	Average	26·9	26·8	99·4	10·8	16·0	0·4	0·5	0·6	0·7	1·3	0·6	3·6	1·4	9·1	33·9

* T.C.T., total carbohydrate test by phenol–sulfuric acid method.

TABLE 52 *Free sugar determinations on bottom sediment samples from
Blue Lake and Clear Lake, Minnesota (Rogers, 1965)*

(mg sugar/g dry sediment)

Lake (sediment type)	Date	Method	Malt-ose	Suc-rose	Lac-tose	Raffinose near Raffin-ose	Gluc-ose	Galac-tose	Xylose	Man-nose
Blue Lake 1 (sapropel)	28 July 1961	Water	Trace	None	None	None	None	None	0·1	None
Blue Lake 2 (sapropel)	28 July 1961	70% EtOH	None	None	None	Trace	None	0·1	2·1	0·8
Clear Lake 3 (copropel)	22 Apr. 1961	Water	None	None	None	None	None	None	0·2	None
Clear Lake 4 (copropel)	22 Apr. 1961	70% EtOH	None	None	None	None	None	None	1·9	None
Clear Lake 5 (marl & copropel)	21 Oct. 1960	Water	None	None	None	Trace	None	None	1·2	1·8
Clear Lake 6 (marl & copropel)	21 Oct. 1960	Water & ultrasonics	Trace	None	None	Trace	1·1	None	0·8	0·9
Blue Lake 7 (sapropel)	21 Oct. 1960	70% EtOH	Trace	Trace	Trace	Trace	None	0·9	None	1·9
Blue Lake 8 (sapropel)	21 Oct. 1960	70% EtOH & ultrasonics	None	Trace	None	Trace	None	None	None	None

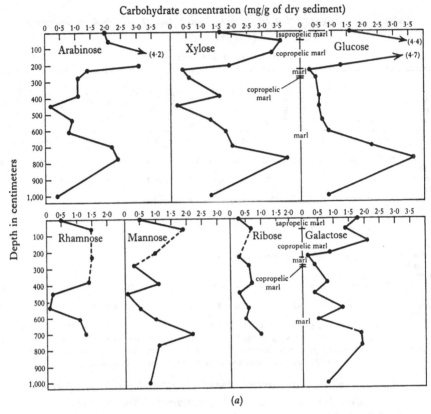

FIGURE 96   Fluctuations of carbohydrate materials with depth in deep Blue Lake Minnesota core (Rogers, 1965).

that their formation is going on faster than their breakdown. In the case of the hydrolyzable, presumably polymeric, carbohydrates, the balance is between recruitment of carbohydrates from sinking organic debris and its incorporation with sediments, and emigration due to slow leakage from resistant preservative mechanism. In the latter case the presence of large quantities of hydrolyzable sugars shows that accumulation and preservative stabilization are dominant.

The individual sugars of the aquatic plants and of the sediments of Blue Lake and Clear Lake exhibit a grouping related to their natural stability to seasonal and depth fluctuations (Rogers, 1965):

> Fairly stable—xylose, glucose, rhamnose and arabinose.
> Moderately stable—ribose, mannose.
> Fairly unstable—galactose.
> Very unstable—glucuronic acid.

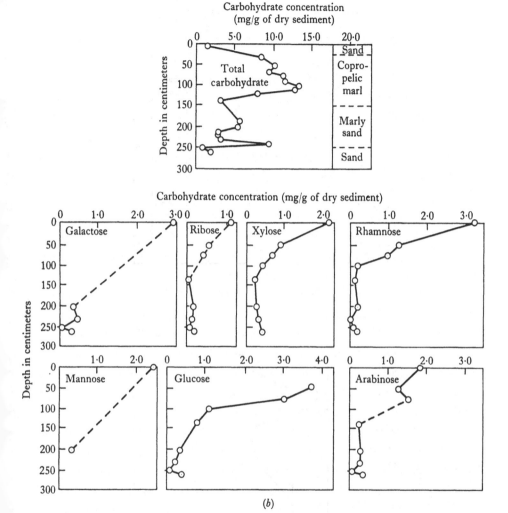

FIGURE 96   Carbohydrate materials in Clear Lake Minnesota core.

Although glucose is one of the monosaccharides most readily attacked by micro-organisms its abundance in the source materials of these lacustrine carbohydrates explains its inclusion as a fairly stable carbohydrate.

The carbohydrate residues of a largely silt and sand lake sediment sequence of Hall Lake from southern Minnesota are differently distributed than in Blue and Clear Lakes (figs. 96, 97). There is a general increase in total carbohydrates in the middle part of the lake sequence that corresponds

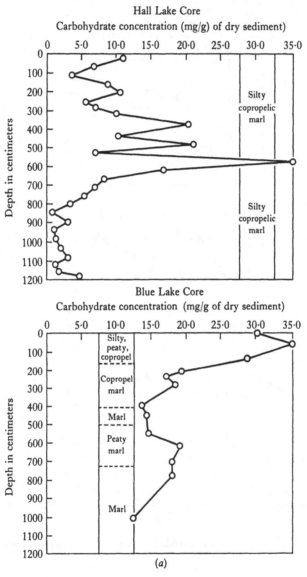

FIGURE 97 Fluctuations of carbohydrate materials with depth in deep Hall and Blue Lake cores, Minnesota. (*a*) Total carbohydrates. (Sample treated with concentrated sulfuric acid, phenol is added and λ_{max} at 490 nm is compared to glucose standard.)

to an increase in organic content. Deeper in the deposit total carbohydrates decrease perhaps due to bacterial degradation. Support for this conclusion lies in the great increase in ribose in the middle of the sediment (fig. 97)

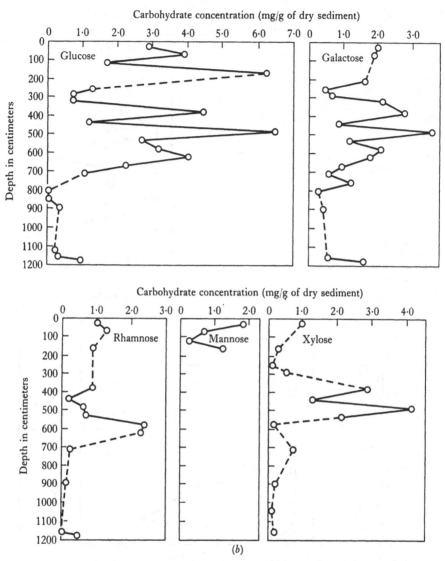

FIGURE 97   Fluctuations of carbohydrate materials with depth in deep Hall Lake
core, Minnesota. (*b*) Individual sugars; M. A. Rogers, analyst.

which suggests a high level of bacterial action at that depth, because ribose
is an important constituent of bacteria. Such relatively large concentrations of ribose would be most easily explained on the basis of bacterial
action.

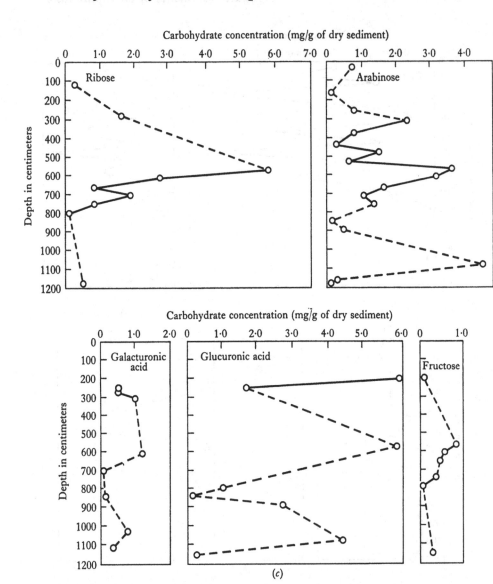

FIGURE 97 Fluctuations of carbohydrate materials with depth in deep Hall Lake core, Minnesota. (*c*) Individual sugars; M. A. Rogers, analyst.

6.3 CARBOHYDRATES OF STABILIZED LOWMOOR PEAT BOGS

The residual monosaccharides in moss peats and underlying lake peats of Rossburg Bog, Aitkin County, Minnesota are shown in fig. 98 (Swain, 1967*b*). The generalized stratigraphy of the bog is shown in figs. 24 and

98 and other properties of the peat are given in table 53. The total carbohydrates were analyzed by the phenol–sulfuric acid method and monosaccharides were studied by sulfuric acid extraction and paper chromatography. Glucose–oxidose and galactose–oxidase enzymatic tests were used
to verify the chromatographic analyses in several samples.

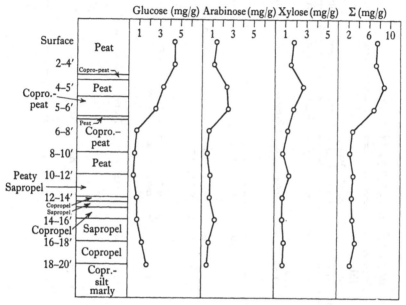

FIGURE 98   Monosaccharide components of Rossburg Bog peat,
Minnesota (Swain, 1967b).

There is an increase in total carbohydrates at 0·7–1·0 ft below the surface
(table 53), in profile Zone II, the cause of which is not completely understood. There evidently is more rapid decomposition of protein as compared
with carbohydrate in the upper few inches of the *Sphagnum* moss peat which
results in relative enrichment of carbohydrates at shallow depths. Somewhat deeper in the moss peat, at 3–4 ft, the process appears to reverse and
carbohydrates breakdown somewhat more rapidly. The variations in total
carbohydrates with increasing depth in the peat probably are caused both
by differences in source as well as in diagenetic conditions.

The individual sugars separated from the peat were glucose, arabinose,
xylose, galactose, mannose and ribose. Mannose is relatively rare in the
upper 2 ft of moss peat, galactose is rare above a depth of 4 ft, and ribose
occurs principally in the copropelic peat and underlying silts. Ribose which
is mainly found in nucleic acids was also identified by Rogers (1965) in

TABLE 53 *Properties of peat from Rossburg Bog, Minnesota*

Zone	pH	Eh mV	N₂ (%)	Ether extracts (%)	Sat. HC (%)	Arom (%)	Asph. (%)	Naphthol	CH abs.	C=O abs.	Pollen zone
II₀, surface *Sphagnum* and forest peat	4	+400	1	2	0·06	0·10	0·35	None	Medium	High	Upper pine zone
IIₘ, *Sphagnum* moss and forest peat to depth of 2 ft	4·2	+400	1	1·5	0·02	0·05	0·30	Present	High	High	
βₚ, moss peat 2–6 ft	4–5	+300 to +400	1–3	1–4	0·02–0·04	0·1–0·3	0·05–2	High	High	Medium to high	
βₚ₋₀₋₈, moss peat with copropel and sapropel layers, 6–12 ft	5–6	+200 to +400	1·5–2	0·4–0·6	0–0·015	0–0·08	0·04–0·5	Medium	Medium	Medium	
β₀₋₈, copropel and sapropel 12–20 ft	6·2–6·8	+150 to +250	2·3	0·2–0·5	0·025–0·085	0·04–0·07	0·09–0·14	Low or absent	Medium	Low	Oak-herb zone/ Pine zone
M zone, marly clay and silt, 20–26 ft	7–7·2	+80 60 +110	1	0·2	0·01	0·02	0·1	None	Medium	Low	Spruce zone
Γ zone, sand at base of deposit, 26–28 ft	7	−10	n.e.	n.e.	n.e.	n.e.	n.e.	None	n.e.	n.e.	

TABLE 53 (cont.)

Zone	Total amino acids °/₀₀₀	Total carbo-hydrates mg/gm	Glucose mg/gm	Arabinose mg/gm	Xylose mg/gm	Pheo-phytin mg/gm	β-carotene	Phenols	Toluene	Pollen zone
II₀, surface Sphagnum and forest peat	75	100	4·5	1·5	2	20	Low or absent	n.e.	n.e.	Upper pine zone
IIₘ, Sphagnum moss and forest peat to depth of 2 ft	200	75	4·5	1·0	1·5	20	Low or absent	n.e.	n.e.	
βₚ, moss peat 2–6 ft	200–300	75–100	2·5–3	2·5	1·5–3	20–25	Low or absent	Present	Absent	
βₚ₋ c₋s, moss peat with copropel and sapropel layers, 6–12 ft	200–250	100–150	0·5–1	0·5–1	0·8–1·2	20–40	Medium	n.e.	n.e.	
βc₋s, copropel and sapropel 12–20 ft	200	30–60	0·8–2	0·8–2	0·8–1	20–40	Medium to high	Absent	Present	Oak-herb zone/ Pine zone
M zone, marly clay and silt, 20–26 ft	2	10–70	n.e.	n.e.	n.e.	0·2–2	None	n.e.	n.e.	Spruce zone
Γ zone, sand at base of deposit, 26–28 ft	n.e.	n.e.	n.e.	n.e.	n.e.	2	n.e.	n.e.	n.e.	

copropelic lake sediments, together with mannose and the other sugars listed above. The sugar identified chromatographically as ribose may be typical of temperate-region eutrophic lake and bog sediments. Glucose is at its maximum concentration in the upper half of the moss peat while

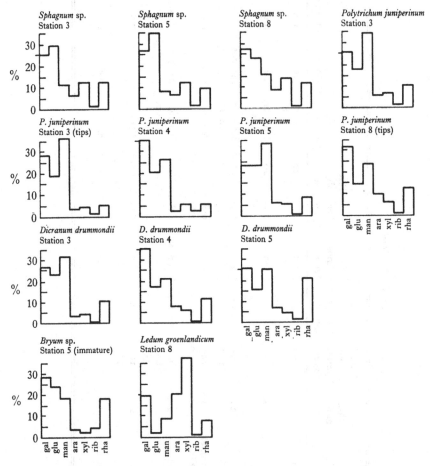

FIGURE 99   Monosaccharide components of plant species from Rossburg Bog peat (Swain, 1967*b*).

xylose and arabinose increase in concentration 4–5 ft below the surface followed by a decrease at greater depths. All three sugars have low and rather uniform concentrations in the lower part of the moss peat and underlying copropelic peat; evidently a point of stability has been reached by the sugars under present conditions in the bog.

Carbohydrates analyses, for monosaccharides, of five species of plants from the surface peat of Rossburg Bog are plotted in fig. 99. Galactose is the most abundant monosaccharide in *Sphagnum* sp., followed by glucose

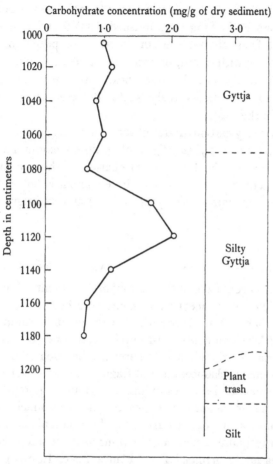

FIGURE 100   Fluctuations of total carbohydrates with depth in portions of a deep Kirchner Marsh core; M. A. Rogers, analyst.

and xylose, with arabinose, mannose and rhamnose occurring in smaller amounts. Both galactose and mannose are present in large and about equal amounts in *Polytrichum juniperinum* followed by glucose, xylose and rhamnose in decreasing amounts. Xylose is abundant, galactose and arabinose are in more or less equal amounts in *Ledum groenlandicum*, an angiosperm, while mannose, rhamnose, glucose and ribose are in small to trace amounts.

Neither mannose nor galactose, important in *Sphagnum* are abundant in the Rossburg peat as a whole. These two sugars are apparently less stable than glucose and xylose in this deposit. A similar relationship was noted by Rogers (1965) in lake sediments from central Minnesota.

In a recent examination by enzymatic methods of the polysaccharides of Rossburg Bog peats F. T. Ting (unpublished data) found that starch is completely absent from all but the surface layer of peat and cellulose decreases very rapidly with depth, disappearing at about 150 cm.

Free sugars extracted from Rossburg peat with water and separated from polymeric components by dialysis also decrease in amount with increasing depth in the peat.

The total sedimentary carbohydrates of the dystrophic peats of Kirchner Marsh, Minnesota (Swain, 1968*b*) (fig. 100) attain a maximum at or near the end of the glacial episode. The monosaccharides of the sequences have not been studied. The high total carbohydrate content is not matched by the amino acid components of the peat (fig. 89) and its origin is not at present understood.

6.4 CARBOHYDRATES OF NON-GLACIAL LAKE DEPOSITS

Almost no study has been made of the carbohydrate contents of non-glacial lake sediments. The diatomaceous and copropelic bottom sediments of Lake Nicaragua contain total carbohydrates in amounts comparable to those of eutrophic lake sediments of Minnesota (table 54) (Swain, 1966*a*). The source materials of these carbohydrates are believed to be diatoms, other algae and bacteria, cladocerans and fragments of terrestrial plants.

Watershed litter in a semi-arid region may contribute important amounts of organic nutrients to depositional lake basins (McConnel, 1968). In Santa Cruz County, south-eastern Arizona, the accumulated litter of a year or more, relatively unleached, may be transported in a few days of the summer rainy season to artificial Pena Blanca Lake. In 1959 this lake received 750 g/m² or more of oak litter, which represents about 329 kg cal/m². Of this about 26% was from carbohydrates, 54% from phenolic compounds, and 20% from unidentifiable compounds. Glucose was the predominant free sugar, followed in much lower concentration by xylose and fructose; galactose and mannose may also have been present. Additional sugars obtained from the water extracts by hydrolysis included xylose, arabinose, glucose and galactose. These were taken to indicate the presence of water-soluble gums and starches. The biomass production in the lake due to oak litter influx in 1959 was an estimated 0·4 g/m² which represents more than 16% of the annual fish harvest for the 1959–61

TABLE 54 *Total carbohydrate content of Lake Nicaragua samples compared to Minnesota lake sediments* (*Swain*, 1966*a*, *Journal of Sedimentary Petrology*, **36**, 522–40)

Station	Description of sediment	Glucose equiv., mg/g of dry sediment
2	Pale grayish brown, pelletal diatomaceous copropel	10
3	Pale tan cladoceran, diatomaceous copropel	10
4	Pale medium gray pelletal diatomaceous copropel	10
7	Pale brownish gray pelletal copropel with gastropods and ostracodes	18
9	Pale brownish gray pelletal diatomaceous copropel	12
10	Pale medium gray diatomaceous copropel	14
11	Pale brownish gray pelletal diatomaceous copropel	17
15	Pale brownish gray, pelletal waxy diatomaceous copropel	18
30	Pale grayish brown pelletal diatomaceous copropel	28
Av.		15
Blue Lake, Minn. surface sediment	Marly sapropel	30–35
Clear Lake, Minn.	Copropel and marl	18–34
Hall Lake, Minn. surface sediment	Silty copropelic marl	11

period. This study brings to mind possibilities for interpretation of highly fossiliferous thin layers of freshwater limestone and other rock types such as are found in the lower Green River Formation of Colorado and Utah and in the Permian Dunkard Group of south-western Pennsylvania.

6.5 CARBOHYDRATES OF FLUVIAL DEPOSITS

As in the preceding case, carbohydrate studies of fluvial sediments are very scarce. The clam-rich sediments of Delta–Mendota Canal are relatively smaller in amount in total carbohydrates than eutrophic lake sediments (table 55, fig. 24*a*) (Swain, 1967*a*). The canal organisms are not particularly high in carbohydrates and relatively little plant debris that floats into the canal is trapped in the sediments.

TABLE 55   *Carbohydrate components of samples from Delta–Mendota Canal, California in* 10^{-4} *g/g of dry sample, total carbohydrate in mg/g* (*Swain*, 1967 *a*)

Sample	Miles	Galact-ose	Glucose	Man-nose & Arabin-ose*	Xylose	Ribose	Rham-nose	Σ	Total carboh.
Corophium spinicorne empty carapaces	4·7	0	0	0	0	0	0	0	22·8
Fredericella sultana and assoc. sed. (– 20 + 80 mesh)	4·7	2·23	1·70	2·34	0·87	0·55	0·54	8·23	6·0
Corophium spinicorne crust (20 mesh)	4·7	11·14	6·80	5·06	1·95	0·72	2·45	28·12	6·4
Corophium spinicorne crust (– 20 + 80 mesh)	4·7	1·51	1·78	2·58	2·06	0·72	1·36	10·01	22·8
Fredericella sultana and assoc. sed. (20 mesh)	4·7	3·27	3·87	2·95	1·19	0·48	1·61	13·37	11·8
Corbicula fluminea live animal shell	4·7	0·52	1·00	1·59	0	0	0	3·11	4·6
Corbicula fluminea live animal	4·7	—	—	Good separation of monosaccharides not obtained					400·0
Sediment assoc. with live *Corbicula* (20 mesh)	4·7	3·86	3·83	6·71	2·28	1·49	1·96	20·13	11·8
Sediment assoc. with live *Corbicula* (– 20 + 80 mesh)	4·7	3·95	6·02	5·54	1·19	0·24	0·67	17·61	7·6

* Both mannose and arabinose are present in the samples examined, but were not separated in the solvent system used for these analyses.

The monosaccharide components of the canal sediments (table 55) indicate a variety similar to that in aquatic plants elsewhere (fig. 95). Glucose and galactose are the principal carbohydrates in the asiatic clam *Corbicula fluminea*.

6.6   CARBOHYDRATES OF NON-MARINE SEDIMENTARY ROCKS AND FOSSILS

Among the few records of analyses of the carbohydrate contents of non-marine sedimentary rocks are those by Palacas (1959), Palacas *et al.* (1960); Vallentyne (1963); Swain (1961*b*, 1963*a*, 1966*b*) and Swain *et al.* (1967).

6.6.1   *Carbohydrates in rock samples and fossils*

The total carbohydrate contents of several non-marine formations of Paleozoic age are in the range 9·6–33 p.p.m. (table 56), and the absorption maxima of the carbohydrate degradation products indicate that furfural (λ_{max} 480 nm) and/or hydroxymethylfurfural (λ_{max} 490 nm) are present in

TABLE 56   *Carbohydrate components of Paleozoic rock samples (Swain, 1966b, Coal Science, Advance in Chemistry Series, No. 55, 10–13)*

Description and interpretation of origin	Formation	Age	Org. C (%)	Glucose eq. (p.p.m.)	Abs. max. (nm)
L Gn–Gr, M–C, argillaceous sandstone (deltaic)	Bald Eagle	U. Ordovician	0·14	23·2	n.d.
M R Br–Gn, M–C, argillaceous ferruginous sandstone (deltaic)	Juniata	U. Ordovician	0·17	19·8	n.d.
W, M–F, angular-subrounded, silty silica cemented sandstone (prodeltaic)	Tuscarora	L. Silurian	0·04	11·2	482, 496 (f + h)
P R Br, silty, slightly calcareous, ferruginous fissile shale	Bloomsburg	U. Silurian	0·03	10·0–32·6	±480 (very weak) (f)
L G–Gn, platy, finely micaceous calcareous slightly fossiliferous shale (tidal flat)	Wills Creek	U. Silurian	0·38	71·6–103·6	482 (f)
M Gr, dense to sublithographic, wavy laminated, brecciated limestone (lagoonal)	Tonoloway	U. Silurian	n.d.	20·0	486 (h)
Gn–Gr, platy silty shale, plant fragments (deltaic)	Harrell	U. Devonian	n.d.	n.d.	n.d.
GN–Gr, flaggy, VF–F sandstone and siltstone; plant fragments (deltaic)	Brailler	U. Devonian	0·11	14·4	480 (very weak) (f)
M Gr and M R–Br, silty shale, Gr and W F–C sandstone, marine fossils (nearshore marine-deltaic)	Chemung (Trimmers Rock)	U. Devonian	n.d.	100·8	484 (very weak) (f + h)
Pale R–Br, silty micaceous shale and F–C felspathic, hematitic sandstone (alluvial)	Catskill	U. Devonian	0·00	21·2	482–492 (f + h)
F–M, L Gn–Gr, angular-subrounded, argillaceous micaceous sandstone, rock fragments (alluvial)	Pocono	L. Mississippian	n.d.	9·6	n.d.
Pale R–Br, silty sandy, micaceous slightly calcareous shale (alluvial)	Mauch Chunk	M.–U. Mississippian	0·00	33·0	482 (f)

Abbreviations: P, pale; Gr, gray; W, white; Gn, green; Br, brown; B, black; M, medium; C, coarse; F, fine; R, red; f, furfural; h, 5-hydroxymethylfurfural. Whether λ_{max} suggests furfural or 5-hydroxymethylfurfural is given in parentheses in column 6.

several of the samples. The average total carbohydrate contents of the red-bed samples (finely divided particulate matter) is higher than that of the marine and non-marine sandstones but lower than that of the marine shales and limestones of the Paleozoic rocks of the area. The carbohydrate components of the non-marine sediments are disseminated terrestrial plant debris, although the organic carbon was too low to be measured in two of the samples.

TABLE 57   *Carbohydrate components of non-marine and marine rock samples (Palacas, 1959)*

Formation	Age	Component sugars				
		Glucose	Galactose	Arabin-ose	Xylose	Rham-nose
Stonehenge*	Ordovician	+	.	(?+)	(?+)	.
Simpson*	Ordovician	+	(?+)	+	+	.
Upper Chambersburg*	Ordovician	+	.	(?+)	.	.
Marcellus*	Devonian	+	(?+)	+	+	.
Woodford*	Upper Devonian	+	.	+	+	.
Des Moines*	Pennsylvanian	+	.	+	+	.
Missouri*	Pennsylvanian	+	.	.	.	(?+)
Leonard*	Permian	+	(?+)	.	+	.
Schuler**	Jurassic	+	.	+	+	.
Green River**	Eocene	+	(?+)	.	(?+)	.
Elko**	Miocene	+	.	.	.	.

The rock was treated with hot 0·5 N sulfuric acid and the sugar identified chromatographically. * marine; ** non-marine.

Palacas (1959) and Palacas *et al.* (1960) studied the carbohydrate components of the non-marine facies of the Upper Jurassic Schuler Formation of Louisiana, the lacustrine Eocene Green River Formation of Utah, and the lacustrine Miocene Elko Formation of Nevada.

The Schuler Formation represents the upper part of the Upper Jurassic Cotton Valley group of the subsurface of the northern Gulf of Mexico region. It includes several thousand feet of both nearshore or non-marine facies of red, green and varicolored shales, sandstones and conglomerates and their offshore equivalents of dark-gray fossiliferous shale, limestone and sandstone. The core sample studied (table 57) consisted of non-calcareous silty shale that has the composition and appearance of sediment deposited under aerobic or oxidizing conditions. No carbohydrates were found in small 50 g core samples of red and green sandstone (Swain, 1961*b*, 1963*a*). One would suppose that the contained organic matter would have been destroyed under these conditions but a 500 g sample analyzed by Palacas was found to contain small amounts of glucose, arabinose, xylose and possibly glycerol. The sediment and contained organic matter were probably deposited rapidly enough to retard reduction of the ferric oxide matter and the decomposition of the organic matter. The recent suggestions as to the *in situ* origin of the hematite red beds such as these (van Houten,

1966) are not supported by the presence of the carbohydrates in the Schuler unless they were introduced post-depositionally.

The Green River Formation of Middle Eocene age represents several thousand feet of lacustrine thin-bedded, organic-rich carbonate rocks (oil shales), sandstones, fissile shales and algal oölitic limestones. The core samples studied for carbohydrates came from shallow-water fossiliferous mollusk-bearing shales in the lower part of the formation. *Elliptio*, a mussel, and *Goniobasis*, a snail, are the principal fossils. The dark color and relatively high organic content of the rock indicate that it formed under negative Eh conditions. Treatment with 12% hydrochloric acid yielded both furfural and 5-hydroxylmethylfurfural which suggest that both pentosans and hexosans are present in the rock. Sulfuric acid hydrolyzates of the rock yielded glucose and possibly galactose. A sample of the Upper Green River Formation, the so-called Mahogany Ledge, was also analyzed for carbohydrates. This differs from the preceding molluscan shale in being a deep-water (hypolimnetic) calcium-carbonate shale containing a large amount of kerogen rather than fossil shells. The Mahogany Ledge yielded glucose and possibly xylose and glycerol. The carbohydrate components of the molluscan shale may have included littoral larger vegetation in addition to algae and animal remains, while those of the kerogen–oil shale were probably planktonic algae for the most part.

The Elko Oil Shale of probable Miocene age occurs in the lower part of the Humboldt Formation of central Nevada. The rock is fissile and platy shale that is lower in carbonate content than the Green River Oil Shale but also is characterized by large amounts of algal-derived kerogen and is probably of similar origin. Small amounts of glucose, and possibly arabinose and xylose were obtained by Palacas from the Elko Oil Shale.

The total carbohydrate contents of specimens of several kinds of nonmarine fossils and of the associated matrix and of matrix alone in other specimens are shown in table 58 (Swain *et al.* 1967). The wide variation in total carbohydrates both in the fossil specimens and in the matrix is related to original carbohydrate content, diagenetic effects, degree of oxidation and mineral content. The high carbohydrate contents of the fossil *Triceratops* and the crocodilian and the contrasting much lower value for the *Lepidodendron* are believed by the writer to result from the protective effect of the mineral matter in the vertebrate specimens and lack of such protection in the more highly carbonized fossil plants.

When Devonian black shale samples were heated in air or in nitrogen atmosphere at 200 °C or more the yields of total carbohydrates was increased (Rogers, 1965). When fossil *Calamites Suckowi* was heated under the same conditions no consistent results were obtained on carbohydrate

TABLE 58 (*a*) *Total carbohydrate content of fossil Reptilia and Plantae and associated matrix and* (*b*) *other geologic samples, in p.p.m.* (*Swain, Rogers, Evans & Wolfe*, 1967)

(*a*)

Species	Age	Location and geological formation	Description	Total carbohydrates	
				Specimen	Matrix
Triceratops sp.	Late Cretaceous	SE¼, Sec. 23T. 24N. R43E. McCone Co. Montana, Hell Creek Formation	Light-brown sandy shale (fluvial, geosynclinal?)	900	384
Crocodilian	Late Cretaceous	Same as preceding	Same as preceding (fluvial, geosynclinal?)	140	161
Cryptozoan sp.	Middle Ordovician	Union Furnace, Pennsylvania, Hatter Formation, algal reef as base	(Littoral, geosynclinal)	70	n.d.
Algal bed	Middle Silurian	Barree, Pennsylvania, McKenzie Limestone, 18–20 ft above base	(Littoral, geosynclinal)	153	n.d.
Lepipodendron gaspianum?	Early Mississippian	Trough Creek State Park, Pennsylvania, Pocono Sandstone, Burgoon Member	Dark gray, micaceous, sandy shale (paludal, geosynclinal)	74	59

(*b*)

Geologic formation	Age and location	Description	Carbohydrate content: range in parentheses	Absorption maximum of colored product nm
Coutchiching (3 samples)	Early Precambrian, Ontario	Dark green biotite slate and phyllite (basinal)	11 (6–14)	490
'Carlim-Lowville' (1 sample)	Early Middle Ordovician Pennsylvania	Dense, dark gray, carbonaceous, pyritic stylolitic limestone; graphitic residue on stylolites (neritic)	37	n.d.
Burket Shale (1 sample)	Late Devonian, Pennsylvania	Black fissile shale, with *Paracardium doris*	17	n.d.
Pocono Sandstone, Burgoon Member (1 sample)	Early Mississipian, Pennsylvania	Coal (paludal)	126	488
Twin Creek Limestone (4 samples)	Middle Jurassic, Wyoming	Light-gray thin bedded limestone (neritic)	31 (14–55)	490
Twin Creek Limestone (9 samples)	Middle Jurassic, Utah	Pale gray-brown sublithographic limestone (neritic)	23 (6–45)	490
Mariposa Shale (3 samples)	Late Jurassic, California	Dark gray micaceous argillite (basinal)	18 (10–30)	n.d.
Twist Gulch Formation (7 samples)	Late Jurassic, Utah	Pale gray soft, calcareous shale (alluvial)	48 (14–88)	484–487
Arapien Shale (1 sample)	Middle Jurassic, Utah	Pale gray calcareous carbonaceous gypsiferous shale (alluvial)	24	486

yields. Apparently oxidative degradation of the carbohydrates occurred before burial of the fossil plant to a greater extent than in the Devonian black shale. In the latter the disseminated carbohydrates are protected by mineral matter with the result that less degradation has occurred and preheating the samples is more effective in increasing carbohydrate yields.

6.6.2 *Carbohydrates of Paleozoic plant fossils*

The carbohydrate contents of 10 species of Paleozoic non-marine plant fossils ranging in age from Lower Devonian to Lower Permian (table 59) comprise (Swain, Bratt & Kirkwood, 1967) mainly glucose, galactose and xylose residues, together with smaller amounts of mannose, rhamnose and arabinose. These monosaccharides were suggested to have occurred in the original plants as cellulose, starch, galactans and xylans. In most of the specimens studied glucose is the principal sugar but in the primitive cordaite *Callixylon* sp. from the Upper Devonian galactose is the predominant sugar which may indicate that galactans were the principal cell-wall constituents in that plant.

Eight additional species of Devonian to Pennsylvanian plants were analyzed for individual sugars (table 60) (Swain, Bratt & Kirkwood, 1968a). In the fossil Psilophytales and Lycopodiales, the general types and ratios of monosaccharides are similar to those of modern ferns. The fossil *Calamites* (Equisetales), by comparison, seem to have more galactose than modern *Equisetum*, which suggests that, instead of cellulose as the main cell wall polysaccharide as in *Equisetum*, the fossil forms had a good deal of galactans.

The carbohydrates of a suite of Upper Carboniferous plant fossils from Radstock, England (table 61) proved to be similar in most respects to those of comparable species from North America (Swain, Bratt & Kirkwood, 1968a). A smooth-trunked *Lepidodendron* with few leaf scars, however, had an excess of galactose over glucose in its carbohydrate residues. Pectic polysaccharides or other galactans may have been relatively higher in this species than in other lycopods studied. Species of Radstock seed ferns, *Alethopteris* and *Annularia* also contain enough galactose and mannose to suggest that galactans and perhaps mannans were present in the cell-wall polysaccharides in these extinct plants. Analyses of additional fossil plants of the Paleozoic yielded similar results (Swain, Bratt, Kirkwood & Tobback, 1969).

Free sugars have been separated from several species of fossil plants and from Silurian limestone (Swain, Bratt, Kirkwood & Tobback, 1969)

TABLE 59 (*a*) *Monosaccharide components of Paleozoic fossil plants in* $\mu g/g$, *and* (*b*) *of living Filicales and Equisetales in* g/g (*Swain, Bratt &* *Kirkwood*, 1967)

(*a*)

Species	Locality and age	Weight of sample (g)	Galac-tose	Glucose	Man-nose	Xylose	Rham-nose	Σ
Zosterophyllum sp.	Gaspé Bay, Que., L. Devonian	10·13	3·00*a* 6·26*b*	4·00*a* 20·00*b*	4·00*a*	o	o	11·00*a* 26·26*b*
Zosterophyllum sp. Another specimen	Gaspé Bay L. Devonian	8·30		2·35*b*	o	o	o	2·35*b*
Lepidodendron cf *L. gaspianum*	Trough Creek Park, Pa., L. Mississippian	3·18	0·12*a*	0·47*a,b*	o	0·95*a*	0·21*a*	1·63*a*
Lepidophloios sp.	Jackson, Ohio, L. Pennsylvanian	11·44	1·17*b*	1·34*b*	n.d.	n.d.	n.d.	2·51*b*
Cordaites sp.	Jackson, Ohio, L. Pennsylvanian	1·48	29·00*b*	37·67*b*	n.d.	n.d.	n.d.	66·67*b*
Cordaites sp. (in coal ball)	Cherokee Co., Kans., M. Pennsylvanian	—	0·07*a*	0·09*a*	o	0·05*a*	0·01*a**	0·22*a*
Sigillaria sp.	Clay, Ky., L. to M. Pennsylvanian	8·42	o*b*	6·17*b*	o	o	o	6·17*b*
Sigillaria cf. *S. approximata* (leaves, 1)	Vance, Pa., L. Permian	85·55	o*b*	3·5	o	o	o	3·5*b*
Sigillaria cf. *S. approximata* (leaves, 2)	Vance, Pa., L. Permian	66·62	0·20*b*	0·98*b*	n.d.	n.d.	n.d.	1·18*b*
Sigillaria cf. *S. approximata* (leaves, 3)	Vance, Pa., L. Permian	108·64	o*b*	o*b*	n.d.	n.d.	n.d.	o*b*
Sigillaria cf. *S. approximata* (leaves, 4)	Vance, Pa., L. Permian	137·57	0.11	0.55	n.d.	n.d.	n.d.	0.66*b*
Sigillaria cf. *S. approximata* (leaves, 5)	Vance, Pa., L. Permian	42.54	o*b*	3.42*b*	n.d.	n.d.	n.d.	3.42*b*
Sigillaria cf. *S. approximata* (leaves, 6)	Vance, Pa., L. Permian	41·70	0·41*b*	1·04*b*	n.d.	n.d.	n.d.	1·44*b*
Sigillaria cf. *S. approximata* (trunk, 1)	Vance, Pa., L. Permian	21·14	tr *a*	tr *a*	o	o	o	tr *a*
Sigillaria cf. *S. approximata* (trunk, 2)	Vance, Pa., L. Permian	11·37	tr? *a* n.d. *b*	tt *a* 1·27*b*	o	o	o	tr *a*
Sigillaria cf. *S. approximata* (trunk, 3)	Vance, Pa., L. Permian	126·73	o*b*	0·35*b*	n.d.	n.d.	n.d.	0·35*b*
Sigillaria cf. *S. approximata* (trunk, 5)	Vance, Pa., L. Permian	82·35	tr *a*	tr *a*	o	o	o	tr *a*
Sigillaria cf. *S. approximata* (trunk, 6)	Vance, Pa., L. Permian	108·05	tr? *a*	tr? *a*	o	o	o	tr? *a*

(b)

TABLE 59 (*cont.*)

Species	Wt of sample (g)	Galac-tose	Glu-cose	Man-nose	Arabi-nose	Xylose	Rib-ose	Rham-nose	Σ
Polypodium virginianum (Rainy Lake, Minn.)	0·217	n.d.	0·124b	n.d.	n.d.	n.d.	n.d.	n.d.	0·124b
Dryopteris cristata (Cedar Creek Bog)	0·061	0·035a	0·1211a	0·002a	0·005a	0·002a	oa	0·005a	0·1375
Dryopteris spinulosa (Cedar Creek Bog)	0·036	0·0663a	0·090a	0·008a	0·004a	0·002a	oa	0·004a	0·171
Equisetum arvense†	0·217	0·028a	0·065a	0·010a	0·011a	0·008a	oa	0·007a	0·129
Equisetum arvense‡	0·217	0·018a	0·051a	0·017a	0·018a	0·011a	oa	0·014a	0·129
Equisitum arvense	0·127	0·022b	0·150b	n.d.	n.d.	n.d.	n.d.	n.d.	—

a, Chromatographic analyses;   b, enzymatic analyses; n.d., not determined.

* Arabinose.   † Developed in amyl alcohol:pyridine:water.

‡ Developed in pyridine:ethyl acetate:water.

(table 62). The origin of the free-sugar components may lie both in indigenous and in introduced materials. The latter is a more plausible explanation of the large amounts of rhamnose and ribose noted in two fossil specimens.

The polysaccharide components of three species of non-marine Paleozoic plants were found to contain (table 63) small amounts of linear α-1 $\rightarrow$ 4 glucopyranose (starch) and linear β-1 $\rightarrow$ 4 glucopyranose (cellulose) on the basis of enzymatic analyses (Swain, Bratt, Kirkwood & Tobback, 1969).

Extracts obtained by refluxing with boiling water of samples of *Rhynia gwynnevaughani* from the Devonian of Scotland and *Calamites duboisi* and *C. Suckowi* from the Upper Carboniferous of Iowa and England, respectively, were dialyzed against distilled water to remove free monosaccharides. The extracts were then treated with α-amylase, β-amylase and cellulase preparations. Small amounts of maltose were produced by the α-amylase reaction suggesting presence of starch. In two of the species glucose was produced by the cellulase reaction suggesting presence of cellulose. Rogers (1965) had previously demonstrated the occurrence in Devonian black shale (Marcellus Formation of Pennsylvania) of a β-1 $\rightarrow$ 3 linked polysaccharide of laminaran type.

TABLE 60 *Monosaccharide components of Paleozoic fossil plants (values in μg of each sugar/g of fossil material)*
(Swain, Bratt & Kirkwood, 1968)

Name	Location	Age	Weight of sample (g)	Glucose	Galactose	Man-nose	Xylose	Other	Totals for a and b and non-resolvable c
Trimerophyton robustius	Cap Aux Os, Quebec	L. Devonian	0·42	a.a.a., 187·7b	a.a.a., 148·0b	a	a	—	335·7b
Callixylon sp.	Escuminac Bay, Quebec	U. Devonian	0·93	a, 0b	a.a., 26·2b	—	—	—	26·2b
Calamites cf. Suckowi (14 samples)	Radstock, England	U. Carboniferous	19·49–153·44 (av.: 71·41)	a.a., 0–2·8b (av.: 0·9b)	5·4a, 0–6·1b (av.: 2·1b)	a	a	a?	0·8–7·4b (av.: 3·1b)
Calamites sp., leaves (3 samples)	Radstock, England	U. Carboniferous	5·55–297·06	a, 0·03–1·5b (av.: 0·6b)	a?, 0·7–5·3b (av.: 2·2b)	—	a?	—	0·8–6·8b (av.: 2·8b)
Calamites sp. (3 samples)	Radstock, England	U. Carboniferous	163·28–254·23	0·07–0·6b (av.: 0·3b)	0·1–0·6b (av.: 0·4b)	n.a.	n.a.	n.a.	0·4–1·1b (av.: 0·7b)
Aborescent lycopod, leaves (2 samples)	Radstock, England	U. Carboniferous	20·91–27·20	1·6–2·1b (av.: 1·9b)	0–3·2b (av.: 1·8b)	n.a.	n.a.	n.a.	1·6–5·3b (av.: 3·7b)
Lepidodendron sp. (3 samples)	Radstock, England	U. Carboniferous	8·5–8·8	2·5–2·8b (av.: 2·4b)	0–2·4b (av.: 1·1b)	n.a.	n.a.	n.a.	0·8–4·8b (av.: 3·5b)
Same as preceding (1 sample)	(Heated 2 h at 175 °C)	—	8·4	1·09b	0b	n.a.	n.a.	n.a.	1·09b
Cordaites sp.	Jackson, Ohio	L. Pennsylvanian	1·48	a.a., 36·7b	a.a., 29·0b	a?	a?	n.a.	65·7b

TABLE 60 (cont.)

Name	Location	Age	Weight of sample (g)	Glucose	Galactose	Mannose	Xylose	Other	Totals for a and b and non-resolvable c
Sigillaria sp.	Clay, Kentucky	L. to M. Pennsylvanian	8·42	a.a., 6·17b	a, ob	—	—	—	6·17b
Matrix and fossil ferns	Wood, Pennsylvania	M. Pennsylvanian	1·87	a.a., 6·1b	a.a., 2·0 b	a?	a?	—	8·1b
Pecopteris cf. *P. vestita*	Garlick mine, Pennsylvania	M. Pennsylvanian	3·46	a.a.	a.a.	a?	a?	—	—
Neuropteris cf. *N. vermicularis*	Garlick mine, Pennsylvania	M. Pennsylvanian	6·30	195·94, c	109·3 a	115 a	—	Unknown oligosaccharide?	420·2 a
Matrix and *Neuropteris* sp.	Beccaria, Pennsylvania	M. Pennsylvanian	5·30	a.a.	a.a.	—	—	—	—
Matrix and *Neuropteris* sp.	Highway 53, Pennsylvania	M. Pennsylvanian	10·67	1·9b	ob	—	—	—	1·9b
Odontopteris schlotheimii	Garlick mine, Pennsylvania	M. Pennsylvanian	1·58	a.a., 102·9b	a, ob	a?	—	—	102·9b
Neuropteris fimbriata	Wood, Pennsylvania	M. Pennsylvanian	6·50	8·7b	ob	—	—	—	8·7b
Sigillaria cf. *S. approximata* (2 samples)	Vance, Pennsylvania	L. Permian	108·1	a, tr.-1·27b	a, o-tr.b.	a?	a?	—	o-1·27b, c

a, Silyl ether derivatives of fossil sugars (a, trace amounts; a.a., moderate amounts; a.a.a., abundant amounts) as indicated on chromatograms;
b, Glucose-oxidase and galactose-oxidase tests;
c, Paper chromatography (indicated in right-hand column if not resolvable); n.a., not examined.

TABLE 61 *Carbohydrates of fossil plants from Radstock, England (in μg/g) (Swain, Bratt & Kirkwood, 1969)*

Species and no. of samples analyzed	Weight of sample (g)	Ignition loss (%)	Total carbon	Galactose	Glucose	Mannose	Arabinose	Xylose	Ribose/ Rhamnose	Σ
Lepidodendron sp. A. 1 *a*, 2 *b*	16·8–55·08	8·1	18·8	0 *a*‡ 0 *b*	1·6 *a* 1·8–4·7 *b**	1·3 *a*	—	—	—	4·2 *a* x
Lepidodendron sp. B. 4 *a*, 2 *b*	8·41–96·40	15·9	13·7	0·2–0·4 *a* av. 1·3 0·2–1·0 *b*†	0–1·5 *a* av. 0·4 1·0–2·0 *b*	0–0·2 *a* av. 0·1	0–0·8 *a* av. 0·4	—	—	0·2–2·9 *a* av. 2·5 *a*
Lepidodendron cf. *loricatum* Arber 2 *a*, 2 *b*	9·9–17·36	8·3	22·1	0–0·7 *a* 0 *b*	0·4–3·8 *a* 1·4–4·4 *b*	0–0·3 *a*	0–0·1 *a*	—	—	0·4–4·9 *a*
Arborescent lycopod leaves 3 *a*, 3 *b*	9·48–27·2	7·5	n.d.	0·6–2·1 *a* av. 1·5 0–3·2 *b* av. 1·8	1·2–2·5 *a* av. 2·0 1·7–2·1 *b* av. 2·0	0·3–0·6 *a* av. 0 5	Present	—	—	2·1–5·2 *a* av. 2 0 *a*
Calamites Suckowi Brongniart 10 *a*, 14 *b*	11·61–147·02	8·0 (av.)	22·6	0–3·1 *a* av. 0·6 *a* 0·6–6·0 *b* av. 2·1	0–3·5 *a* av. 0·9 *a* 0–2·6 *b* av. 1·0	0–0·3 *a* av. 0·04 *a*	0–0·2 *a* av. 0·002 *a*	0–1·0 av. 0·02 *a*	0–1·0 (ri) av. 0·1 *a*	0–0·1 *a* av. 1·7 *a*
Calamites sp., leaves 4 *a*, 3 *b*	5·55–297·06	7·8 (av.)	9·5 (av.)	0·1–0·8 *a* av. 0·4*a* 0·6–5·3 *b*	0·1–2·5 *a* av. 0·8 *a* 0·03–1·5 *b*	—	0–0·9 *a* av. 0·2 *a*	0–0·5 *a* av. 0·1 *a*	0–0·7 *a* av. 0·2 *a*	0·2–5·4 *a* 1·7 *a*
Calamites sp. 2 *a*, 3 *b*	163·28–254·23	7·6	n.d.	0–0·5 *a* 0·1–0·6 *b* av. 0·4 *b*	0·5–0·9 *a* 0·07–0·6 *b* 0·3 *b*	0–0·1 *a*	—	—	0–0·05 (ri) 0–0·22, (rh)	0·5–1·5 *a*
Annularia sphenophylloides Zenker 1 *a*, 2 *b*	69·07	n.d.	14·8	0 *a* 0 *b*	0·5 *a* 0–0·2 *b*	0·4 *a*	0·4 *a*	0·6 *a*	—	2·0 *a*
Alethopteris sp. 1 *a*, 1 *b*	31·75	8·4	15·1	2·8 *a* 1·1 *b*	1·7 *a* 1·9 *b*				—	4·5 *a*

* The presence of D-glucose is verified by the glucose-oxidase test.
† The presence of D-galactose is strongly indicated by the galactose-oxidase test.

TABLE 62 *Free sugars in fossil plants and Silurian limestone (in µg/g) (Swain, Bratt, Kirkwood & Tobback, 1969)*

Species and sample number	Age	Gal	Glu	Man	Ara	Xyl	Rib	Rha	Σ
Rhynia gwynnevaughani	Devonian	0 a 0 b	0·03 a 0·02 b	0·09 a		0·09 a	0 a	0 a	0·11 a
Calamites Suckowi	Pennsylvanian	1·26 a 2·00 b	0·63 a 0·17 b	0·48 a		0·56 a	8·11 a	6·95 a	17·99 a
C. Suckowi (different specimen)	Pennsylvanian	22·44 b	1·06 b	n.d.	n.d.	n.d.	n.d.	n.d.	—
C. duboisi	Pennsylvanian	0 b	0·31 b	n.d.	n.d.	n.d.	n.d.	n.d.	—
Lepidophloios laricinus	Pennsylvanian	2·11 a 14·09 b	0·86 a 0·43 b	0·68 a		1·42 a	13·34 a	8·89 a	25·30 a
Keyser Limestone	Silurian	2·66 b	0·07 b	n.d.	n.d.	n.d.	n.d.	n.d.	—

a, Chromatographic analyses; *b*, enzymatic analyses; n.d. not determined.

TABLE 63 *Polysaccharide components of non-marine Paleozoic plants (Swain, Bratt, Kirkwood & Tobback, 1969)*

Species and sample no.	α-Amylase (glu µg/g)	β-Amylase (glu µg/g)	Cellulase (glu µg/g)	Polysaccharides indicated by enzymatic activity
Rhynia gwynnevaughani	0 a 0 b Trace maltose a	0 a 0·89 b	0 a 0 b	Linear α-1 → 4 glucopyranose units (starch)
Calamites Suckowi	0·20 a* 0·16 b	0·18 a* 0 b	0·20 a* 0·16 b	Linear α-1 → 4 glucopyranose units (starch); Linear β-1 → 4 glucopyranose (cellulose)
C. duboisi	0 a 0 b† Trace maltose a	0 a 0 b†	0 a 0 b†	Linear α-1 → 4 glucopyranose units (starch) suggested by maltose

a, Chromatographic analyses for glucose following reaction with indicated enzyme; *b*, enzymatic analyses for β-D-glucose.

* Small amounts of mannose, arabinose and ribose (~ 0·1–0·2 µg/g) were present on chromatograms of extracts treated with indicated enzyme.

† Small to moderate amounts of D-galactose (0·5–5·1 µg/g) detected by galactose–oxidase in extracts treated with indicated enzyme.

6.7 SUMMARY

The monosaccharide components of a suite of aquatic plants include eight common sugars. Glucose is the commonest of the plant sugars and is also one of the most stable, therefore it is the one most likely to be found in associated sediments. Other sugars that show a fairly high degree of stability in accumulating sediments are xylose, rhamnose and arabinose. In decreasing order of stability in lacustrine sediments are: ribose, mannose, galactose and glucuronic acid.

As aquatic plants decay and their carbohydrate residues accumulate in the sediments a reduction of one to two orders of magnitude in total carbohydrates occurs with respect to the original plant material. The neutral to slightly alkaline lacustrine environments favor the higher values of preservation. Despite the fact that individual sugars are more unstable in alkaline than in acid solutions, in polymeric form the sediment carbohydrates are somewhat more stable in alkaline than in acid sediments. The negative Eh values characteristic of alkaline sediments are suggested as responsible for the reducing conditions that favor this preservation.

Free sugars in water of a Minnesota eutrophic lake include glucose, mannose, sucrose and xylose, whereas galactose is scarce or absent. This difference may result from the lower solubility of galactose in water.

Monosaccharide components recovered by acid hydrolysis from homogeneous lake sediments typically show a slow downward decrease probably due to bacterial degradation. In non-homogeneous lake sediments the carbohydrate components reflect prior intervals of increased or decreased organic productivity, and thus may have stratigraphic value. The increased concentrations of ribose at depth in lake sediments may be of bacterial origin. In general, the particular monosaccharides of lake sediments may differ appreciably in type as well as in amount from those of aquatic source plants.

The monosaccharides of living bog-forming plants in a cool-temperate peat bog are predominantly glucose and xylose in the angiosperms, and galactose, glucose, mannose and xylose in the bryophytes. Neither galactose nor mannose is abundant in the underlying peat whereas glucose and xylose are the main sugars. The first two sugars are apparently less stable than the second two in this environment.

The sparse data on carbohydrates of lake sediments of non-glacial regions indicate similarity in amount to those of eutrophic glacial-lake sediments.

Glucose and galactose are the principal carbohydrates of canal-sediment accumulations in California, but total carbohydrates are less than those of eutrophic lake sediments. The more oxidizing environment of the canal and perhaps more rigorous degradation of the particulate organic matter in the source areas are suggested as accounting for the low carbohydrate values. Other carbohydrate studies of fluvial material appear generally to be lacking.

The total carbohydrates of continental or nearshore red shales of middle and early Paleozoic age of the Appalachian Mountains are higher than in associated sandstone and lower than in marine shales. Here the tendency to oxidize the sedimentary organic matter in the alluvial red sand environment is somewhat superceded by the relatively abundant sources of particulate organic matter.

Red and green alluvial shales of the Jurassic of the United States also contain carbohydrate residues including glucose, arabinose, and xylose. If these red beds originated by oxidative weathering *in situ* as suggested by some authorities, it is questionable whether the carbohydrates would have been preserved. Rapid deposition of red lateritic sands and particulate organic debris seems to be a more plausible explanation of their origin.

The carbohydrate components of the lower eutrophic part of the Eocene Green River Shale are different from those of the upper meromictic facies of the Green River. It cannot be determined whether these variations are caused by differences in source material or by other causes.

Studies of the carbohydrates of Paleozoic plant fossils have shown that glucose, presumably from structural cellulose, is the main sugar in most cases, but exceptions occur. Fossil *Calamites*, an Equisetales, appears to have more galactose than glucose which suggests that galactan rather than glucose as in modern Equisetales, may have been the main structural polysaccharide in those primitive forms. Similarly some forms of *Lepidodendron*, a Lycopodiales, seem to have had structural galactans rather than glucans in important amounts.

Free sugars were found in some of the Paleozoic plant fossils but the occurrence of large amounts of rhamnose and ribose at least suggests that these were introduced post-depositionally as they would not be expected to be abundant in the living plant.

Polysaccharides identified in Paleozoic fossil plant specimens are based on enzymatic treatment of fossil material freed of free sugars. Small amounts of linear α- and β-1 $\rightarrow$ 4 glucopyranose strongly suggest the presence of starch and cellulose respectively in some specimens. Furthermore, β-1 $\rightarrow$ 3 linked polysaccharide of laminaran type has been identified by M. A. Rogers in Devonian black shale. Thus it appears that the carbohydrate structures of some of the Paleozoic plants at least was similar to that of modern plants, although other polysaccharides of larger land plants, particularly xylans, have not yet been specifically determined.

7 Organic pigments of non-marine deposits

7.1 TYPES OF ORGANIC PIGMENTS

Three kinds of organic pigments are most likely to be found in aquatic plants, animals and sediments (Orr & Grady, 1957; Orr *et al.* 1958): (1) chlorinoid pigments, including chlorophylls and porphyrins, (2) carotenoids, including the orange pigments carotene and xanthophylls, and (3) flavoproteins, the yellow pigments and related N-heterocyclic substances.

Several other natural organic pigments may be present in small amounts in aquatic materials (Swain, Paulsen & Ting, 1964): (1) flavonoids or anthoxanthins and anthocyanins, the coloring matter of flowers and fruits; (2) napthoquinones and anthroquinones, the coloring matter of tree bark and echinoids; (3) bile pigments; (4) pterins, the yellow pigments of insects, fish scales and animal urine.

7.2 PIGMENTS OF NON-MARINE AQUATIC SOURCE ORGANISMS

7.2.1 *Chlorinoid pigments*

The chlorophyll of living plant material is principally chlorophyll *a*, which is $C_{55}H_{72}MgN_4O_5H_2O$, the fundamental structure of which is that of four pyrrole (C_4H_5N) nuclei joined in a complex called a porphin nucleus. If substituent groups are present in the eight-positions of the pyrrole nuclei a porphyrin is formed. Chlorophyll is a magnesium chelate complex of the porphyrin, in which two hydrogen atoms have saturated one double bond of the pyrrole group (i.e. a dihydroporphyrin) and which is esterified with the long chain alcohol phytol, $C_{20}H_{39}OH$. Degradation of chlorophyll in the aquatic environment typically involves a two-step process in which magnesium is released and a free chlorin pigment, pheophytin *a*, is formed; this in turn may be recomplexed with another metal to form a porphyrin. In some instances the magnesium may be replaced directly by another metal. The change from chlorophyll *a* to pheophytin *a* involves replacement of Mg by two H_2 atoms, one cause of which in sediments is thought to be the acidic digestive fluids of herbivores feeding on the phytoplankton and other aquatic vegetation.

Other changes proceeding from chlorophyll to petroleum porphyrins are shown on p. 247.

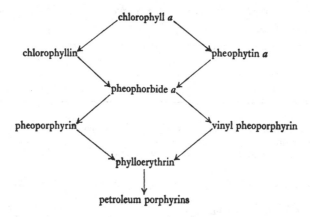

The change of pheophytin to pheophorbide is by means of hydrolysis of the phytol group which may occur in digestive tracts of animals and in sediments.

Chlorophyll *b* differs from chlorophyll *a* in that an aldehyde (CHO) group is in the same position (3-position) occupied by a methyl (CH₃) group in chlorophyll *a*. Chlorophyll *a* is about 2·5 times as abundant in

TABLE 64   *Chlorophyll content of marine and brackish water plants from Chesapeake Bay, Virginia (Swain, Paulsen & Ting, 1964)*

Species	Station no.	Chlorophyll (mg/g)
Chlorophyta		
Enteromorpha linza (Linn.)	1	2
E. prolifera (Muller)	1	3·5
Rhodophyta		
Gracilaria verrucosa (Hudson)	1	3
Ceramium strictum (Kutzing)	1	3
Agardhiella tenera (J. Ag.)	1	4
Tracheophyta		
Najas flexilis (Willd.)	1	23
Zostera marina (Linn.)	1	1·5
Ceratophyllum echinatum (Gray)	2	2
Ruppia maritima (Linn.)	4	16
R. maritima (Linn.)	11	13
Myriophyllum spicatum (Linn.)	14	10
M. spicatum (Linn.)	6	2
Potamogeton sp.	6	10
Potamogeton sp.	9	25
Av.		~ 8

TABLE 65   *Chlorophyll content of marine plants from Dillon Beach area,*
California (Swain, Paulsen & Ting, 1964)

Species	Locality	Date collected	Chloro-phyll (mg/g)
Chlorophyta			
Enteromorpha intestinalis (Linn.)	Dillon Beach, California	Mar. 1962	4
Ulva lobata (Kutzing)	Dillon Beach, California	Mar. 1962	5
Codium Setchellii (Gardner)	Dillon Beach, California	May 1962	10
Phaeophyta			
Desmarestia munda (Setchell & Gardner)	Dillon Beach, California	Mar. 1962	250
Laminaria sinclairii (Areschoug)	Dillon Beach, California	Mar. 1962	22
Macrocystis pyrifera (Linn.)	Tomales Bay, California	Mar. 1962	8
Pelvetia fastigata (J. Agardh)	Bodega Head, California	Apr. 1962	29
Fucus furcatus (C. Agardh)	Bodega Head, California	Apr. 1962	7
Postelsia palmaeformis (Ruprecht)	Bodega Head, California	Apr. 1962	6
Nereocystis luetkeana (Mertens) stipe partly decomposed	Dillon Beach, California	Apr. 1962	3
Nereocystis luetkeana (Mertens) blade partly decomposed	Dillon Beach, California	Apr. 1962	2
Dictyoneurum californicum	Tomales Point, California	Apr. 1962	147
Laminaria Andersonii (Eaton)	Tomales Point, California	Apr. 1962	1
Cystoseira osmundacea (Menzies) washed ashore	Dillon Beach, California	Apr. 1962	7
Alaria marginata (Postels & Reprecht)	Bodega Head, California	May, 1962	104
Egregia Menziesii (Turner)	Dillon Beach, California	Mar. 1962	44
Rhodophyta			
Porphyra lanceolata (Setchell & Hus)	Dillon Beach, California	Mar. 1962	9
Cryptopleura violacea (J. Agardh)	Dillon Beach, California	Mar. 1962	7
Lithothrix aspergillum (Gray)	Dillon Beach, California	May 1962	104
Bossea Gardneri (Manza)	Bodega Head, California	May 1962	4
Farlowia mollis (Harvey & Bailey)	Dillon Beach, California	May 1962	20
Iridophycus flaccidum (Setchell & Gardner) purplish color	Tomales Point, California	May 1962	1
Iridophycus flaccidum (Setchell & Gardner) greenish color	Tomales Point, California	May 1962	0·5
Gastroclonium coulteri (Harvey)	Tomales Point, California	May 1962	21
Gigartina corymbifera (Kutzing)	Dillon Beach, California	May 1962	3
Rhodomela larix (Turner)	Dillon Beach, California	May 1962	10
Av. of Algae			~ 33

TABLE 65 (*cont.*)

Species	Locality	Date collected	Chloro-phyll mg/g
Tracheophyta			
Zostera marina Linn. (nearshore)	Tomales Bay, California	30 Mar. 1962	256
		30 Mar. 1962	96
		10 Apr. 1962	131
		16 Apr. 1962	134
		20 Apr. 1962	166
		26 Apr. 1962	150
		1 May 1962	66
		5 May 1962	125
		10 May 1962	286
		15 May 1962	215
		20 May 1962	172
		28 May 1962	225
Zostera marina (offshore)	Tomales Bay, California	6 Apr. 1962	102
		10 Apr. 1962	160
		25 Apr. 1962	132
		5 May 1962	145
		10 May 1962	126
Average of *Zostera*			~ 158
Average of all plants			~ 38

leaves of green plants as is chlorophyll *b*. Chlorophyll *c*, *d* and bacterio-chlorophyll are other relatively uncommon chlorophylls.

The chlorophyll and related pigments of a variety of freshwater, brackish water and marine larger plants show wide variations (tables 64–66). There is considerable seasonal variation in the freshwater chlorinoid pigments and the maximum concentration is not in the summer months in all cases (Paulsen, 1962; Swain, Paulsen & Ting, 1964).

The chlorinoid pigments identified from absorption spectra of 90% acetone extracts of the plants are as follows:

Chlorinoid Pigments	*Non-Chlorinoid Pigments*
1 Chlorophyll *a*	8 Carotenes and xanthophylls
2 Chlorophyll *b*	not studied further
3 Chlorophyll *c*	9 Aromatic hydrocarbons
4 Pheophytin *a*	10 Unknown compounds
5 Bacteriochlorophyll ?	
6 Chlorophylline *a* ?	
7 Chlorophylline *c*	

TABLE 66   *Chlorophyll content of aquatic plants from*
Coos Bay, Oregon (Swain, Paulsen & Ting, 1964)

Species	Station no.	Description of locality	Chloro-phyll content
Chlorophyta			
Enteromorpha intestinalis (Linn.)	6	Haynes Inlet, mudflat in water 0–6 ft deep at low tide	31
Ulva lobata (Kutzing)	9	Russel Point, mudflat in water 2–5 ft deep	8·5
E. tubulosa (Kutzing)	15	Cape Arago, sand beach, rocky coast	10
Phaeophyta			
Egregia Menziesii (Turner)	15	Cape Arago, sand beach and rocky coast in water up to 25 ft deep	113
Fucus furcatus (C. Agardh)	10	South of Glasgow, mudflat in water 0–5 ft	19
Pelvetiopsis limitata (Setchell)	12	Kentuck Slough, north-west part of inlet mudflat flanking salt marsh, water 3 ft deep	11
Nereocystis luetkeana (Mertens)	15	Cape Arago, sand beach and rocky coast water up to 25 ft deep	4·1
Rhodophyta			
Porphyra lanceolata (Setchell & Hus)	15	Cape Arago, sand beach and rocky coast	11
Plocamium violaceum (Farlow)	3	Sunset Bay State Park, sand beach, rocky shore	11
P. violaceum (Farlow)	15	Cape Arago, sand beach, rocky coast	2
Gracilaria Sjoestedtii (Kylin)	14	Hanson's Landing, mudflat, water 0–1 ft deep at low tide, subject to effluent contamination	38
Tracheophyta			
Zostera marina (Linn.)	9	Russell Point, mudflat, in water 2–5 ft deep	21
		Av.	25

Chlorophyll *a* predominates over the other chlorinoid pigments in all the plants. It is identified by a prominent absorption peak at 432–435 nm, a smaller peak at 661 nm and other smaller peaks (fig. 101). Chlorophyll *b* (fig. 101) has principal absorption maxima at 456–460 nm and 643–649 nm, as was found in the following freshwater species studied by Swain, Paulsen & Ting, (1964): *Ceratophyllum demersum, Myriophyllum exalbescens, Najas guadalupensis, Nymphaea odorata* and *Potamogeton illinoiensis*. It was also noted in the spectra of extracts of several marine species. Chlorophyll *b* is characteristic of the Chlorophyta algae, but was not found in all the specimens of this family studied.

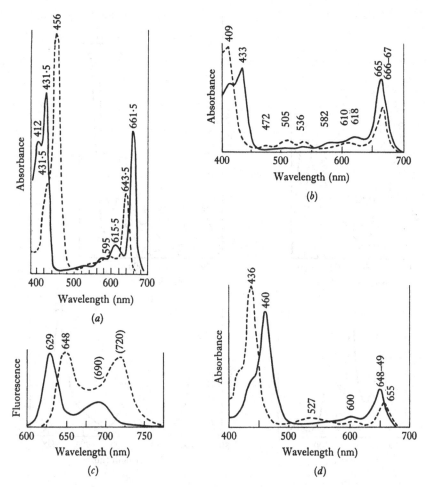

FIGURE 101   Absorption spectra of chlorinoid pigments (Swain, Paulsen & Ting, 1964). (*a*) chlorophyll *a* (solid line) and *b* (dashed line) in dry acetone (Harris & Zscheile, 1943); (*b*) chlorophyll *a* (solid line) and pheophytin *a* (dashed line) in 80% acetone; pheophytin determined in presence of oxalic acid (Vernon, 1960); (*c*), chlorophyll *b* (solid line) and pheophytin *b* (dashed line) in 80% acetone (Vernon, 1960); (*d*) fluorescence spectra of chlorophyll *c* and pheophytin *c* in ether (French *et al.* 1957).

Chlorophyll *c* (fig. 102) has absorption maxima at 629 and at 690 nm; it was found only in the marine specimens studied (*Zostera marina*, *Fucus furcatus* and *Postelsia palmaeformis*). Chlorophyll *d* (fig. 102) with λ_{max} at 400, 445 and 696 nm was not found in any of the specimens examined.

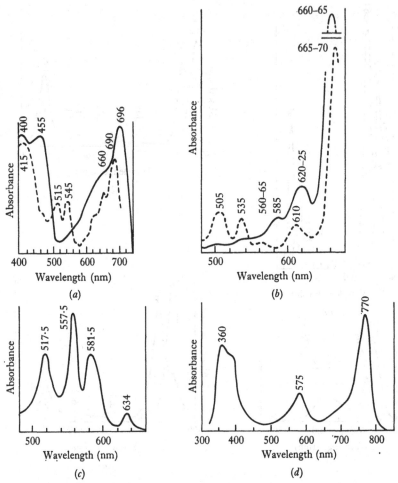

FIGURE 102 Absorption spectra of chlorinoid pigments (Swain, Paulsen & Ting, 1964). (*a*) chlorophyll *d* (solid line) and pheophytin *d* (dashed line) in methanol (Manning & Strain, 1943); (*b*) bacteriochlorophyll in dry acetone (Goedheer, 1958); (*c*) methyl chlorophylline *a* (solid line) and methyl pheophorbide *a* (dashed line) showing effect of magnesium (in the chlorophyllin) on the spectrum, in dioxane (Rabinowitch, 1951); (*d*) phylloerythrin in dioxane (Stern & Wenderlein, 1935).

Pheophytin *a*, the first degradation product of chlorophyll *a*, has λ_{max} at 409 and 666–667 nm, together with smaller peaks at 472, 505, 536 and 610 nm. It is widely distributed in aquatic plant species analyzed by Swain, Paulsen & Ting (1964) especially in older parts of the specimen and in winter collections. A paper chromatogram of pigments from *Elodea canadensis* (fig. 115) shows a combination of chlorophyll *a* and pheophytin *a*. The rapid degradation of chlorophyll *a* to pheophytin *a* under laboratory conditions even when air and light are nearly excluded is shown by the fact that several fresh plant extracts when dried and redissolved in *n*-heptane had spectra of pheophytin rather than chlorophyll.

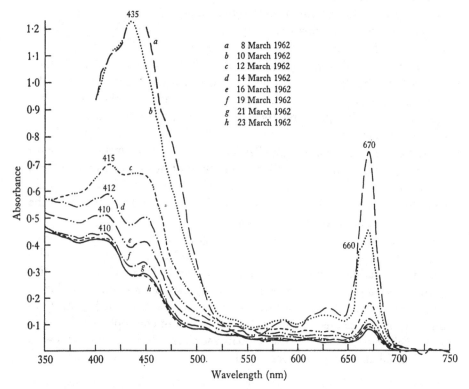

FIGURE 103 Pigment extraction of *Egregia Menziesii* (Swain, Paulsen & Ting, 1964). Absorption spectra of 90% acetone extracts of *Egregia Menziesii* from Dillon Beach, California, prepared from sample standing in sea water in laboratory over a two-week period, showing gradual change from chlorophyll *a* to pheophytin *a*.

The absorption spectra of pigment extracts of a marine alga *Egregia Menziesii* show a change from chlorophyll *a* to pheophytin *a* (fig. 103) when the plant was allowed to stand in seawater in the laboratory over a 2-week period. The change is rapid at first but approached equilibrium at the end of 2 weeks. The final absorbance of the pheophytin was about $\frac{1}{10}$ that of the original chlorophyll. By contrast, when chlorophyll is treated with oxalic acid, a typical procedure to make pheophytin, a reduction in absorbance of $\frac{1}{4}$ to $\frac{1}{3}$ results (Zscheile & Comer, 1941). Microbial decomposition probably accounts mainly for the greater reduction noted in the naturally decayed sample, but irreversible photobleaching of chlorophyll may also be involved (Livingston, 1949). On the basis of limited data it was tentatively concluded that a unit of pheophytin represented in geochemical samples represents about 10 units of original chlorophyll, although this may vary with the environment to some extent.

The separate determination by spectrophotometry of chlorophylls *a*, *b* and *c* and of carotenoids in 90% acetone extracts of water samples was discussed by Richards (1952); Richards & Thompson (1952) and by Creitz & Richards (1955). In all such extracts other pigments may be present that absorb at wavelengths close to those at which the chlorophylls absorb, so that the spectrophotometric determinations of such mixtures is not truly quantitative.

Studies by Gillbricht (1952), Banse (1957) and Kozminski (1938) have shown that much of the so-called chlorophyll of deeper marine and fresh bodies of water, while appreciable in amount, is from detritus rather than from living cells. An estimate was made that ash-free dry plankton of the Bay of Kiel contained 4–12% chlorophyll.

In detailed studies of the chlorinoid pigments of oak, hazel and aspen leaves on a seasonal basis Sanger (1968) found that chlorophyll *a* and *b* occurred in ratios of about 2·5 : 1 or 3 : 1 per unit leaf area during the active growing period (June–September) in oak leaves; about 2 : 1 in hazel leaves and less than 2 : 1 in aspen leaves. In fallen oak leaves a small quantity of pheophytin *a* remained as a degradation product of chlorophyll *a*. In aspen and hazel leaves, however, little or no measurable chlorinoid pigments were found in fallen leaves, October–April.

7.2.2 *Carotenoid pigments*

Carotenoid pigments of living plants fall into several chemical classes (Dunning, 1963): (1) carotenes ($C_{40}H_{56}$), (2) carotenols ($C_{40}H_{56}O$, $C_{40}H_{56}O_2$, $C_{40}H_{56}O_3$ and $C_{40}H_{56}O_4$), (3) ketones ($C_{40}H_{54}O$, $C_{40}H_{50}O_2$), (4) hydroxycarbonyl compounds ($C_{40}H_{60}O_6$) and (5) carboxylic compounds ($C_{20}H_{24}O_4$, $C_{27}H_{38}O_4$). Of these the following would seem to be of geochemical importance:

(1) α-, β-, γ-*Carotene* ($C_{40}H_{56}$), found in most plants, absorption maxima in petroleum ether at 423, 448 and 478 nm; 426, 452 and 484 nm; and 431, 462 and 495 nm, respectively, and in carbon disulfide at 477 and 490 nm; 450, 485 and 520 nm; and 463, 496 and 534 nm respectively.

(2) *Rhodopurpurin* ($C_{40}H_{56}$), found in purple bacteria, absorption maxima in carbon disulfide at 479, 511 and 550 nm.

(3) *Rhodopol* ($C_{40}H_{56}O$) found in purple bacteria, absorption maxima in CS_2 478, 508 and 547 nm.

(4) *Luteol* ($C_{40}H_{56}O_2$), found in most plants, absorption maxima in CS_2 440, 474 and 506 nm.

(5) *Rhodoviolascin* ($C_{42}H_{60}O_2$), found in purple bacteria, absorption maxima in CS_2 496, 534 and 573 nm.

(6) *Myxoxanthin* ($C_{40}H_{54}O$), found in blue-green algae, absorption maximum in CS_2 at 488 nm.

(7) *Rhodoxanthin* ($C_{40}H_{50}O_2$), found in tree leaves, absorption maxima in CS_2 at 491, 525 and 564 nm, and in petroleum ether at 458, 489 and 524 nm.

(8) *Fucoxanthin* ($C_{40}H_{60}O_6$), found in brown algae, absorption maxima at 445, 477 and 510 nm.

(9) *Azafrin* ($C_{27}H_{38}O_4$), found in plant root, absorption maxima at 419, 446 and 476 nm.

(10) *Bixin* ($C_{24}H_{28}O_4$), found in certain plants.

In addition to those listed, Vallentyne (1956) has reported *lycopene* ($C_{40}H_{56}$), an acyclic tetraterpene, from post-glacial lake sediments, although this pigment is usually restricted to fruits. In the sediments it may have formed from carotene by opening of the terminal rings.

Of the substances listed above α- and β-carotene are the most abundant and stable in the geochemical environment. The absorption spectra of carotenoid pigments generally consist of three absorption bands, a middle principal band with two smaller bands on either side. Thus in the lists above the middle value is the main absorption band, except where only two are listed in which case the first number is the main band.

The carotenoid pigments of oak, hazel and aspen leaves were found by Sanger (1968) to comprise β-carotene, lutein, violaxanthin, neoxanthin and other unresolved xanthophylls. All but the xanthophyll lutein, degraded below measurable amounts by the end of the growing period (October–November), lutein continued to remain at reduced levels in the fallen oak leaves. In fact lutein formed 80–90 % of carotenoid pigments in fallen oak leaves, with β-carotene forming the remainder. In hazel leaves although both carotenoids and chlorinoid pigments are richer in the growing leaf than in oak leaves, both types disappear entirely in the fallen leaves. The pigments in aspen leaves suffer a similar fate.

7.2.3 *Flavinoid pigments*

The flavinoid pigments and related heterocyclic substances of several species of freshwater, brackish water and marine plants are listed in tables 78, 79. Characterization of these was by means of paper chromatography, color of fluorescence in ultraviolet light and wavelength of absorption maxima (Swain, Paulsen & Ting, 1964). Additional studies of these substances in organisms that are important geochemically should be made.

TABLE 67 *Average amount of sedimentary chlorophyll degradation products (SCDP units/g ignitable matter) for samples greater than 8·5 m deep from six Connecticut lakes (Vallentyne & Craston, 1956)*

Data for two 8·5 m samples from Pataguanset Lake are included. Standard deviations are listed for the SCDP averages. Data for oxygen deficit and seston chlorophyll are from Deevey (1940).

Lake	Number of samples	Average SCDP (units/g ignitable matter)	O_2 deficit (mg/cm^2/day)	Seston chlorophyll (mg/m^3)
Pocotopaug	3	46·9 ± 17·7	11·9	6·37
Bashan	5	49·9 ± 12·0	12·8	2·08
Rogers	10	68·7 ± 9·2	31·0	2·87
Pataguanset	4	80·7 ± 7·0	20·2	4·07
Linsley Pond	9	119·6 ± 24·0	41·8	15·25
Ball Pond	7	120·6 ± 28·6	18·2	5·54

7.3 ORGANIC PIGMENTS IN EUTROPHIC LAKE SEDIMENTS OF GLACIATED REGIONS

7.3.1 *Chlorophyll and related pigments*

The chlorinoid pigments of eutrophic lakes of several different types in Connecticut ranged from 46·9 to 155·0 sedimentary chlorophyll degradation product units (SCDP) per gram of ignitable matter (Vallentyne & Craston, 1957) (table 67). The lakes represent three types: (1) shallow lakes, (2) deep lakes in which amount of SCDP was related to or (3) was independent of depth. Standard deviations in SCDP measurements ranged from 5·4 to 28·6. In the shallow lakes (2–4 m water depth) sedimentary chlorophyll per gram of ignitable matter seems independent of depth (table 68). In surface water of Dooley Pond a large amount of seston chlorophyll, 58·4 mg/m^3, had been recorded (Deevey, 1940), but this was not matched by high SCDP values in the mud. In another shallow pond, Beseck Lake, seston chlorophyll was only 13·56 mg/m^3 but SCDP was about the same as that in Dooley Pond.

TABLE 68   *Ignitable matter and sedimentary chlorophyll degradation products (SCDP units/g ignitable matter) for Ekman dredge samples from Dooley Pond, Connecticut (Vallentyne & Craston, 1957), Reproduced by permission of the National Research Council of Canada from the Canadian Journal of Botany* **35,** *35–42*

(The standard deviation is listed for the average.)

Depth (m)	Ignitable matter (%)	SCDP (units/g ignitable matter)
2	16·4	69·0
2	16·4	61·0
2·5	16·4	58·0
2·5	17·5	81·2
3	18·3	45·3
3	20·1	61·2
3	19·3	53·8
3·5	18·7	44·4
4	21·4	56·6
4	20·4	64·7
	Av.	59·5 ± 10·9

TABLE 69   *Ignitable matter and sedimentary chlorophyll degradation products (SCDP units/g ignitable matter) for Ekman dredge samples from Lower Linsley Pond, Connecticut (Vallentyne & Craston, 1957 loc. cit.)*

Depth (m)	Ignitable matter (%)	SCDP (units/g ignitable matter)
3	53·7	51·8
3·5	29·2	97·9
4·5	22·5	83·1
5·5	47·7	78·6
6·5	27·7	140·0
7	41·3	91·7
9	35·0	104·0
9	33·5	102·0
10	40·6	121·1
10	23·7	166·0
11	33·9	118·5
12	37·6	105·0
12·5	39·7	155·0
13	33·9	118·0
13	35·0	86·6

In the deeper lakes that show increases of SCDP with increasing depth, for example, Lower Linsley Pond (table 69) the sediments also have higher contents of ignitable matter than in the shallow ponds, although the standing phytoplankton crops are lower than in the shallow ponds.

TABLE 70 *Ignitable matter and sedimentary chlorophyll degradation products (SCDP units/g ignitable matter) for Ekman dredge samples from Rogers Lake, Connecticut*

(The standard deviation is listed for the average.)

Depth (m)	Ignitable matter (%)	SCDP (units/g ignitable matter)
10	31·2	63·0
11	32·1	59·4
11·5	32·8	56·5
12	32·0	65·4
12	29·2	65·5
13	27·0	65·5
16	30·4	74·9
17	32·7	74·0
18	30·4	90·4
18	31·5	72·0
		Av. 68·7±9·2

The deep lakes that show no appreciable change in SCDP with depth have intermediate values of ignitable matter (table 70). Ignitable matter is also low in the shallow water of the deep lakes.

The explanation for the observed variations in SCDP in the Connecticut eutrophic lakes seems to lie in the greater breakdown rate of pigments in the shallow as compared to the deep water.

Several species of aquatic plants (table 63) and associated sediments from an alkalitrophic lake and a eutrophic–alkalitrophic lake in Minnesota contained the following pigments (Swain, Paulsen & Ting, 1964) in 90% acetone extracts:

Chlorophyll *a*   Chlorophylline *c*?
Chlorophyll *b*   Carotenes and xanthophylls?, not
                                      studied in detail
Chlorophyll *c*   Hydrocarbons and related compounds
                                      other than carotenes
Pheophytin *a*
Bacteriochlorophyll?   Unknown compounds
Chlorophylline *a*?

Chlorophyll *a* predominates over the other chlorinoid pigments in the living plants. It was identified from a prominent absorption peak at 432–435 nm, a lesser peak at 665–667 nm and other smaller peaks as shown in fig. 104. Absorption spectra of several species of lake plants are shown in fig. 104 in which chlorophyll *a* is the principal constituent. Chlorophyll *b* has prominent absorption peaks in acetone and acetone water at 456–460 nm and at 643–649 nm. Chlorophyll *b* was detected in extracts from

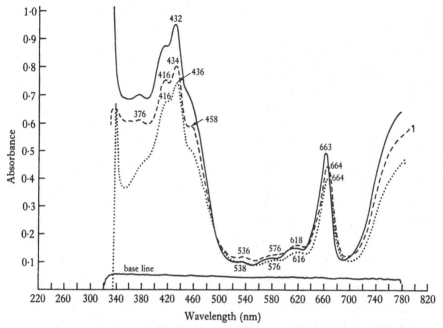

FIGURE 104   Visible absorption spectra of 90% acetone extracts of aquatic plants from Blue Lake, Minnesota (Swain, Paulsen & Ting, 1964). Solid line, *Myriophyllum exalbescens* Fern., collected 22 April 1961; dashed line *Ceratophyllum demersum* Linn., collected 29 December 1960; dotted line *Elodea canadensis* Michx., collected 29 December 1960. Chlorophyll *a* is the principal constituent, as shown by absorption at 416, 432–436, 536, 616–618 and 663–664 nm. A trace of chlorophyll *b* or a carotenoid is suggested by the shoulder at 458 nm in *Ceratophyllum*.

the following freshwater plants, although it should be present in most given plants:

> *Ceratophyllum demersum* from Blue Lake, Minnesota
> *Nymphaea odorata* from Blue Lake, Minnesota
> *Myriophyllum exalbescens* from Clear Lake, Minnesota
> *Najas guadalupensis* from Clear Lake, Minnesota
> *Potamogeton illinoiensis* from Clear Lake, Minnesota

Chlorophyll *c* has absorption maxima at 629 and 690 nm. It was not found in the freshwater plants studied. Chlorophyll *d*, characterized by absorption maxima at 400, 445 and 696 nm with a shoulder at 660 nm was not found in any of the specimens studied. Pheophytin *a* is present in most of the plant specimens and sediments examined. It has absorption peaks at 409 and 666–667 nm and smaller peaks at 472, 405, 536 and 610 nm.

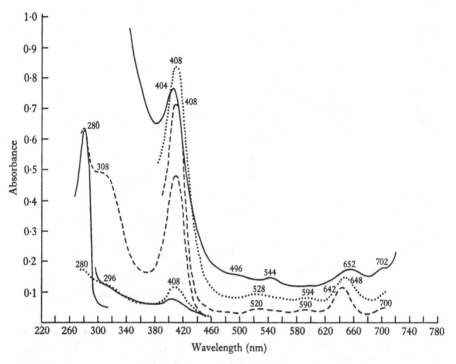

FIGURE 105 Pigment extracts from freshwater plants. Chromatographic fractions remaining adsorbed on alumina after elution with heptane, benzene, pyridine and methanol of benzene, etc., extracts of lake plants. Solid line *Chara* sp., Clear Lake, Minnesota, collected 16 October 1960 (in heptane). Dashed line *Nymphaea odorata* (Ait.) (in acidified EtOH) Blue Lake, Minnesota, collected 16 October 1960. Dotted line *Ceratophyllum demersum* Linn., collected 16 October 1960 (in acidified EtOH). Substances probably represented include pheophytin *a*, pheophytin *b* (λ_{max} 562 nm), pheophytin *c* (λ_{max} 648 nm), pheophytin *d* (λ_{max} 544, 702 nm), and some aromatic substances with absorption peaks below 400 nm.

Some of the living plant samples contained both chlorophyll *a* and pheophytin *a*, particularly the older parts of the plant or winter specimens (fig. 105). Specimens that contain more than a small amount of pheophytin cannot be used for evaluation of the chlorophyll content. Pheophytin *d*, chlorophylline *a* and *c* and bacteriochlorophyll were detected spectrally in several samples.

TABLE 71   *Chlorophyll* a *in plants from Blue Lake and Clear Lake, Minnesota (on dry weight ash-free basis) based on spectral data at* 665–670 nm *(Swain, Paulsen & Ting,* 1964)

Species	Date	Carbon (%)	Nitrogen (%)	Chloro-phyll (mg/g)
Chlorophyta				
Algae (Blue Lake)	29 Dec. 60	n.d.	n.d.	3
Tracheophyta				
Ceratophyllum demersum	29 Dec. 60	43·34	2·25	113
L. (Blue Lake)	11 Feb. 61	43·60	1·59	30
	22 Apr. 61	41·22	1·53	5
	28 Jul. 61	44·93	2·58	48
Elodea canadensis	29 Dec. 60	43·20	1·41	12
Michx. (Blue Lake)	11 Feb. 61	37·68	2·74	12
	22 Apr. 61	38·22	2·38	12
	28 Jul. 61	41·57	2·84	110
Nymphaea odorata Ait. (Blue Lake)	28 Jul. 61	46·11	2·78	58
Najas guadalupensis	29 Dec. 60	42·74	2·38	17
(Spreng.) (Clear Lake)	11 Feb. 61	39·54	2·57	23
	22 Apr. 61	38·29	2·07	21
	28 Jul. 61	37·28	1·57	18
Myriophyllum	29 Dec. 60	45·01	2·88	26
exalbescens Fern.	22 Apr. 61	44·14	2·84	20
(Clear Lake)	7 June. 61	n.d.	n.d.	28
	28 Jul. 61	n.d.	n.d.	54
Typha latifolia	29 Dec. 60	44·75	0·44	2
Linn. (Clear Lake)	22 Apr. 61	47·69	1·18	0·8
	28 Jul. 61	42·29	n.d.	30
Potamogeton illinoiensis	22 Apr. 61	n.d.	1·18	12
Morong (Clear Lake)	28 Jul. 61	n.d.	n.d.	49
Av.	29 Dec. 60 (6 samples)			∼ 29
	11 Feb. 61 (3 samples)			∼ 22
	22 Apr. 61 (6 samples)			∼ 12
	28 Jul. 61 (7 samples)			∼ 52
	all samples			∼ 29

Seasonal variation studies of chlorophyll *a* in several plant species are shown in table 71. *Ceratophyllum demersum*, although it had the largest amount of chlorophyll also had the greatest seasonal variation. That photosynthesis continues when under the lake ice cover in early winter is shown by the high chlorophyll content of *C. demersum* and several other species. The general levels of chlorophyll content are fairly high in mid-winter even under a fairly heavy snow cover on the ice. Low levels of chlorophyll in spring collections of *Ceratophyllum* and *Typha latifolia* are probably caused by bacterial decay, as the plant samples were partly degraded when collected, as was the winter sample of algae. The low nutritive levels of the lake waters in the early spring are reflected in the low chlorophyll values in those samples.

The chlorinoid pigments from the sediments of Blue Lake, Minnesota are relatively high at the sediment surface but decrease rapidly downward (fig. 106). The change is probably related to sediment type which changes from calcareous sapropel downward to calcareous copropel and then into purer marl. The shallow water environment (3 m) is apparently not ideal for the preservation of pigments. The water temperature is as high as 16 °C at the bottom of the lake in August and dissolved oxygen was found in the bottom waters; a fairly high percentage of surface light probably reaches the bottom.

The length of the shoreline of a lake is related to its general productivity (Welch, 1952). This can be expressed as the development of the shoreline, defined as the ratio of the length of shoreline to the length of the circumference of a circle of area equal to that of the lake. In Blue Lake and through other Minnesota lakes the length of shoreline and development of shoreline is as follows (Paulsen, 1962):

Blue Lake, Isanti Co., Minn.	9,817 m	2·47;
Lake Itasca, Clearwater Co., Minn.	22,720 m	3·68;
Long Lake, Clearwater Co., Minn.	5,050 m	1·74;
Upper Red Lake, Beltrami Co., Minn.	93,284 m	1·25.

The chlorinoid pigments and a rough indication of the carotenoid pigments in short sediment cores in the four lakes is shown in fig. 106. Some correlation can be seen between pigment content and shoreline development. Welch (1952) also suggested that a lake basin which is steep-sided, such as Long Lake, will have lower productivity because decomposable material is removed to the deep part of the lake where it becomes less accessible as food for other organisms.

A longer core of Blue Lake sediments (fig. 107) shows that the lake underwent an early period of eutrophication that was followed by an interval of decreased deposition of organic marl and an increase in marl before the beginning of the present cycle of productivity. The relatively large amount of pheophytin in the lower part of the sediment shows that this substance can persist for thousands of years relatively unchanged. The pheophytin content of the Blue Lake sediment is expressed both as parts per million of dry sediment and as per cent of H_2O_2-determined organic matter. Both methods show an earlier episode of eutrophication not shown

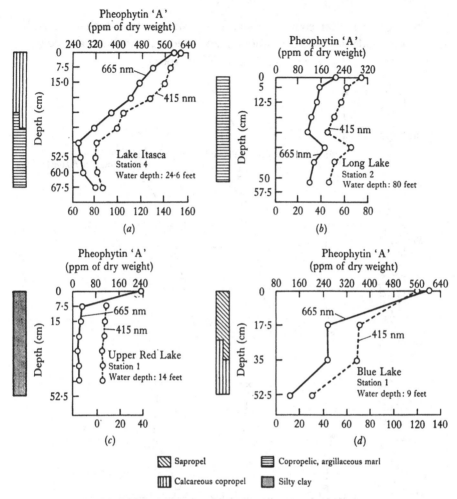

FIGURE 106 (a) Sedimentary pigments in 90% acetone extracts of core samples from Lake Itasca, Minnesota, station 4. Numerical difference between curves at 415 and 665 nm represents mainly carotenoid pigments, whereas curve at 665 nm is mainly pheophytin. Values are on a dry weight, ash-free basis. (b) Sedimentary pigments in 90% acetone extracts of core sample from Long Lake, Minnesota, station 2. Numerical difference between curves at 415 and 665 nm represents mainly carotenoid pigments, whereas curve at 665 nm is mainly pheophytin. Values are on a dry weight, ash-free basis. (c) Sedimentary pigments in 90% acetone extracts of core sample from Upper Red Lake, Minnesota, station 1. Numerical difference between curves at 415 and 665 nm represents mainly carotenoid pigments, while curve at 665 nm is mainly pheophytin. Values are on a dry weight, ash-free basis. (d) Sedimentary pigments in 90% acetone extracts of core sample from Blue Lake, Minnesota, station 1 (Swain, Paulsen & Ting, 1964). Numerical difference between curves at 415 and 665 nm represents mainly carotenoid pigments, while curve at 665 nm is mainly pheophytin. Values are on a dry weight, ash-free basis.

by the organic matter content alone. The residual monosaccharides and the amino acids show similar though not identical increases at 700 cm depth (Rogers, 1965; Swain, Paulsen & Ting, 1964).

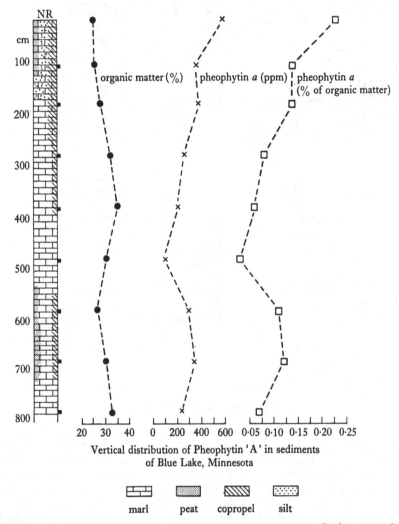

Vertical distribution of Pheophytin 'A' in sediments
of Blue Lake, Minnesota

marl   peat   copropel   silt

FIGURE 107   Sedimentary pigments in 90 % acetone extracts of a long core from Blue Lake, Minnesota (Swain, Paulsen & Ting, 1964). Values are on a dry weight, ash-free basis.

As a further observation on chlorinoid pigments in Lake Itasca, Minnesota, sediments, the pheophytin gradually decreases with depth in sediments that contain nearly the same amount of organic matter. The decrease may result from its use by mud-eating organisms as food or it may be due in part to dark-reaction photochemical effects between pheophytin in solution and heterocyclic organic compounds (Pariser, 1950; Livingston, 1949).

The rather low content of pheophytin in the sediments of Long Lake, Minnesota may seem anomalous when compared to pigment data from Lake Itasca and Blue Lake. The sediments are a uniform copropelic argillaceous marl in Long Lake. Conditions for preservation in 80 ft of water (low temperature, lack of light and generally low oxygen levels) are fairly good. In this lake, however, R. O. Megard (personal communication) has found oxygen in the profundal waters throughout much of the year. The reasons for the low pigment content are that the lake has lower productivity of phytoplankton and aquatic vegetation than Lake Itasca or Blue Lake, persistent, though small oxygen content, and greater depth of water that allows more time for the chlorinoid pigments to degrade before reaching the bottom.

The extracts from Lake Itasca, Long Lake, and Blue Lake sediments had a small peak near 750 nm which suggests the presence of bacterio-chlorophyll. The Red Lake sediment extracts contained chlorophyll *a*, unlike the other group of Minnesota lakes discussed here. Red Lake sediment acetone extracts had no absorption at 750 nm but had a broad peak at 970–1000 nm that is due to unknown background absorption.

Gorham (1960) has studied the relationship between chlorophyll degradation products and sediment-facies in five English lakes. The lakes included: (1) two (Wastwater and Ennerdale Water) which are large, deep, have oxygen in the hypolimnion throughout the period of stratification, are low in calcium bicarbonate and potash and have sparse phytoplankton; (2) a similar lake (Windermere, North Basin) but with moderate amounts of lime and potash; (3) a smaller and shallower lake (Esthwaite Water), but also stratified, in which the hypolimnion becomes completely deoxygenated near the mud surface, the water is richer in lime and potash, and there is a large phytoplankton crop; and (4) a small, shallow presumably unstratified pond (Priest Pot), richer in lime and potash than the others and with a very dense phytoplankton crop. The physical and chemical properties of the lakes (table 72) and the chlorophyll-derivatives, sulfur, and carbon contents of the sediments (table 73) show a definite relationship to productivity. It appears that the chlorophyll derivatives (measured spectrophotometrically at 667 nm) may be a more sensitive index of productivity than the carbon content of the sediments.

Gorham found that living *Melosira* (diatom) contained about 250 units of chlorophyll/gm of carbon (1 unit = optical density of 0·1 in a 1-cm cell when dissolved in 100 ml of 90% acetone), and that after 3–4 days and conversion to pheophytin the absorbance declined to about half the original

TABLE 72 *Physical and chemical properties of some*
English lakes and their waters (Gorham, 1960)

	Area km²	Max. depth m	Total salts	Na	K	Ca	Mg	HCO₃	Cl	SO₄	NO₃
							(m-equiv/l)				
Wastwater	2·9	79	0·31	0·15	0·01	0·09	0·07	0·05	0·16	0·10	0·006
Ennerdale Water	2·9	45	0·33	0·17	0·01	0·09	0·07	0·05	0·18	0·10	0·005
Windermere North Basin	8·2	67	0·51	0·16	0·01	0·27	0·07	0·17	0·17	0·14	0·018
Esthwaite Water	1·0	16	0·75	0·20	0·02	0·41	0·12	0·31	0·20	0·20	0·013
Priest Pot	0·01	4·5	1·09	0·25	0·08	0·62	0·15	0·66	0·23	0·19	0·002

TABLE 73 *Chlorophyll derivatives, total carbon and total*
sulfur in English lake muds (Gorham, 1960)

	Carbon	Sulfur	Chlorophyll derivatives units/g	Ratio of optical densities
	% dry weight		dry weight	410:350 nm
Wastwater	7·2	0·18	0·21	0·7
Ennerdale Water	6·6	0·19	0·24	0·9
Windermere North Basin	8·7	0·37	0·83	1·0
Esthwaite Water	11·6	0·61	1·37	1·4
Priest Pot	18·9	1·22	6·88	1·8

value. As noted elsewhere in this chapter, however, the value would very probably continue to decline and reach approximate equilibrium at about 10^{-1} of the original value with longer time (Swain, Paulsen & Ting, 1964).

In further study of the chlorophyll-derivative distribution in sediment cores of Esthwaite Water and Ennerdale Water (Gorham, 1961) organic carbon, sulfur and pheophytin exhibited parallel trends (fig. 108). In Ennerdale Water the maxima of the three constituents were not reached until the lake was quite far along in its history of fertility while in Esthwaite Water eutrophication and resulting maxima in the three components set in much earlier.

An interesting study of the relationship of chlorophyll-derivatives to carotenoids in Esthwaite Water sediments was made by Fogg & Belcher (1961). They extracted the sediments at 10-cm intervals to a depth of 440 cm which marked the base of the lake sediment sequence. The chlorophyll degradation products are plotted in fig. 109, and the epiphasic and hypophasic carotenoid contents of the sediments are shown in fig. 110. The epiphasic carotenoids were characterized by paper and column chromatography and found to consist of α- and β-carotene and two more polar pigments not identified. The hypophasic carotenoids represented mainly

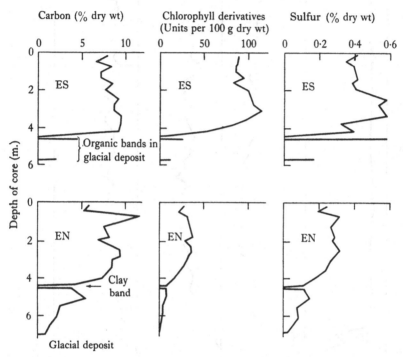

FIGURE 108   The distribution of carbon, chlorophyll derivatives, and sulfur in cores from Esthwaite Water (ES) and Ennerdale Water (EN), England. (Gorham, 1961). Reproduced by permission of the National Research Council of Canada from the *Canadian Journal of Botany* **39**, 333–338.

one component which behaved chromatographically like lutein (a xanthophyll) (fig. 111). The most polar epiphasic component resembled one (R_1) recognized by Andersen & Gundersen (1955) from interglacial gyttja in Denmark. The nature of the pigments did not change appreciably throughout the core. The pigments are believed to have been largely derived from vegetation in existence when the sediments formed and that little if any *in situ* bacterial synthesis has occurred. There seems to have been very little vertical movement of the pigments because they are strongly adsorbed on the sediment particles not soluble in non-polar solvents but soluble in polar solvents, and because their variations appear to some extent to reflect ecologic changes of the past. There does not seem to have been much change in the amounts of sedimentary chlorophyll and of epiphasic carotenoids vertically in the core but the ratio of sedimentary chlorophyll

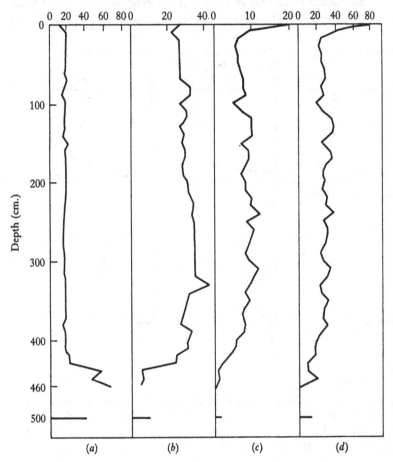

FIGURE 109 Variation with depth in a core of Esthwaite Water (England) sediments of (*a*) per cent dry matter, (*b*) per cent loss in dry weight on ignition, (*c*) chlorophyll degradation products, units/g dry weight, (*d*) chlorophyll degradation products, units/g loss on ignition (Fogg & Belcher, 1961).

to hypophasic carotenoid increases with depth probably owing to progressive degradation of lutein. If the age of the lake sediments is about 10,000 years, based on a pollen zone date at 440 cm of 8,300 years B.C., the half-life of lutein in this deposit is of the order of 20,000 years.

The authors (Fogg & Belcher, 1961, p. 137) suggest that comminuted deciduous leaves have supplied a large part of the organic matter in the

Esthwaite Water sediments owing to the large proportion of lutein, although in Connecticut lake sediments phytoplankton seems to have been the main source (Vallentyne & Craston, 1957).

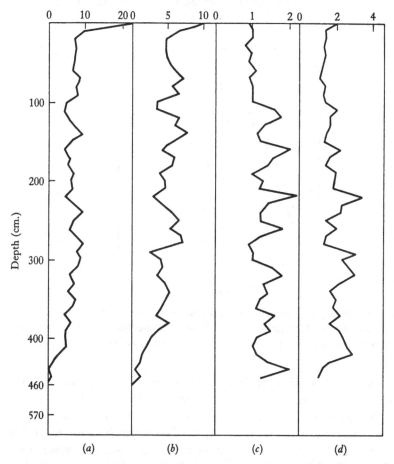

FIGURE 110 Variation with depth in a core of Esthwaite Water (England) sediments of (a) epiphasic carotenoids, units/g dry weight, (b) hypophasic carotenoid, units/g dry weight, (c) the ratio of chlorophyll degradation products to epiphasic carotenoids, (d) the ratio of chlorophyll degradation products to hypophasic carotenoid (Fogg & Belcher, 1961).

The variations in ratios of sedimentary chlorophyll to epiphasic carotenoids is nearly constant at 0·963 ± 0·028 from the surface ooze down to 100 cm, but rises to peaks of about 2 at 120, 160, 220, 260, 320, 370–390 and 440 cm; the intervening basic value being about 1·0. These regular fluctuations are very likely not of random origin but their cause is not known; variation in nitrogen nutrition levels, such as in nitrate concentration or in rates of decomposition (aerobic *v.* anaerobic) might be relevant. A statistical analysis of the periodicity of the fluctuations in chlorophyll degradation product to epiphasic carotenoid ratio (D. E. Barton, in Fogg & Belcher, 1961) indicated periods averaging 53·3 cm.

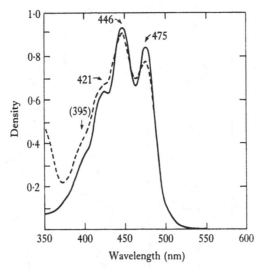

FIGURE III   Absorption spectra of α-carotene from carrots (solid line) and α-carotene from the mud of Little Round Lake, Ontario (dotted line). Solutions adjusted to approximately equal density. Solvent: *n*-hexane (Vallentyne, 1956).

The complexing of nickel with pheophytin-derived porphyrin pigments in Alberta, Canada lake sediments has been described by Hodgson *et al.* (1960). The sediments of North Cooking Lake, 22 miles south-east of Edmonton, Alberta consist of about 3·7 m of copropel, beneath which lies 0·6 m of blue-black clay, 3·6 m of water-laid silt and sand that rests in turn on glacial till. The lake waters are slightly alkaline, pH 8·1, and the underlying copropel, sand and till are pH 7·4, 7·6–8·6 and 6·7, respectively. The chlorophyll-derived pigments extracted with 90% acetone decrease downward from the copropel to the clay and from the clay to the silt (table 74). The principal pigment extracted throughout was pheophytin *a*, but in the silt, and to a lesser extent above, a nickel-complexed porphyrin was present. By experiment the pheophytin in the copropel was found to complex with nickel, and it was postulated that aquatic plants concentrated the nickel and during decomposition yielded it to the copropel and associated interstitial water. The present writer has not noted metal-complexed porphyrins in lake sediments of Minnesota but S. R. Silverman (personal

TABLE 74 *Chlorinoid-pigment content of North Cooking Lake sediments* (*Hodgson* et al. 1960)

Sampling point	10	9	8
Distance from shore (m)	15	70	225
Thickness of sampled section (m)	1·4	1·3	2·6
Pigment concentration (p.p.m. dry wt)			
Sample 1 gyttja (copropel)	300	280	350
Sample 2 gyttja	250	140	310
Sample 3 gyttja	—	72	360
Sample 4 gyttja	—	78	280
Sample 5 blue clay	—	—	110
Sample 6 silt	—	—	2

communication) has observed such material in Holocene sediments at several localities.

The pigment contents of three types of soils in a cool-temperate deglaciated region (Minnesota) were examined by Sanger (1968). The chief pigment in all three soil humus layers was pheophytin *a*, followed by smaller quantities of pheophorbide *a* and chlorophyllide *a*, and much smaller amounts of lutein and *β*-carotene. The oak forest humus had up to 168 mg/g pheophytin *a* in the L soil zone and less than half that amount in the underlying F and H layers. The spruce–cedar woodland humus contained somewhat less pheophytin, up to 120 mg/g, and the prairie soil humus contained much less, 25 mg/g, than the oak forest humus. The relative resistance of pheophytin *a* in surface humus layers is shown by this evidence, as is the relatively unfavorable condition for pheophytin accumulation in prairie as compared to the other two soil types.

As both fallen leaves and soil humus layers are relatively low in both chlorinoid and carotenoid pigments (Sanger, 1968) the chief contribution of terrestrial pigments to lake sediments may be by storm-borne green leaves that settle into the lake sediments. Experiments that Sanger performed in Cedar Creek Bog, Minnesota, indicate that green oak leaves lying on and buried in the peaty swamp sediment undergo gradual degradation of the chlorinoid pigments. The chlorophyll derivatives accumulating in the leaves are mainly pheophorbide *a* and *b* and pheophytin *a*. Chlorophyllide *a* is present throughout the degradation period in amounts of less than 10% of the total chlorophyll derivatives. The pathway of degradation of the original chlorophyll thus seems to be mainly by the pheophytin rather than the chlorophyllide pathway. The pheophorbides show maximum amounts of 10–15 mg/cm^2 at around 40–80 days of leaf-degradation and gradual subsequent decline to levels of 1–2 mg/cm^2 after 320 days. Pheophytin *a* attains a maximum of about 5 mg/cm^2 at 20 days for leaves lying on the mud surface and 20–40 days for buried leaves; after 80 days the pheophytin levels have dropped to 1–2 mg/cm^2 or less.

7.3.2 *Carotenes and related pigments*

The carotenoid pigments of several eutrophic post-glacial lake deposits were found by Vallentyne (1956) to include α- and β-carotene, myxoxanthin, rhodoviolascin and lycopene.

The carotenoids consist of about 80 colored lipid-type compounds that have the following common characteristics: (1) 40 carbon atoms per molecule, (2) a system of ethylene-type double bonds of which most are conjugated (alternating with single bonds), and (3) well-defined light absorption maxima in the range 400–600 nm. Two generalized groups of carotenoids are recognizable on the basis of number of hydroxyl (OH) groups per molecule and phase separation between 90% methanol and petroleum ether: (1) the carotenes which for the most part contain no OH groups and which pass to the upper petroleum ether phase are termed epiphasic; (2) the xanthophylls which contain two hydroxyl groups and which rest in the lower methanol phase are termed hypophasic. Several carotenoids with one OH group are separable into a hypophase with 95% methanol following preliminary epiphase partition in 90% methanol.

Carotenes tend to predominate slightly over xanthophylls in sediments but exceptions occur (Muraviesky & Chertok, 1938).

Of the twenty or more carotenoids separated from the surface sediments only β-carotene and rhodoviolascin have been definitely identified (Savinov *et al.* 1950; Karrer & Koenig, 1940). Trask & Wu (1930) first recorded carotenoids in surface sediments, and other older papers include those by: Baudisch & von Euler (1934), Klimov & Kazakov (1937), Fox (1937), Lederer (1938), Beatty (1941), Fox & Anderson (1941), Fox *et al.* (1944), Phinney (1946), Andersen & Gundersen (1955), and Züllig (1955, 1956).

The carotenoids of three eutrophic lakes in Connecticut and Ontario are shown in tables 75, 76 and figs. 111, 112. The pigments of Lower Linsley Pond, Connecticut and Little Round Lake, Ontario are dredge samples of surface sediments, those of Bethany Bog and Upper Linsley Pond, Connecticut are of cores. The α- and β-carotene and lycopene would be expected to occur in these and other lacustrine sediments because they are widely distributed in the plant kingdom. Myxoxanthin or echinenone is found only in blue-green algae among fresh-water organisms, and this group of plants is well represented in Linsley Pond. Rhodoviolascin occurs

TABLE 75   *Carotenoid fractions from column chromatography of an extract from Lower Linsley Pond mud, Connecticut (Vallentyne, 1956)*

(An asterisk * indicates too low a concentration for a reliable determination of the absorption maxima.)

Fraction	Color	Abs. max. in CS_2 nm	Remarks
1	Green	—	Sedimentary chlorophyll degradation products
2a	Very faint pink	*	Unidentified
2b	Faint yellow	*	Unidentified
2c	Orange-red	570, 527, 499	Rhodoviolascin? (abs. max., 473, 534, 496). Could be torulene
3a	Yellow	509, 482, 451	Compound A; properties do not agree with any known carotenoid
3b	Very faint pink	*	Unidentified
3c	Orange-yellow	Single band, 488–496	Myxoxanthin? (abs. max., single band at 488).
4a	Yellow	510, 482, 451	= 3a
4b	Faint pink	*	= 3b
4c	Orange-yellow	Single band, 488–494	= 3c, myxoxanthin?
5a	Orange-yellow	Single band, 488–496	= 3c = 4c, myxoxanthin?
6a	Orange-yellow	547, 510, 483	Lycopene? (abs. max., 548, 507, 477)
7a	Orange	518, 489 (452)	β-carotene (abs. max., 520, 485, 450)
7b	Yellow	510, 479, (452)	α-Carotene (abs. max., 509, 477). The shoulder at 452 may indicate contamination with β-carotene.

TABLE 76   *Chemical analysis of a sediment profile from Upper Linsley Pond, Connecticut (Vallentyne, 1956)*

(Pollen zone nomenclature according to Deevey (1940). Depth refers to that below water surface. Mud–water interface at depth of 6·0 m.)

Depth (m)	Pollen zone	Ignition loss (%)	Nitrogen (%)	Pigment units/g dry weight				
				Carotene	Compd. B	Compd. A	Myxoxanthin	SCDP
6·5	C3	69·6	3·3	4·9	1·4	+	+	n.d.
7·2	C2	62·3	2·2	3·3	1·2	−	+	41·2
7·9	C2	61·8	2·8	2·1	0·6	−	−	48·1
8·7	C1–C2	60·4	2·7	1·1	0·6	−	−	36·2
9·4	C1	59·1	2·1	5·3	0·9	+	+	33·7
10·3	C1	60·0	2·2	2·1	0·2	−	−	37·6
11·0	B–C1	59·8	2·4	3·9	1·2	+	+	39·7
11·7	A	20·0	0·8	1·9	1·0	+	+	15·5
12·2	A	26·0	1·2	2·0	0·8	+	+	20·0
13·0	A	3·7	0·1	0·9	0·1	+	+	0·8

n.d. = not determined.

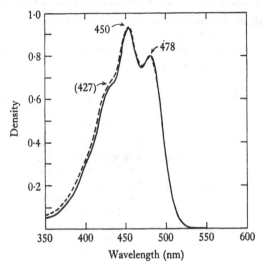

FIGURE 112 Absorption spectra of *β*-carotene from carrots (solid line) and *β*-carotene from the mud of Little Round Lake (dotted line). Solutions adjusted to approximately equal density. Solvent: *n*-hexane (Vallentyne, 1956).

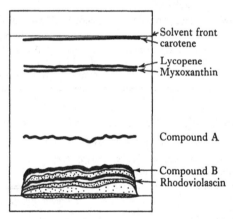

FIGURE 113 The characteristic appearance on a paper chromatogram of an acetone extract of a core sample of lake sediment (Vallentyne, 1956). Epiphasic carotenoid zones shown in black. Sedimentary chlorophyll degradation products shown as stippled zones. Portions of xanthophylls not shown.

in purple bacteria, and although these have not been searched for in the mud of Linsley Pond they are at least a likely source of the pigment.

The *α*- and *β*-carotenes are not separable on one-dimensional paper chromatograms (fig. 113). The total *cis*- and *trans*-carotene distribution in Bethany Bog sediments is shown in fig. 114. The high content of carotenoids between 7 and 13 m represents the mesotrophic stage of the lake and the lower content from 0 to 7 m represents the dystrophic stage in which there was a marked decrease of blue-green algae that provided a principal source of the carotenes. Vallentyne suggested that a small part of the

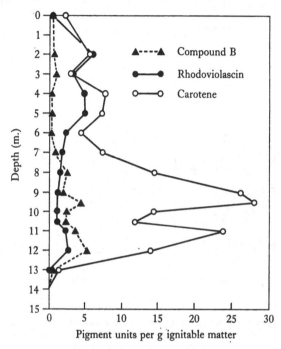

FIGURE 114   Compound B, rhodoviolascin, and carotene in a sediment core from Bethany Bog, Connecticut. Concentrations expressed as pigment units/g ignitable matter. The values for the 9·5 and 10·5 m samples are included on the assumption that the ratios of ignition loss to N are constant from 9 to 11 m (Vallentyne, 1956).

sedimentary carotenes might have been synthesized by micro-organisms in the sediments citing the work of Snow & Fred (1926) in which colored bacteria were cultured from lake sediments.

The sediments of Long Lake, Minnesota being studied for chlorinoid pigments contained appreciable amounts of carotenoid pigments in acetone extracts. These were separated by paper chromatography. The dried extract was dissolved in carbon disulfide and spotted in a series of 10 lambda spots $\frac{1}{8}$ in apart on Whatman No. 1 filter-paper. The chromatogram was developed in petroleum ether for 8 h with the results shown in fig. 115c. The bands were cut out, redissolved in CS_2 and the absorption spectrum of each was determined in the 350–700 nm range (fig. 115c).

Organic pigments of non-marine deposits

Several of the maxima correspond closely to published maxima of carotenoid pigments (table 77). The absorption maximum at 425–430 nm that appears in most bands of the chromatogram is common to the carotenoids listed in table 77. The maximum at 542–44 nm on band 1 of the chromatogram is also common to those same carotenoids and also to α- and

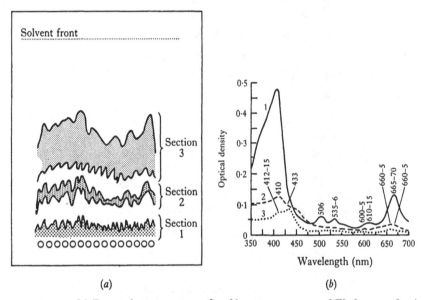

(a)　　　　　　　　　　　　(b)

FIGURE 115 (a) Paper chromatogram of 90 % acetone extract of *Elodea canadensis*, Blue Lake, Minnesota, developed in petroleum ether. (b) Absorption spectra of the three bands; curve no. 2 probably represents both pheophytin and carotenoids; curve no. 2 is mostly chlorophyll *a*.

β-carotene and echinone. The peak at 505–510 nm in bands 1, 2 and 4 is characteristic of β-carotene and lutein. Generally similar results were obtained by Andersen & Gundersen (1955) who extracted carotenoid pigments from interglacial Danish gyttja. Carotenoid pigments were also extracted from sediments of English lakes (Fogg & Belcher, 1961).

In a recent study of carotenoid pigments from upland sources and associated lowmoor sediments Sanger (1968) found that leaf pigments of upland forest litter supplied a small amount of carotenoid to the swamp accumulations.

As was noted in the discussion of chlorinoid pigments Sanger found that green leaves brought to the lake during storms contribute much of the terrestrial pigment matter in lake sediments. The carotenoid pigments of oak leaves lying on, and buried in, the bottom sand of a pond in Cedar

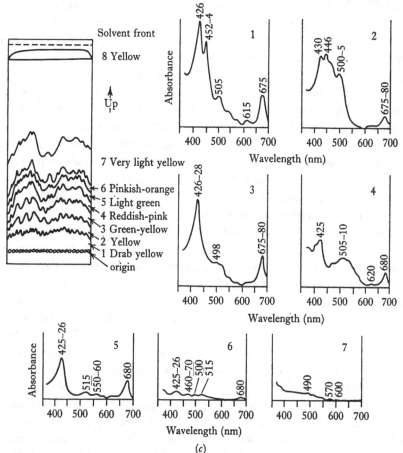

(c)

FIGURE 115   (c) A similar chromatogram and group of spectra of extract of sediment sample from Long Lake, Minnesota; interpretation of spectra in table 77 suggests bands represent mainly carotenoid pigments, rather than chlorinoids as in *Elodea* preparation, some chlorinoid absorption represented by peak at 675–680 nm (Swain, Paulsen & Ting, 1964).

TABLE 77   *Absorption maxima of carotenoids which have peaks corresponding to bands of the chromatogram (fig. 115)*

Solvent: carbon disulfide, Karrer & Jucker (1950) (Swain, Paulson & Ting, 1964).

Carotenoid	Formula	Maxima	(nm)
Dihydro-β-carotenone (derivative of β-carotene)	$C_{40}H_{58}O_4$	426	455
Aurochrome (derivative of β-carotene)	$C_{40}H_{56}O_2$	428	457
Auroxanthin	$C_{40}H_{56}O_4$	423	454
Apo-3-norbixinal methyl ester (derivative of bixin)	$C_{18}H_{22}O_3$	427	455
Dihydronorbixin (derivative of bixin)	$C_{24}H_{30}O_4$	428	454
Dihydrobixin (derivative of bixin)	$C_{25}H_{32}O_4$	428	454

Organic pigments of non-marine deposits

Creek Bog, Minnesota, include β-carotene, violaxanthin, neoxanthin, lutein and 'lutein tail'. The latter is a chromatographic mixture that may contain lutein esters and other components. Lutein and β-carotene, the most important carotenoids in the leaves, show declines from around 0·3 to 0·4 mg/cm^2 each during the first 40 days of leaf-degradation to around 0·1–0·2 mg/cm^2 between 80 and 320 days. The greatest decrease seemed between 40 and 80 days. The lutein tail shows an increase then a decrease between 80 and 320 days, evidently responding to degradation and esterification? of the lutein. Violaxanthin is unstable and disappears at about 25 days. Neoxanthin decreases from about 0·1 mg/cm^2 to about 0·02 mg/cm^2 after 80 days but thereafter stays fairly constant.

The sediments at depths of 200–225 cm in Little Round Lake, Ontario are rich in bacterial carotenoids (Brown, 1968). These substances, extracted and crystallized from the sediment are spheroidenone and lesser amounts of spheroidene (pigment Y) and 2-ketospirilloxanthin (P-518). These are associated in the sediments with algal carotenoids, chlorophyll derivatives, and bacteriophytin. The bacterial carotenoids are represented by three stereoisomers of spheroidenone and two of spheroidene. Thus, cis-, trans-isomerization of the compounds seems to have taken place in the sediments. The P-518 pigment was found only in the trans- form. The absorption maxima of the cis- Y pigment (and thus their means of identification) are: 344·5, 423, 446 and 478 nm; of trans- Y pigment 426, 453 and 486 nm; of cis-spheroidenone 366, (455), 475 and 505 nm; of trans-spheroidenone (460), 583 and 513 nm; and of P-518 pigment 492, 517 and 554 nm. Experiments with senescent cultures of *Rhodopseudomonas spheroides* indicated that these purple photosynthetic bacteria produced the carotenoids in question. Growth most reasonably occurred in some aerobic environment in the lake rather than in the hypolimnion and it appears there was very little post-depositional alteration of the pigments.

This discovery adds to the list of carotenoid pigments known to occur in lake sediments: myxoxanthin (echinone) and crystalline β-carotene from Lake Biserovo, U.S.S.R. (Lederer, 1938), rhodoviolascin (spirilloxanthin) from Lake Nakuru, Africa (Karrer & Koenig, 1940); α- and β-carotene in lake sediment extracts from Connecticut, Ontario and California (Vallentyne, 1956, 1957); lutein from Esthwaite Water, England (Fogg & Belcher, 1961); myxoxanthophyll from Swiss lakes (Züllig, 1961); oscilloxanthin from McKay Lake, Ontario (Brown & Coleman, 1963); and the pigments from Long Lake, Minnesota referred to above.

7.3.3   *Flavinoids and related pigments*

Only a few reports have been made of the occurrence of flavinoid and other heterocyclic pigments in sedimentary materials (Swain & Venteris, 1964; Swain, Paulsen & Ting, 1964).

Several fluorescent substances believed to represent flavinoid, as well as flavonoid? and indole compounds were separated on paper chromatograms and characterized by paper chromatography from Cedar Creek Bog peat, Minnesota (Swain & Venteris, 1963). The flavinoid and possible indole acid substances in the upper and middle Cedar Creek peat are related in distribution and amount to the organic contents of the sediments. In the deeper, less organic parts of the bog the flavinoids are more scarce but their slight increase at depths of 26–27 ft conforms to a rise in organic contents of the sediments at that depth. The absence of flavinoid substances in the deeper part of the bog despite the appreciable total organic content of the sediments suggest that they have undergone degradation. The recurrence of moderate amounts of fluorescent substances in the deepest samples possibly is due to their introduction by underground waters circulating through the porous sands beneath the bog.

Flavoproteins are affected by, and are involved in, oxidation as well as, but to a lesser extent by anaerobic processes. In an oxygen-poor, but not completely oxygen-less, sedimentary environment such as that of Cedar Creek Bog, some flavinoid compounds would have a chance of formation by metabolic activity of mud-dwelling organisms and of preservation of these products. The observed antiseptic properties of this and other bog accumulations (Swain & Prokopovich, 1954; Swain, 1968b) point to a very low level of bacterial activity in the middle and upper parts of the bog, and provide a reasonable explanation for the presence of flavinoid substances. The general lack of flavinoids in the copropelic marls of the lower bog, except in the basal layer referred to above, may indicate that phenols and other preservative substances had not yet formed in the early history of the bog.

TABLE 78   *Fluorescent substances in hydrochloric acid extracts of lake sediments and peat*

Riboflavin, riboflavin phosphate and other derivatives of yellow enzymes or flavoproteins, including myricetin, lumiflavin and flavineadenine dinucleotide (FAD)

Derivatives of indole and tryptophan, including kynurenine and hydroxyanthranilic acid

Alkaloid derivatives, possibly including tertiary nicotinyl compounds

Porphyrins, such as etioporphyrin

TABLE 79 *Paper chromatographic analysis of flavinoid and other fluorescent organic substances in acid hydrolyzates of lake samples from Minnesota, Nevada, and Louisiana; in several solvents (Swain & Venteris, 1964)*

Lake and station	Type of sediment	Solvent	Color of fluores-cence	R_F	Con-centra-tion p.p.m.	Identity of compound
Blue Lake, Minnesota	Sapropelic marl	BuOH:HAc:H₂O	Yellow	0·14	20	Flavinoid pigment?
Budd Lake, Minnesota	Copropel	BuOH:HAc:H₂O	Yellow	0·24–0·26	8–16	Riboflavin
Budd Lake, Minnesota	Copropel	Phenol:water	Yellow	0·06–0·13	8–16	Flavinoid pigment?
Budd Lake, Minnesota	Copropel	Collidine	Yellow	0·09	6	Flavinoid pigment?
Cedar Creek Bog Lake, Minnesota	Copropelic marl	BuOH:HAc:H₂O	Yellow	0·02	8	Flavoprotein?
Cedar Creek Bog Lake, Minnesota	Copropelic marl	BuOH:HAc:H₂O	Yellow	0·07	8	Riboflavin phosphate
Cedar Creek Bog Lake, Minnesota	Copropelic marl	Phenol:water	Yellow	0·14	6	Flavinoid pigment?
Cedar Creek Bog Lake, Minnesota	Copropelic marl	Phenol:water	Yellow	0·26	20	Flavinoid pigment
Cedar Creek Peat	Forest and sphagnum peat	Phenol:water	Yellow	0·13	6	Riboflavin phosphate
Cedar Creek Peat	Forest and sphagnum peat	Phenol:water	Yellow	0·34	4	Flavinoid pigment
Cedar Creek Peat	Forest and sphagnum peat	Phenol:water	Yellow	0·39	6	Flavinoid pigment?
Lindstrom Lake, Minn.	Copropel	BuOH:HAc:H₂O	Yellow	0·26–0·28	16	Riboflavin
Lindstrom Lake, Minn.	Copropel	Collidine	Yellow	0·14	10	Flavinoid pigment?
Lindstrom Lake, Minn.	Copropel	BuOH:HAc:H₂O	Yellow	0·42	8	Lumiflavin?
Lindstrom Lake, Minn.	Copropel	BuOH:HAc:H₂O	Yellow	0·56	10	Myricetin?
Rainy Lake, Minn.	Silty clay	BuOH:HAc:H₂O	Blue	< 0·1	—	Flavine-adenine dinucleo-tide?
Rainy Lake, Minn.	Silty clay	BuOH:HAc:H₂O	Yellow	0·24	20	Riboflavin
Pyramid Lake, Nevada	Sapropelic silt	BuOH:HAc:H₂O	Yellow	0·11–0·16	16–20	Flavinoid pigment?
Pyramid Lake, Nevada	Sapropelic silt	BuOH:HAc:H₂O	Blue	0·16	50	Flavineadenine dinucleotide?
Pyramid Lake, Nevada	Sapropelic silt	BuOH:HAc:H₂O	Yellow	0·27–0·28	20	Riboflavin
Catahoula Lake, La.	Reddish brown clay	BuOH:HAc:H₂O	Blue	0·075	20	Flavineadenine dinucleotide?
Catahoula Lake, La.	Reddish brown clay	BuOH:HAc:H₂O	Bright yellow	0·16	30	Flavinoid pigment?
Catahoula Lake, La.	Reddish brown clay	BuOH:HAc:H₂O	Blue	0·11–0·13	10–20	Unknown
Catahoula Lake, La.	Reddish brown clay	BuOH:HAc:H₂O	Yellow	0·23–0·28	8–30	Riboflavin

TABLE 80   *Other fluorescent substances in acid hydrolyzates of bog and lake sediments in various solvents (Swain & Venteris, 1964)*

Location	Type of sediment	Solvent	Color of fluorescence	R_F	Concentration p.p.m.	Identity of compound
Cedar Creek Bog 1–2 ft	Copropelic marl	BuOH:HAc:H$_2$O	Golden	0·30–0·84	30–40	Indolic or flavinoid pigment?
Cedar Creek Bog 6–7 ft	Peaty copropel	BuOH:HAc:H$_2$O	Blue	0·50	16	Kynurenine
Cedar Creek Bog 14–15 ft	Copropelic marl	BuOH:HAc:H$_2$O	Yellow	0·50	16	Myricetin or lumiflavin?
Cedar Creek Bog 38–39 ft	Sideritic marl	Collidine	White	0·23	30	Tertiary nicotinyl compound?
Cedar Creek Bog 38–39 ft	Sideritic marl	Collidine	White	0·33	16	Tertiary nicotinyl compound
Cedar Creek Bog 41–42 ft	Sideritic sand	Collidine	White	0·25	20	Tertiary nicotinyl compound
Cedar Creek Bog 41–42 ft	Sideritic sand	BuOH:HAc:H$_2$O	Blue	0·25	20	Unknown
Clear Lake, Minnesota	Marl	BuOH:HAc:H$_2$O	Golden	0·58	60	Myricetin
Budd Lake, Minnesota	Copropel	Phenol:water	Bluish white	0·76–0·81	20–30	Tryptophan derivative, possibly hydroxyanthanilic acid?
Budd Lake, Minnesota	Copropel	Collidine	Yellow	0·51	8	Unknown flavinoid pigment
Lake of the Woods, Minnesota	Copropelic silt	BuOH:HAc:H$_2$O	Blue	0·16	20–24	Unknown alkaloid? compound
Prior Lake, Minnesota	Copropelic marl	Collidine	White	0·14	20	Tertiary nicotinyl compound?
Green Lake, Minnesota	Copropel	BuOH:HAc:H$_2$O	Yellow	0·82	30	Unknown flavinoid pigment
Bison rib from Minneapolis peat bog	Peat	BuOH:HAc:H$_2$O	Yellow	0·58	6	Myricetin?
Bison rib from Minneapolis peat bog	Peat	BuOH:HAc:H$_2$O	Pink	0·09	8	Etioporphyrin or related compound (red in visible light

Organic pigments of non-marine deposits

The distribution and general amounts of flavinoid and other hetero-aromatic fluorescent substances in lake sediments of glaciated regions (tables 78–80) indicate their widespread occurrence in freshwater aquatic sediments. The substances of this sort that can be separated from aquatic plants are more variable and difficult to identify than those from associated sediment samples. The sediments, however, seem to contain relatively larger amounts of flavinoid compounds compared to the total organic matter than the associated plants, probably as a result of bacterial degradation of the plant material in the sediments.

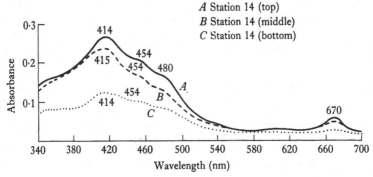

FIGURE 116 Visible absorption spectra of 90 % acetone extracts of Lake Nicaragua sediments (Swain, 1966a, *Journal of Sedimentary Petrology*, **36**, 522–40).

7.4 ORGANIC PIGMENTS IN LAKE SEDIMENTS OF
UNGLACIATED REGIONS

Flavinoid pigments including riboflavin and riboflavin-phosphate were detected by spectrophotometry and paper chromatography in peats from Dismal Swamp, Virginia (Swain & Venteris, 1964; Swain, Paulsen & Ting, 1964). No indole acid or tryptophan derivative was noted in these acidic peats, whereas the more alkaline peats of Cedar Creek Bog appear to contain some of these substances. Although the evidence is inconclusive it may be that the acid environment favors more rapid degradation of the indole compounds either through chemical or biochemical processes.

The chlorinoid pigments of the sediments of a tropical lake, Nicaragua, were extracted with 90% acetone from short Phleger- and Davis-sampler cores (table 81) and proved to be predominantly pheophytin *a* (Swain, 1966a). Carotenes and probably xanthophylls were also present. The upper parts of the cores typically are higher in pheophytin than the lower part of the core, but exceptions occur that do not seem to be related simply to variations in organic content. Some of the cores are marked by relatively

TABLE 81 *Chlorinoid pigments of sediments from Lake Nicaragua (Swain, 1966 a, Journal of Sedimentary Petrology 36, 522–40)*

Station no.	Position in core	Description of sediment	Pheophytin *a*, p.p.m.
35	Top	Light-brown ash and peat, pumice fragments	5·9
35	Middle	Same as above.	5·2
35	Bottom	Pale gray, reddish and greenish silty, peaty sand and gravel	3·7
36	Top 6 cm	Pale gray to tan very diatomaceous copropelic silt	21·3
36	Middle 13 cm	Same as above, less diatomaceous	24·4
36	Bottom	Pale tannish gray, diatomaceous ashy silt and silty clay	13·6
41	Top 6 cm	Pale gray to tan ashy silty, slightly diatomaceous copropel	17·1
41	Middle 14 cm	Same as above, more diatomaceous, cladocerans	35·4
41	Bottom 8 cm	Same as above, less silty, more diatomaceous and copropelic	41·1
45	Top 8 cm	Medium grayish brown copropelic silt, common *Botryococcus*	72·6
45	Middle 11 cm	Pale gray silty diatomaceous copropel	45·7
45	Bottom 4 cm	Pale gray waxy, silty, copropelic, very diatomaceous clay	38·2
47	Top 10 cm	Pale gray, very diatomaceous copropelic clay with *Botryococcus*	69·5
47	Middle 13 cm	Same as above, *Botryococcus* uncommon	2·8
47	Bottom 14 cm	Pale gray silty and finely sandy copropelic clay	6·3
48	Top 4 cm	Pale gray very diatomaceous copropelic clay	7·7
48	Middle 11 cm	Pale gray very diatomaceous copropelic clay or diatomite	70·4
48	Bottom 8 cm	Pale gray waxy, silty copropelic diatomaceous clay	50·6
51	Top 9 cm	Fine-grained tan carbonaceous diatomaceous, silty angular sand	31·9
51	Middle	Pale grayish brown finely sandy silt and silty sand, diatomaceous	13·6
51	Bottom	Pale gray reddish and greenish silty peaty sand and gravel; igneous rock fragments predominantly	12·5
51	Davis core 7 cm	Fine- to coarse-grained argillaceous and silty peaty sand	9·0
52	Davis core 11 cm	Light-grayish green sandy clay and fine- to medium-grained angular to rounded sand; many rock grains	8·9
53	Top 5 cm	Pale gray very diatomaceous clayey silt	24·8
53	Middle 14 cm	Pale gray ashy very diatomaceous, copropelic silt	22·6
53	Bottom	Pale gray very diatomaceous silt	25·1

uniform but low pheophytin content, owing to low total organic content at nearshore localities. Others, at offshore localities, particularly, show marked decrease, presumably degradative, in the deeper part of the sediment (fig. 116). A pronounced increase in pheophytin a few centimeters below the surface in apparently undifferentiated sediments occurs in

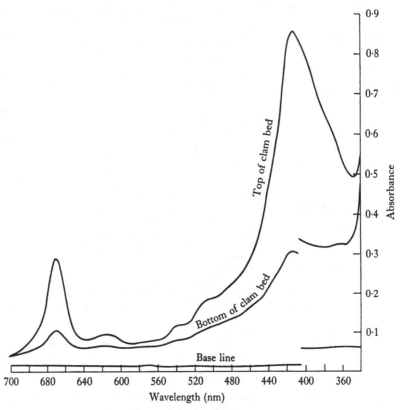

FIGURE 117   Visible absorption spectra of 90% acetone extracts of shells of *Corbicula fluminea* from Delta–Mendota Canal, California (Swain, 1967*a*).

core 41. This increase has also been noted in a few instances in Minnesota lakes and at present is unexplained. The total pheophytin content of sediments of Lake Nicaragua is comparable to that of Minnesota eutrophic lake sediments when one assumes the average organic content of the Nicaragua sediments to be 8–10%.

Yellow-fluorescing pigments of flavinoid character judging from their absorption spectra were separated from acid extracts of Lake Nicaragua sediments by paper chromatography. The material amounted to only about 1×10^{-4} μmol riboflavin/gm of sediment.

7.5 ORGANIC PIGMENTS IN FLUVIAL DEPOSITS

The chlorinoid pigments of sediment accumulations and associated organisms in Delta–Mendota Canal, California (Swain, 1967a) consist only of pheophytin *a*. The chlorophyll *a* that provided the source of the pheophytin occurs in the canal waters in diatoms and other phytoplankton, some of which enter the canal from San Joaquin River. Diatoms should contain chlorophyll *c* but this was not detected in the samples studied (fig. 117). The sediment is considerably richer in chlorinoid pigments than are most of the organisms in the canal with the exception of the sponges (*Spongilla* sp.). In these there are rather large concentrations of chlorinoid pigment that is possibly due to presence of symbiotic algae. Green pigmented algae also occur in the periostracum of the asiatic clam (*Corbicula fluminea*). In the sand sediments the richest concentrations of pheophytin *a* are in the surficial part of the sediment. At the bottom of the sedimentary accumulation which at the time of sampling varied from a few inches to several feet in thickness on the canal floor, the pheophytin content was highly erratic in amount but was appreciably lower than at the surface of the sediment.

7.6 ORGANIC PIGMENTS IN NON-MARINE SEDIMENTARY
        ROCKS

The study of organic pigments in non-marine sedimentary rocks has mainly dealt with oil shales, bituminous rocks, asphalts and mineral waxes (Triebs, 1934a, b, 1935a, b, 1936; Dhéré, 1934; Dhéré & Hradil, 1934; Glebovskaya & Volkenshtein, 1948; Moore & Dunning, 1955; Blumer, 1950, 1952; Sugihara & McGee, 1957). Most of the pigments in oil shales and bituminous rocks are porphyrins (table 88). These substances are typified by four strong spectral absorption bands at or near 620, 565, 535 and 500 nm, in addition to which there is strong absorption near 400 nm (the Soret band).

Very little can be said about the geological aspects of these porphyrin occurrences other than that plant chlorinoid pigments probably were the source of most of the rock porphyrins particularly if they have the 'phyllo' spectrum (fig. 118) (Dunning, 1963). The 'etio' spectrum which suggests the hemoglobin series of porphyrins may indicate an animal origin, but this series also occurs in some plants.

Carotenoid and flavinoid pigment structures have not been found in sedimentary rocks, nor have flavones or anthocyanins. The latter two types of substance decompose to phloroglucinal and salicylic and pyrocate-

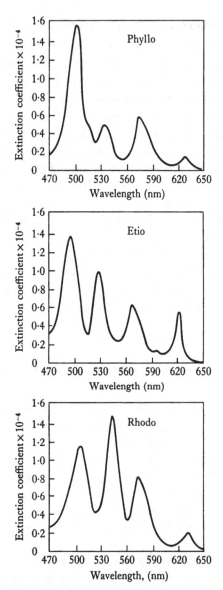

FIGURE 118   Absorption spectra of porphyrins (Dunning, 1963).

chuic acids. These in turn may have been the source of cresols, xylenols, phenols and naphthols that occur in coal tars and other geochemical materials.

TABLE 82   *Porphyrins in oil shales, bituminous rocks
and coal (Dunning, 1963)*

Porphyrin	Characteristics	Occurrence	Reference
Desoxophyllerythrin	R_F in CCl_4–C_8H_{18} (70:30) (methyl ester) 0·15; 'phyllo' type spectrum (fig. 118)	Oil shale, cannel and boghead coal	Triebs, 1934a, 1935b
Etioporphyrin	λ_{max} (and absorbance) in dioxane: 621 nm (0·39), 567 nm (0·462), 528 nm (0·712), 496 nm (1·0)	Oil shale	Triebs, 1934a
Mesoporphyrin?	λ_{max} (and absorbance) in dioxane: 620 nm (0·376), 567 nm (0·459), 528 nm (0·69), 496 nm (1·0)	Oil shale	Triebs, 1934b
Vanadium porphyrin complex	λ_{max} of this group of porphyrins 570–573 nm and 533–537 nm in dioxane or pyridine	Bituminous rocks, asphalts and mineral waxes, boghead coal and torbanite	Triebs, 1934b
Iron–porphyrin complex	λ_{max} of this group of FeOH porphyrins 587, 536 nm in dioxane	Bituminous rocks, asphalts and mineral waxes, coals	Triebs, 1934b
Nickel–porphyrin complex; porphyrin component, desoxyphylloerythin	λ_{max} of this group of porphyrins 550–557 nm, 514–516 nm in pyridine or dioxane	Gilsonite	Sugihara & McGee, 1957
Deuteroetiophorphyrin	'Etio'-type spectrum (fig. 118).	Coal	Triebs, 1935b

7.7   PRODUCTIVITY STUDIES IN LAKES BY MEANS OF PIGMENT ANALYSES

Analyses of pigment extracts of plankton have been developed to study plankton populations and productivity (Richards, 1952; Richards & Thompson, 1952; Parsons & Strickland, 1963; Duxbury & Yentsch, 1956; Strickland & Parsons, 1965). The procedures used by Strickland & Parsons (1965) are modifications of several earlier methods and are briefly discussed here. The methods are described for seawater but apparently are applicable equally well to freshwaters (Talling, 1966; R. O. Megard, personal communication). The three chlorophylls (*a*, *b* and *c*) commonly found in planktonic algae, and a collective value for carotenes and xanthophylls are recorded. If Myxophyceae are present in the sample phycobilin pigments extracted may interfere with the values other than those for chlorophyll *a*.

Chlorophyll *a* and *b* are prepared (Parsons & Strickland, 1963) by chromatographic separation of a 90% extract of a mixture of grasses and

TABLE 83　*Absorption coefficients of chlorophyll* a, b *and* c *(Parsons &*
Strickland, 1963, *Journal of Marine Research*, **21** (3), 151–71.)

Wavelength (nm)	Chlorophyll		
	a	*b*	*c*
430	*	*	*
450	*	*	*
480	1·2	28·0	4·7
510	2·1	3·1	1·8
580	10·4	9·3	11·5
630	13·9	16·4	19·5
645	21·8	54·0	4·3
665	89·0	6·3	0·7

* Indicates values not determined.

clover. The λ_{max} and ratio of blue-to-red peak heights for *a* are 430 and
1·31; for *b* these values are 455 and 643 nm and 2·85.

Plant xanthophylls, peridinin and fucoxanthin, are prepared from 90%
acetone extracts of the dinoflagellate *Amphidinum carteri* and the seaweed
Sargassum muticum respectively. The extracted xanthophylls and chlorophyll *a* are transferred to a hexane epiphase dried under nitrogen taken up
in ethyl ether and ligroine is added. The mixture is reduced in volume and
cooled to −20 °C; the orange precipitate is filtered off and redissolved in
ethyl ether; hexane is added and the mixture segregated by column chromatography on powdered sugar; development is with hexane followed by
N-propanol in increasing proportions, in hexane. The earlier elutions
remove chlorophyll *a* and some carotenoids while the final elution separates fucoxanthin from its isomers; after further solution and reprecipitation
crystalline fucoxanthin is obtained (λ_{max} and values of *E* are 430 nm (73·1),
450 nm (88·0), 480 nm (72·5) and 510 nm (19·9). The *E* values represent
specific absorption coefficients (Log I_0/I) of pigments in 90% acetone.

Peridinin is chromatographed and purified in a similar manner (Parsons
& Strickland, 1963) and has the following λ_{max} and *E* values: 430 nm
(69·0), 450 nm (77·0), 480 nm (83·8), 510 nm (49·1), 580 nm (0·8).

Specific absorption coefficients of chlorophylls *a*, *b* and *c* are shown in
table 83.

The following equations were developed by Strickland & Parsons (1965)
for estimating chlorophyll concentration in seawater:

$$C \text{ (chlorophyll } a) = 11 \cdot 6 \, E_{650} - 1 \cdot 31 \, E_{645} - 0 \cdot 14 \, E_{630}, \qquad (103)$$

$$C \text{ (chlorophyll } b) = 20 \cdot 7 \, E_{645} - 4 \cdot 34 \, E_{650} - 4 \cdot 42 \, E_{630}, \qquad (104)$$

$$C \text{ (chlorophyll } c) = 55 \, E_{630} - 4 \cdot 64 \, E_{665} - 16 \cdot 3 \, E_{645}. \qquad (105)$$

Where E_{665} nm, etc. represents extinction values at indicated wavelength in 10-cm cells after correcting for a blank.

To find the concentration of chlorophyll in a liter of water (V), using acetone extracts of ml volume (v), in a cuvette of 10 cm pathlength (L) the following equation is used:

$$\text{mg chlorophyll } (a, b \text{ or } c)/\text{m}^3 = \frac{C\,(a, b, c)\,v}{LV}. \tag{106}$$

Where $C\,(a, b, c)$ is the concentration of the chlorophyll concerned, is determined from preceding equations.

The equation used to estimate plant carotenoids in water, as milli-specific pigment units is:

$$\text{m.sp.u.}/\text{m}^3 = \frac{7\cdot6\,(E_{410} - 1\cdot49\,E_{510})\,v}{LV}. \tag{107}$$

Where E is the optical density at the indicated wavelength, L is the lightpath of the cuvette in cm, v is the volume of the acetone extract in ml, and V is the volume of seawater filtered in liters. If the estimation is of plant carotenoids without regard to nature of the crop, the factor $7\cdot6$ $(E_{480}-1\cdot49\,E_{510})$ is used in the right-hand side of the equation. If the crop is predominantly Chlorophyta or Cyanophyta $4\cdot0\,E_{480}$ is used and if predominantly Chrysophyta or Phaeophyta $10\cdot0\,E_{480}$ is used.

An example of the application of the study of chlorophyll contents of the waters to assessment of productivity in two lakes is taken from the work of Talling (1966). The two lakes studied are Windermere, England and Lake Victoria, East Africa. Some characteristics of the two lakes are given in table 84 and outline maps are shown in fig. 119. The average concentration of phytoplankton production and of photosynthetic activity are of similar magnitude in the euphotic zone (fig. 120), but winter minimum of solar radiation accounts for the low seasonal values in Windermere. The seasonal variation in total population density is greater in Windermere than in Lake Victoria. There is a seasonal major diatom maximum in both lakes, but for different reasons. In Windermere the seasonal maximum of *Asterionella formosa* Hass is related to availability and decline of silica. The maximum of *Melosira Nyassensis victoriae* O. Müller is related to incidence in June of cooling and mixing south-east trade winds, the blooming occurring in August.

The photosynthetic productivity per unit area (termed $\Sigma\,n\,P$ by Talling) is much higher in Victoria than in Windermere during the diatom maximum. The equation for determining $\Sigma\,n\,P$ is:

$$\Sigma\,n\,P = \frac{n\,P_{\text{max.}}}{1\cdot33\,k_{\text{min.}}}\,\ln\frac{I_0'}{0\cdot5\,I_k}. \tag{108}$$

TABLE 84   *Some characteristics of Lakes Victoria and Windermere (North Basin) (Talling, 1966)*

	Lake Victoria	Windermere
Latitude	$\frac{1}{2}$ °N–2$\frac{1}{2}$ °S	54 °N
Area (km²)	66250	8·05
Maximum depth (m)	79	64
Mean depth (m)	~ 40	25
Day length (h)	11.9–12.2	7.4–17.2
Incident solar radiation (cal/cm²/year)	154,000	61,750
Minimum extinction coefficient, k_{min} (ln units/m)	0·16–0·33	0·31–0·51
Total ionic concentration (m-equiv./l)	1·05 (±0·03)	0·48–0·54
Alkalinity (m-equiv./l)	0·90–0·94	0·15–0·19
NO_3.N (μg/l)	< 5–11	70–430
PO_4.P (μg/l)	< 5–28	< 0·1–3·7
SiO_2 (mg/l)	2·9–4·6	0·01–1·8
Total Fe (μg/l)	< 10	50

Where $E\,n\,P$ is the integral rate of photosynthesis per unit area (units are mgO_2/m^2 h)

n, is population density (units are mg chlorophyll a/m^3.)

$P_{max.}$, is the light-saturated rate of photosynthesis per unit of population (units are mg O_2/mg^2 h).

$k_{min.}$, (ln units/m) is the minimum value over the spectrum of the vertical extinction coefficient, applicable here to green light.

I'_0, is the subsurface light intensity, and is equal to the surface intensity I_0, corrected for surface loss.

I_k, is the light intensity which measures the beginning of light saturation of photosynthesis.

This equation may lead to overestimation of the productivity of surface waters, because it does not take into account the inhibitory effects on productivity of too much light. Recent studies have taken into account this possible discrepancy (R. O. Megard, personal communication).

The higher $\Sigma\,n\,P$ values in Lake Victoria are believed by Talling to be due principally to high and probably temperature-dependent rates of photosynthesis at light-saturation per unit of population (P_{max}). The high Victoria productivity rates seem to be nearly the same throughout the year, and this fact coupled with low attenuation of light and steady popu-

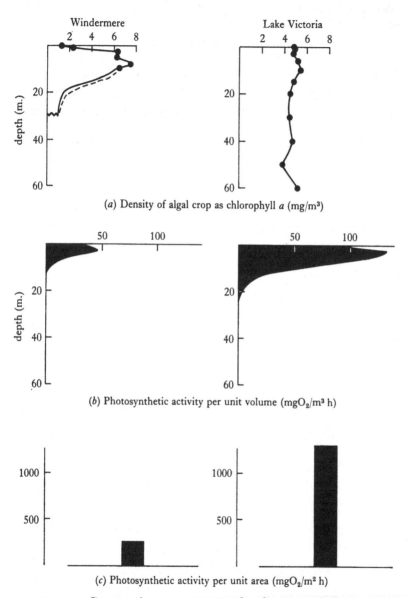

(a) Density of algal crop as chlorophyll *a* (mg/m³)

(b) Photosynthetic activity per unit volume (mgO₂/m³ h)

(c) Photosynthetic activity per unit area (mgO₂/m² h)

FIGURE 119 Comparative measurements for the seasonal diatom maxima in Windermere (14 May 1959) and Lake Victoria (3 August 1961) of (a) depth profile of population density (*n*, as mg chlorophyll *a*/m³), (b) depth profiles of photosynthetic activity (*nP*, as mgO₂/m³ h), and (c) derived estimates of photosynthetic activity per unit area (Σ *nP*, as mg O₂/m² h) (Talling, 1966).

lation) densities resulted in an exceptionally high estimate for annual photo-synthetic productivity.

Although climatic factors—variation in solar radiation for Windermere, and the wind regime for Victoria—are dominant in photosynthetic productivity in these cases, in other freshwaters other mechanisms may exert control. For example in the Gebel Aulia Reservoir on the White Nile and in the nearby Blue Nile, flow rate, is dominant (Rzoska *et al.* 1955; Prowse & Talling, 1958).

8 Organic acids in non-marine deposits

8.1 ORGANIC ACIDS IN AQUATIC SOURCE ORGANISMS

Organic acids include the following types: aliphatic normal carboxylic acids (i.e. formic, acetic) and their esters; fatty acids, both saturated (i.e. palmitic) and unsaturated (i.e. oleic); aromatic carboxylic acids (i.e. benzoic); polycarboxylic acids of saturated dicarboxylic (i.e. oxalic), unsaturated dicarboxylic (i.e. maleic); hydroxydicarboxylic (i.e. tartaric); hydroxytricarboxylic (i.e. citric); and keto-acids (i.e. pyruvic) types.

The generalized distribution of organic acids in plants has been recently summarized by Stevenson (1967). The commonest organic acids in plants are those involved in the tricarboxylic acid cycle: pyruvic, citric, *cis*-aconitic, isocitric, oxalosuccinic, α-ketoglutaric, succinic, fumaric, valic and oxaloacetic. The following aliphatic acids are common in plants: formic, acetic, propionic, butyric, oxalic, glycolic, lactic and tartaric. Fruits and leaves are frequently high in citric and malic acids, succinic and fumaric acids occur in smaller amounts, and oxalic acid in salt form is abundant in some plant leaves.

The acids synthesized by bacteria such as *Acetobacter* are principally formic (HOOC), acetic (CH_3COOH), propionic (CH_3CH_2COOH), butyric ($CH_3CH_2CH_2COOH$) and occasionally lactic ($CH_2CHOH—COOH$), that is simple volatile or semi-volatile types. Fungi, on the other hand produce mainly non-volatile acids such as oxalic and citric; the latter especially by *Aspergillus niger*:

$$\begin{array}{ccc}
\begin{array}{l}
\text{COOH} \\
| \\
\text{CH}_2 \\
| \\
\text{HOC—COOH} \\
| \\
\text{CH}_2 \\
| \\
\text{COOH} \\
\text{Citric acid}
\end{array}
&
\begin{array}{l}
\text{HOOCCOOH} \\
\text{Oxalic acid}
\end{array}
&
\begin{array}{l}
\text{COOH} \\
| \\
\text{CH} \\
\| \\
\text{CH} \\
| \\
\text{COOH} \\
\text{Fumaric acid}
\end{array}
\end{array}$$

Fungi also produce fumaric, succinic, malic and lactic acids; the latter especially by *Rhizopus*.

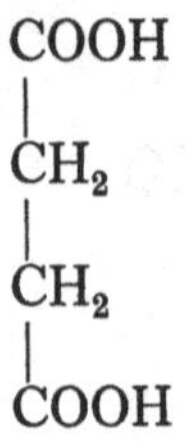

COOH   COOH   COOH
| | |
CH$_2$ HOCH   C=O
| | |
CH$_2$ CH$_2$ HOCH
| | |
COOH   COOH   HCOH

Succinic acid 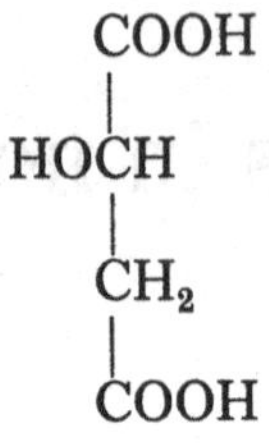   Malic acid |
               HCOH
               |
               CH$_2$OH

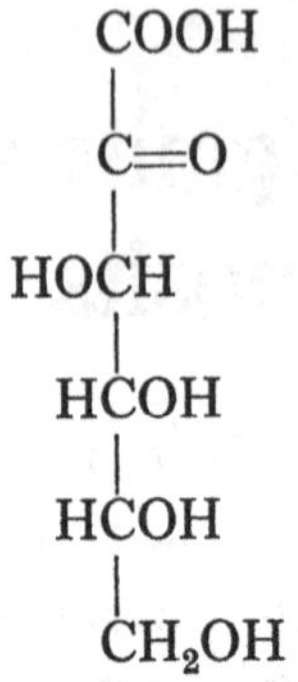

2-Keto-glucuronic acid

Both fungi and bacteria produce sugar acids such as gluconic, glucuronic, galacturonic and 2-ketoglucuronic acid, the latter particularly in bacteria associated with rock surfaces and soil microhabitats.

Aromatic acids are represented in higher plants, and include: cinnamic *p*-coumaric, caffeic, ferulic, melilotic, protocatechuic and gallic.

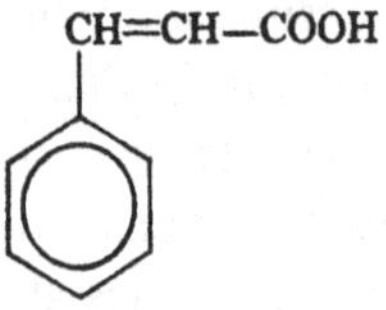

CH=CH—COOH COOH   COOH

Cinnamic acid 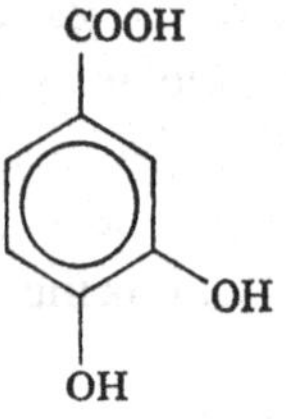 Protocatechuic acid 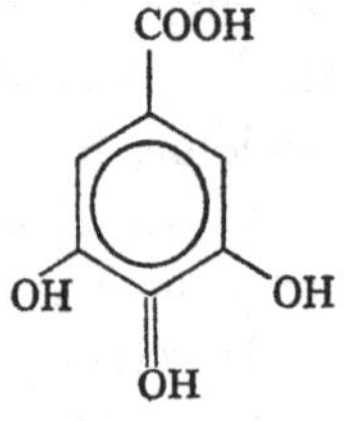 Gallic acid

Micro-organisms such as white-rot fungi may produce phenolic acids, as vanillic, ferulic, syringic, *p*-hydroxybenzoic, and *p*-hydroxycinnamic have been found in solutions decomposed by this fungus from lignin.

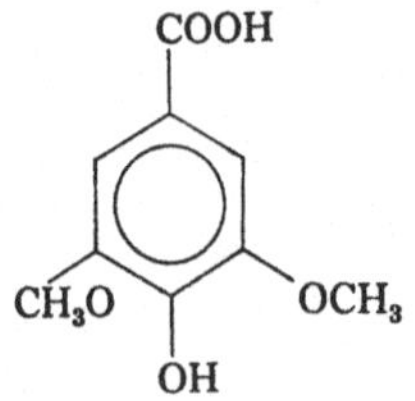

COOH   CH$_2$COOH

Syringic acid 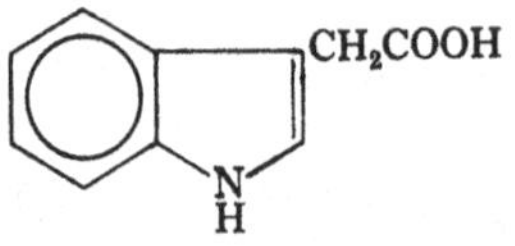   Indoleacetic acid

Carboxylic acid groups are found in plant growth regulators (auxins) such as indoleacetic acid.

Lichen acids have as a basic building block phenylcarboxylic acid, orsellic acid or β-orsellic acid. The depside structure is found in about 30 lichen acids and the depsidone structure in about 15 lichen acids.

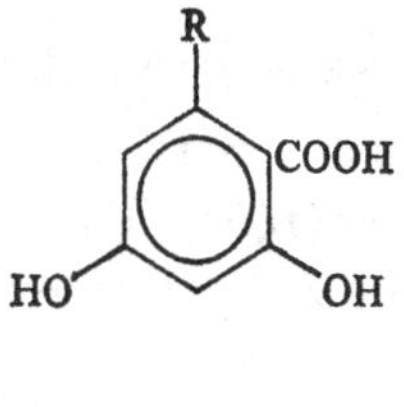

Orsellic acid 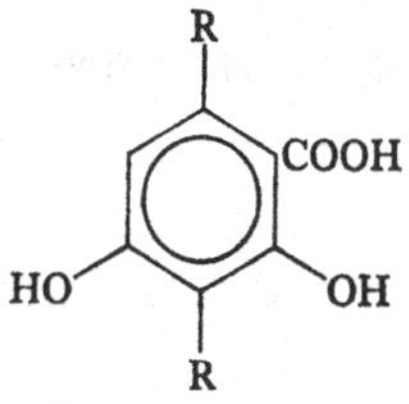   β-Orsellic acid

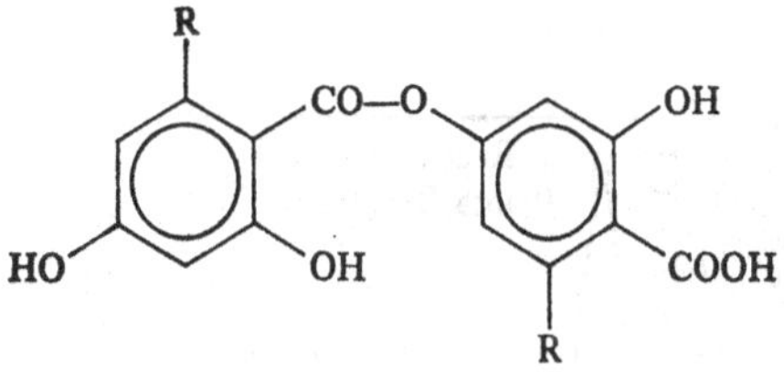

Depside structure 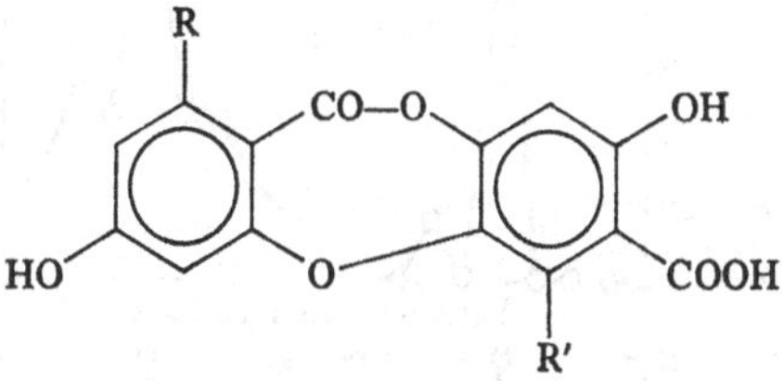   Depsidone structure

Lichens also produce fatty straight chain and lactonic acids, such as caperatic and protolichesteric, respectively.

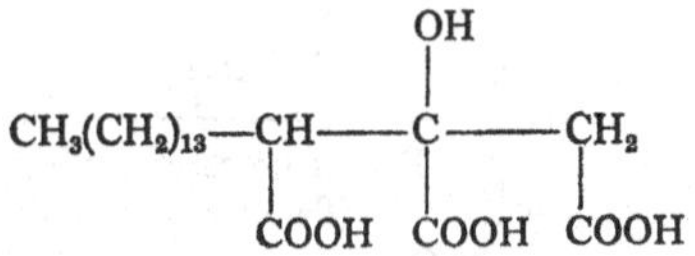

Caperatic acid 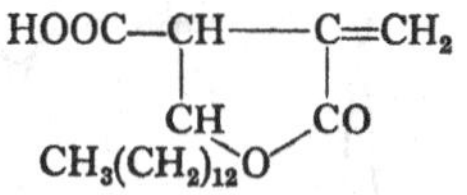   Protolichesteric acid

The organic constituents of *Phylloglossum drummondi* Kunze, a presumed lycopod are of interest because of the primitive nature of the plant (White *et al.* 1967). By means of extractions and paper chromatography the following results were obtained: (1) vanillin, *p*-hydroxybenzaldehyde, *p*-hydroxybenzoic, vanillic and syringic acids were obtained by alkaline oxidation of extracted wood meal with copper oxide; (2) protocatechuic, *p*-hydroxybenzoic, vanillic, ferulic and syringic acid and other phenols were obtained by alkaline hydrolysis of ethanol insoluble residues; (3) sucrose, glucose and fructose in aqueous or ethanolic extracts were, as in *Lycopodium*, the main sugars; (4) alkaloids were detected in small amounts, mainly unidentifiable, but different from the 15 authentic alkaloids of *Lycopodium clavatum*. Whereas the *Lycopodium*-group of lycopods yield syringic acid in ethanolic extracts or on lignin degradation, as well as syringaldehyde, *Phylloglossum* in which these components are scarce or absent is more closely related to *Huperzia* group lycopods.

8.2 ORGANIC ACIDS IN NON-MARINE AQUATIC SEDIMENTS

The fatty acids of recent sediments, sedimentary rocks and of petroleum reservoir waters were analyzed by Cooper (1962). The acids were extracted from the sediments and rocks with sodium hydroxide in methanol under

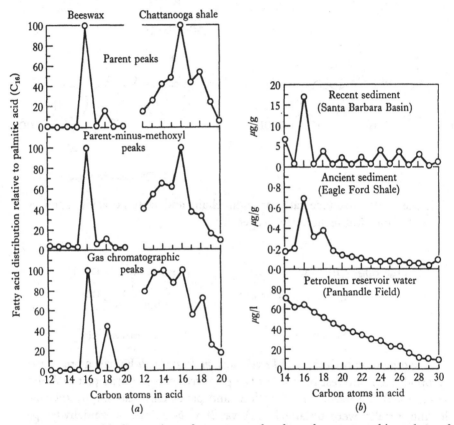

FIGURE 120 (*a*) Comparison of mass spectral and gas chromatographic analyses of fatty acids from beeswax and Chattanooga shale (Devonian) (Cooper, 1962). (*b*) Comparison of the distribution of fatty acids in a recent sediment, an ancient sediment, and in water from a petroleum reservoir (Cooper, 1962).

reflux conditions. Water was added to the extract which was then concentrated by distillation, and washed with carbon tetrachloride to remove non-acidic substances. The sample was divided equally and stearic acid was added to one-half as an internal standard. Each fraction was acidified, the acids were extracted with carbon tetrachloride, and the solvent was evaporated. Esterification of the acids in the residue was carried out with boron trifluoride in methanol and straight chain esters were separated by urea adduction. The adducts were decomposed with water, and the esters were taken up in benzene and separated by gas chromatography. A standard mixture of esters of C_8, C_{10}, C_{12}, C_{14}, C_{16}, C_{18}, C_{20} fatty acids was chromatographed for comparison, and mass spectra were obtained of the fatty acid esters from the Chattanooga Shale and from beeswax. Semi-quantitative results were obtained for these extractions. The recent sediments were found to contain fatty acids having odd numbers of carbon atoms as well as those having even numbers (fig. 120*b*), whereas modern organic matter is typically low in odd-numbered fatty acids (fig. 120*a*). Sediments and seawater do resemble modern organisms in having predominantly palmitic and stearic acids.

Even-numbered acids predominate over odd-numbered acids in all samples studied by Parker and there is a relative increase in abundance of odd- to even-numbered acids with increasing age of the sedimentary rock and of formation water. A possible mechanism for this change with increasing age is that a reactive intermediate is formed on decarboxylation of a fatty acid. Such intermediates could react to give a mixture of odd-numbered paraffinic hydrocarbons. Bray & Evans (1961) found an increase in even-numbered paraffinic hydrocarbons with increasing age in sedimentary rocks; whereas in modern organic material odd-numbered paraffins predominate. Thus, the change in carbon-number preference index (CPI) numbers of paraffinic hydrocarbons and fatty acids with progressive age may be parallel processes. The odd-numbered fatty acids produced by decarboxylation could react further to form even-numbered acids and paraffins. Under conditions which favored acid formation over paraffin formation an equilibrium condition would be reached in which odd- and even-numbered paraffins and acids would show no preference.

8.3 ORGANIC ACIDS OF NON-MARINE SEDIMENTARY ROCKS

The Green River Oil Shale has yielded a variety of aromatic carboxylic acids which include the following (Haug *et al.* 1967): mono-, di- and trimethyl benzoic; mono-, di- and trimethyl propanoic; mono- and dimethyl butanoic; indanoic; tetrahydronaphthoic and naphthoic acids. A large sample (5·3 kg) of the Green River shale from Parachute Creek, Colorado was pulverized and extracted ultrasonically with benzene + methanol. From the extract (55 g, hexane soluble), free acids (0·4 g) were extracted with 1 N aqueous sodium hydroxide. Phenols were extracted from a saturated sodium bicarbonate solution and the acids remaining were esterified with BF_3 and MeOH. The esters were separated by gas chromatography (3%

SE-30 on 80/100 mesh Aeropak and 3% HIEPF 8 BP on 80/100 mesh Gas Chrom. Q, Applied Sci. Corp., State College, Pa.). The collected gas chromatographic fractions were identified by mass spectrometry, examples of spectra of which are shown in fig. 121. The authors suggest that some

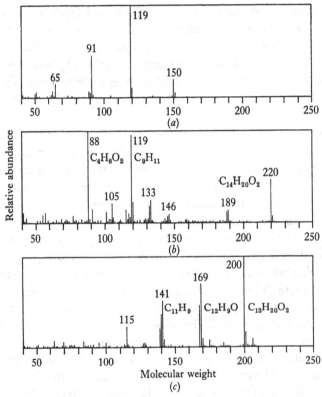

FIGURE 121 Mass spectra of (a) methyl m- or p-methyl-benzoate; (b) methyl 2-methyl-4-(dimethylphenyl) butanoate; (c) methyl methylnaphthoate (Haug, Schnoes & Burlingame, 1968).

of the aromatic acids from the Green River shale may have been derived from cyclic terpenoid precursors. In related papers dicarboxylic and keto acids (Haug *et al.* 1967) and kerogen acids (Burlingame & Simonet, 1968) have been identified in the Green River Shale). Normal, *iso*, *anteiso*, and isoprenoid acids have also been found in the Green River Formation (Abelson & Parker, 1962; Lawlor & Robinson, 1965; Leo & Parker, 1966; Eglinton *et al.* 1966 *a*; Ramsay, 1966; Douglas *et al.* 1968; Haug *et al.* 1967).

Isoprenoid fatty acids, including norphytanic (2,6,10,14-tetramethyl-pentadecanoic acid, $C_{19}H_{38}O_2$) and phytanic (3,7,11,15-tetramethylhexadecanoic acid, $C_{20}H_{40}O_2$) were extracted from Green River Oil Shale (Eglington *et al.* 1966*a*). A 200-g sample of a shale core from Sulfur Creek, Colorado, after pulverizing and acid digestion was extracted ultrasonically with benzene + methanol (1:1). The total free acids were isolated from the extract and converted to their methyl esters (methanol + hydro-

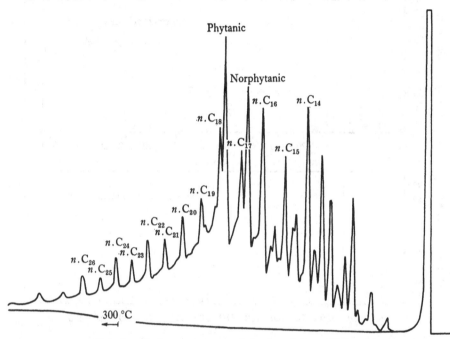

FIGURE 122   Gas–liquid chromatogram of the free fatty acid fraction (as methyl esters) of a sample from the Green River Shale, Colorado. Conditions: 3 m × 3 mm column containing 1 per cent SE-30 on Gas Chrom P, temperature programmed from 150 to 300 °C at 8 °C/min. Flow rate, 20 ml/min. The base-line signal for injection of solvent above is also shown (Eglinton *et al.*, 1966*a*, *Science*, **153**, pp. 1133–4, fig. 1).

chloric acid). After preparative thin-layer chromatography the esters were fractionated on a gas chromatographic column (fig. 122). Authentic isoprenoid acid esters were added to enhance the GLC peaks in the cases of phytanic and norphytanic acid esters. The labeled fractions in fig. 122 were identified by combined gas chromatography–mass spectrometry, Mass spectra of two isoprenoid acids are shown in fig. 123. The isoprenoid acids probably derived from phytyl side chain of chlorophyll, although there is a possibility they may have arisen from the lipids of halophilic bacteria (Kates *et al.* 1965).

It has been found that carbon-balance oxidation techniques can be used to study the chemical structure of complex materials such as coal, peat, and oil shale (kerogen) (Bone *et al.* 1930; Bone *et al.* 1935). Oxidation of the organic material with a boiling aqueous solution of alkaline potassium premanganate is carried out until no additional permanganate is consumed.

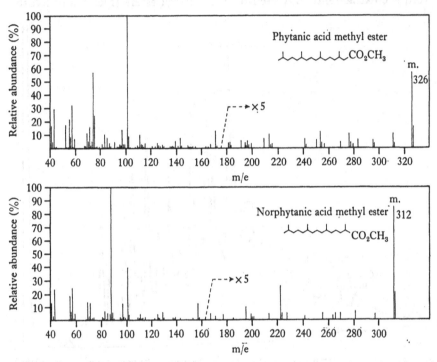

FIGURE 123   Mass spectra of isoprenoid fatty acids (as methyl esters) isolated from the Green River Shale. Spectra obtained by combined gas chromatography–mass spectrometry (2 m × 4 mm column containing 6 per cent SE-30 on gas chrom Q; temperature, 190 °C and scan time 2 sec) on fractions trapped from a Versamid column (3 m × 6 mm column containing 3 per cent Versamid on gas chrom Q) (Eglinton *et al.* 1966, *loc. cit.*, fig. 2).

The distribution of the organic carbon in the carbon dioxide, volatile acids, oxalic acid, non-volatile oxalic acids, and residue is determined.

According to Randall *et al.* (1938) who used the carbon-balance method to determine the structure of approximately 70 known compounds, carbo-hydrates, aliphatic acids, dinuclear aromatic ethers, and non-condensed aldehydes produced only carbon dioxide and oxalic acid by this technique. On the other hand, condensed and non-condensed aromatic hydrocarbons, condensed and non-condensed aromatic compounds containing a carbonyl group, aromatic carboxylic acids and condensed heterocyclic oxygen compounds either were immune to oxidation or produced large yields of non-volatile, non-oxalic acids, mostly benzenoid acids. It was further found that reduced cyclic hydrocarbons and alkyl substituted aromatic compounds were more susceptible to oxidation than were aromatic compounds. Aromatic compounds substituted with a hydroxyl or a methoxy group were found to be susceptible to oxidation, in contrast to which carbonyl group-substituted aromatic compounds are stabilized against oxidation. The rate of oxidation of unsubstituted aromatic carboxylic acids is similar to that of the corresponding hydrocarbons.

Using the carbon-balance oxidation procedure Robinson *et al.* (1963) studied the products of the Mahogany Ledge Oil Shale of the Green River Formation (Eocene) from Rifle, Colorado. Inasmuch as the oxidation products of this non-marine deposit include organic acids of various types the results of Robinson's *et al.* experiments are discussed here.

Samples of the raw oil shale and of concentrates of kerogen from the oil shale were oxidized by alkaline potassium permanganate, air, nitric acid, oxygen and ozone. When samples of oil shale, concentrated kerogen and of kerogen thermally extracted with tetralin were oxidized by alkaline potassium permanganate, large yields were obtained for the first two samples of CO_2 and oxalic acids, and only small amounts of volatile acids and non-volatile, non-oxalic acids. The tetralin-extracted kerogen yielded insolubles, oil, resins and wax that oxidized in different ways. Analyses of several types of known materials by the same methods showed that monocyclic terpenes and non-condensed cyclic ketones produced amounts of non-volatile, non-oxalic acids similar to those from kerogen of the Green River Oil Shale. Polycyclic terpenes, cyclic alcohols, a cyclic alkene and natural products yielded larger amounts of non-volatile, non-oxalic acids.

In order to study intermediate products, controlled stepwise oxidation of kerogen concentrates of the Green River Oil Shale was carried out by successive extractions of a larger quantity of concentrate than used for uncontrolled oxidation. The filtrates from each oxidation step were combined and acidified with hydrochloric acid; water-soluble precipitated acids, removed by filtration, were dried. The dried acids were extracted with ether and ethyl alcohol (Fraction I), the acid filtrate (Fraction II) was evaporated and dried, and a black viscous liquid separated as the volume was reduced. The residue was extracted with ether (Fraction III) and then with ethyl alcohol (Fraction IV). All four resulting regenerative humic acid fractions were purified by electrodialysis, in a 3-compartment cell. The largest of the four fractions was No. II, a lustrous light-brown resinous-looking solid (36·6%). Fraction IV (30·4%) was a dark-brown to black viscous liquid. Attempts to reduce the precipitated humic acids without resorting to high temperatures were only partly successful. Reaction of the humic acids with Raney nickel in boiling methyl alcohol solution yielded acetic acid, C_6–C_9 pyridines, toluene, xylene, ethyl benzene, C_9 aromatics, paraffins and olefins.

Permanganate oxidation products of raw Green River Oil Shale were converted to *N*-butyl esters which were fractionated by distillation (Robinson *et al.* 1963). In the lower-distilling fractions using infrared and mass spectral analysis and X-ray diffraction of crystalline benzylamide derivatives, separately or in combination, esters of the following acids were identified: oxalic, succinic, glutaric, adipic, malonic, pimelic and suberic. The higher-distilling fractions and residue of the butyl-esters had empirical-formula carbon numbers ranging from C_{15} to C_{36} and appear to represent a different series of acid-esters than the normal alkane dicarboxylic acids of the low-distilling fractions.

Partition chromatography of the oxidation products on silicic acid fractionated dicarboxylic acids of the alkane series from malonic to pimelic. Further study of the oxidation products consisted of reducing their *n*-butyl esters to alcohols using an ether solution of lithium aluminium hydride; converting the alcohols to iodide by potassium iodide and phosphorus pentoxide in 85% phosphoric acid; and reducing the iodides to hydrocarbons by zinc and hydrogen chloride gas. Distillation of the hydrocarbons; chromatography into paraffin, aromatic and polar oils, urea adduction of the paraffin oils and waxes, and mass spectral studies resulted in recognition of a variety of hydrocarbons. These included normal paraffins and cycloparaffins, the former ranging from C_{18} to C_{36}, as well as mono-, di-, tri- and tetra-nuclear aromatic oils. Indanes and tetralins were the principal aromatic hydrocarbons. The resin fractions of the hydrocarbon analyses had weak infrared absorptions in the 13·7–13·9-μm region which suggests that there are few carbon chains with more than four methylene groups; strong absorption in the 7·3-μm region indicates considerable terminal methyl groups; absorption in the aromatic region was weak. Thus the resins are believed to be mainly saturated heterocyclic structures. These and cycloparaffinic structures are predominant in the Green River kerogen.

Whereas coals, under similar conditions of oxidation yield mostly benzene–carboxylic acids, the Green River kerogen yields mainly carbon dioxide and oxalic acid as noted above, thus indicating a fundamental structural difference from coal.

8.4 SUMMARY

Relatively few studies have been made of organic acids in sediments and sedimentary rocks, compared to those for some other organic constituents. Modern organic matter contains a predominance of even-carbon-numbered fatty acids, but Holocene sediments seem to contain variable amounts of odd-carbon-numbered fatty acids. Furthermore, there is a relative increase in odd-numbered acids with increasing age of the sediments. The formation of reactive intermediate products during decarboxylation of the fatty acids in the natural environment may account for the phenomenon. A parallel increase in even- as compared to odd-carbon-numbered paraffinic hydrocarbons may also be involved in the same process.

Aromatic carbocyclic acids and isoprenoid fatty acids that have been extracted from the Eocene Green River Oil Shale of freshwater origin are thought to have been derived from cyclic terpenoid precursors and the phytyl alcohol side chain of chlorophyll, respectively.

When Green River Oil Shale kerogen is subjected to oxidative degradation the yield is principally carbon dioxide and oxalic acid. When stepwise oxidation and analysis of the shale kerogen was carried out, a variety of carboxylic acids and other products were obtained in relatively small amounts. The much larger yields of benzene–carboxylic acids from coal deposits indicates a fundamental difference in structure from that of the Green River kerogen. Although the treatment of the organic matter by this method is too rigorous for most environmental and biogeochemical studies the type of information obtained would be useful for the determination of source material origin in sedimentary rocks.

9 Organic geochemistry of non-marine humus

9.1 INTRODUCTION

Various aspects of the geochemistry of humus were discussed in other parts of the book. The literature on humus up to the early 1960s was reviewed (Swain, in Breger, 1963 a), and some additional developments were more recently described (Swain, 1968, 1969). Because the discussions of humus in other places in this book has dealt mainly with the part it plays in aquatic sediments of various kinds, the present chapter will consider principally terrestrial soil humus. In so far as possible, some geological aspects of soil humus will receive first consideration.

The immense literature on soil humus, numbering in thousands of references has been summarized in several treatises of which those of Waksman (1938), Bear et al. (1955), Kononova (1961), McLaren & Peterson (1967) and Black (1968) can be mentioned.

9.2 THE SOIL PROFILE

The soil profile is of importance in humus geochemistry because of its effect on distribution of soil organisms, water, oxygen and soluble mineral salts. The A, or leached zone of the normal profile is characterized by high porosity aeration, low to moderate water content depending on water-holding clays and humus, active micro-organisms and soil fauna, and reduced soluble minerals. The B, or enriched zone, has reduced porosity due to deposition of soluble minerals leached from above and transported from elsewhere, moderate to high-water content, aerobic conditions grading to anaerobic, i.e. less than 3×10^{-6} M oxygen concentration (Greenwood, 1961) in many soils, and accumulation of anaerobic decomposition products such as fatty acids. The C zone of unaltered bedrock typically is gradational with the overlying B zone and is marked by much reduced moisture content except in very porous bedrock and by greatly reduced microbe and soil fauna content.

9.3 SOIL ORGANISMS AND ENZYMES

The soil microflora and microfauna are relatively sparse in newly deposited sediments (Webley et al. 1952), such as in the bare sand of a sand dune. As one proceeds to sediments stabilized to a progressively greater degree

TABLE 85   *Soil enzymes released by faunal and floral catabolism of plant cytological structure (Briggs & Spedding, 1963; McLaren & Peterson, 1967)*

Name of enzyme	Reaction catalyzed
Urease	Urea → ammonia + carbon dioxide
Amylases	Starches → sugars
Glycosidases	Glycosides → sugars + aglycones
Asparaginase	Asparagine → aspartate + ammonia
Aspartate–alanine transaminase	Aspartate + pyruvate ⇌ alanine + oxalacetate
Catalase	$2 H_2O_2 → 2 H_2O + O_2$
Invertase	Sucrose → glucose + fructose
Proteases	Proteins → peptides
Dehydrogenases	Reduced substrate → oxidized substrate
Glutamate–alanine transaminase	Glutamate + pyruvate ⇌ alanine + α-ketoglutarate
Glycerophosphatase	Glycerol phosphate → glycerol + phosphate
Inulase	Inulin → fructose and fructose oligosaccharides
Leucine–alanine transaminase	Leucine + pyruvate ⇌ alanine + α-ketosiovalerate
Nuclease	Purines, etc. → ammonia + ketopurines (etc.)
Peroxidase	Substrate + H_2O_2 → oxidized substrate + H_2O
Phosphatase	Organic phosphates → compound + orthophosphate
Polyphenol oxidase	Polyphenols + O_2 → quinones + H_2O
Tyrosinase	Tyrosine + O_2 → orthoquinones + H_2O

by vegetative growth the bacterial count rises from a few thousand to millions per gram and a 'soil' can be considered to have developed. Soil microfauna are especially active in the oxygenated A zone, while the microflora are the dominant agent of degradation at greater depths (Kononova, 1961). Both fauna and floral catabolism of the plant cytological structure in soil release enzymes some of which are listed in table 85 after Briggs & Spedding (1963). Thus, many of the organic residues in terrestrial soils can be accounted for by enzyme activity. Other soil organic compounds are components of soil organisms themselves as well as the residues of deposited organic matter in various stages of decomposition.

9.4   SOIL ORGANIC MATTER

Vallentyne (1957) summarized knowledge of soil organic matter up to that time and more recent treatises (McLaren & Peterson, 1967) have added to the list.

Organic geochemistry of non-marine humus

9.4.1 *Humic acids and related substances*

Kukharenko and associates have devoted much study to hymatomelanic acids, the alcohol-soluble portion of humic acid. Kukharenko (1948) believes that hymatomelanic acids are simpler forms of humic acid. A sample formula of hymatomelanic acid from highmoor peat would be $C_{42}H_{47}O_4$ $OCH_3(OH)_4(COOH)_2$, M.W. 804, and of the comparable humic acid $C_{68}H_{53}O_4OCH_3(OH)_4(COOH)_4$. According to Kukharenko & Ekaterinina (1965) hymatomelanic acids have an easier oxidizability with higher consumption of $KMnO_4$ than do humic acids; the oxidation products of the former fractions contain aromatic acids and more CO_2. The oxidation of humic acids increases in the order: weathered lignite, unweathered lignite, and weathered lean coal. The yield of aromatic acids increases and of CO_2 decreases in a similar way. Hymatomelanic acid contains more hydro-aromatic rings than humic acid in compact lignite, weathered lignite and low grade bituminous coal. The absorption bands in i.r. range of the spectrum are better defined than those of humic acid, but conjugated double bonds are less intense.

The ease with which exchangeable cations adsorbed in soils, clays, humus and roots are released by various solutions was studied (Nagata, 1957). The natural materials were treated with calcium acetate solution at pH 7 to saturate them with calcium. Following this the materials were leached with chloride solutions of Na, K, Mg, H, Al and Fe, each amount of which was equivalent of the Ca retained. Exchangeably leached Ca was used as a measure of ease of release of each base. In an equal mixture of clay and soil

$$Na > K > Mg > Ca > H > Al;$$

clay alone, $Na > K > Mg > Ca > H{=}Ca > Al;$

humus, $Na > K > Mg > Ca > Al = Fe = H;$

powdered roots, $Na > K > Mg > Ca > Al > H.$

Aluminum salts when applied in dilute solution reacted as an exchangeable base. The processes described would seem to be applicable to formation of underclays and related sedimentary deposits.

Desert alluvial soils with their low humus content show a decrease in humus with depth (Pershina & Bykova, 1959). The ratio of fulvic to humic acids in such soils varies from 1·33 to 4·73 and the humic substances are not only unstable but do not occur as free humus or in combination with mobile sesquioxides.

The humic acid content of *Cyperus papyrus* in swamps of Uganda increases during decomposition and accumulation of the plants in peat deposits (Visser, 1964). The presence of clay minerals appears to favor the formation of the humic acids.

An examination of the reactions of zinc with several organic compounds including humic acid showed that the metal was adsorbed as the bivalent species by solid phase model compounds in contact with 0·1 M zinc acetate solution. With respect to carboxylic acids and acidic phenols, zinc behaved as a weaker acid than calcium and stronger than barium. The zinc retention was dependent on pH and the limiting pH for a given site was dependent on acidity, type of exchange site and the nature of the cation. Humic acid had three more types of sites that were capable of retaining zinc. The zinc adsorbed on some of the humic acid sites could be desorbed only by 0·1 N or stronger HNO_3.

The maximum viscosity of humic acid solutions is in the concentration range 0·46–0·04 g/100 ml, and is lower in humic acids from A-type than from B-type soils (Kumada & Kuawanura, 1965). The maximum viscosity appeared in earlier stages of humification and decreased later. The acidic radicals of the humic acid are mainly carboxyls with 10–20% of the radicals being phenolic OH. The carboxyl radicals were, in the samples studied, 4·72–2·28 m-equiv/g more in A type (in which radicals increased with humification) than in B-type humic acids and titration curves suggested that two types of carboxyl groups were present. The phenol OH radical was about 1 m-equiv/g and was not dependent on type or degree of humification of the humic acids.

Experiments on the formation of humic substances from lignin by Flaig (1964) indicate that humic acid forms as follows: polymerizates are formed simultaneously with degradation of the lignin side chains. As decomposition continues, cross-linking of polymers through side chains decreases and linking of rings increases. Increased demethylation enables condensation with nitrogen-containing compounds to occur.

The studies of the metabolism of model compounds related to humus have shown (Ladd, 1964) that *Achromobacter* isolated from soil will oxidize *n*-(*o*-carboxyphenyl) glycine. This compound was selected because it contains an aromatic ring bonded directly to the nitrogen of an α-amino acid. The model compound is in this case metabolized via anthranilate, 5-hydroxyanthranilate, and 2,5-dihydroxybenzoate. The rate of oxidation of the *n*-(*o*-carboxyphenyl) glycine is inhibited by *n*-methylanthranilate and salicylate, the latter inhibits conversion of anthranilate to 5-hydroxy-anthranilate.

Humic acids prepared from *Cephalosporium gordoni* (fungus), moor soil, black soil (brown and gray humic acids), moor drainage water and brown coal all contained free radicals (Kleist & Muecke, 1966). There were more free radicals in gray than in brown humic acids.

Analyses of brown humus show by methoxy content and presence of phenolic groups that humus contained lignin structure (Johnston, 1964). One of the first events in decomposition of lignin from Maize roots is the loss of methoxy groups, probably from the 2,6-dimethoxylphenol part of lignin. Demethylation and the formation of carboxyl groups accompany the decomposition.

The shrinkage in peats on drying is considered to be caused mainly by the humic acid (Ivanov, 1958) content of the peat, as this component showed 77–78% shrinkage, like that of the peat. Lignin on the other hand showed only 26% shrinkage.

With respect to trace element cations in the humate fraction of Indian soils, exchangeable copper, manganese and nickel were present in amounts of 493, 361 and 352 m-equiv/l or g of soil, respectively (Basu, *et al.* 1961). The amounts of fixed cations on humates, was less than that fixed on clays. The order of ion adsorption was $Zn \geq Mn < Ni < Ca$, and the order of release by exchange with H^+ of Mg^{2+} was $Mn > Ni > Ca$. H^+ was more effective than Mg^{2+} in releasing ions.

In their examination of the humic acids of *Cephalosporium gordoni* (fungus), Aurich *et al.* (1963) identified eighteen amino acids and five unidentified ninhydrin-reacting spots by 2-dimensional paper chromatography. The intramycelial pigment rubacene of this fungus is believed to be represented in the humic acid preparations.

Cobalt forms complex compounds with humic and related acid preparations of peat and soil (Agapov, 1966), over a wide pH range. The humic acids form complexes in highly acidic media whereas fulvic acids are inactive in these media. Exchanges of the Co with Fe and Al are intensified at higher pH values.

The composition of the mineral portion of soil humic acid representing a range of 0·98–46·50% ash was as follows: SiO_2 56·02–73·63%, $Fe_2O_3 +$ Al_2O_3 8·52–23·68%, P_2O_5 0·44–9·79%, CaO 0–0·20%, K_2O 0·85–3·72%, Cu 0·004–0·12%. No correlation between soil type and mineral composition of the humic acids was observed.

The equivalent weight of humic acid from peat was studied over a 52-day period (Pommer & Breger, 1960*a, b*). The humic acid contained 57·15% C and 4·43% H, moisture and ash free. The equivalent weight of the humic acid 30 min after preparation, as determined by discontinuous titration was 119. Between 3 and 52 days the equivalent weight increased from 144 to 183. By use of infrared spectra it could be shown that the change in composition with time probably resulted from condensation, with loss of carbonyl groups. The carbonyl absorption band at 5·85 μm was lost when the humic acid was treated with acid or alkali, and the aliphatic C—C single bonds at 3·45, 3·52, 6·68 and 6·95 μm were decreased. This suggests that there was conversion into polynuclear ring systems by alkali action or into phenols by acid action. The pH of the acid solution increased on standing, while in the alkaline solutions the pH decreased.

Humic acid can be decomposed by *Penicillium luteum* and *Polystictus* sp. that were isolated from soil (Latter & Burges, 1960). The same activity was shown by some of the basidiomycetes.

Paper electrophoresis was applied to the separation of fulvic acid and humic acid of krasnozem, chernozem and sod-podzolized soils and recognizable fractions were obtained (Kononova & Tutova, 1961). Spectral and X-ray investigations of humic and fulvic acids of chernozem and podzolic soils showed the fundamental structure of these acids (Kasatochkin *et al.* 1964). It consists of a flat aromatic condensed ring with functional groups of carboxyl, phenolic and alcohol groups as well as simple aromatic ethers and hydrocarbon chains. The nitrogen occurs in cyclic and in aliphatic forms. The carboxyl groups may be transformed into anionic form by interaction of the humic acid with barium hydroxide, sodium hydroxide and calcium. The writers relate the structure of peat humic substances to that of brown coal and oxidized hard coal.

Continuous paper electrophoresis of humic acids of various soil types shows a group of substances in two zones, gray and brown (Kaurichev *et al.* 1960). In ultraviolet light the brown zone shows white luminescence while the gray zone is dark (probably representing inorganic salts). The electrophoretic patterns are different for different soils, but apparently could not be resolved further.

The separation of humic acid preparations by gel diffusion on Sephadex columns resulted in fractions that had differing 'color quotients' (absorbance at 470 nm/absorbance at 610 nm) (Kleinhempel & Heike, 1965). Slowly diffusing fractions ('oligomers') had higher color quotients that appear to represent a higher content of carboxyl and arylether groups as determined by infrared spectroscopy. The degree of polymerization of soil- and fungus-humic acids can be estimated from the color quotient.

In addition to amino acids, humic acid preparations were found to yield non-hydrolyzed phenolic acids and aldehydes such as 3-methoxy-4-hydroxybenzoic acid (Jaquin, 1960). The latter compound is also a chemical decomposition product of lignin. Repeated solubilization and precipitation followed by paper chromatography were used to separate the components.

Organic geochemistry of non-marine humus

Infrared analyses of humic substances from a variety of geological materials showed the following 'order of aromaticity' (Ishiwatari & Hanya, 1965): peat > bay sediment > sedimentary rock > lake sediment. Absorption peaks at 1640 and 1530 cm^{-1} are believed due to peptide-like bonds in humic acids from bay and lake sediments.

Paper chromatography and paper electrophoresis were used to separate soil humic acid into four fractions (Hayashi & Nagai, 1959). Four fractions were separated on an Al_2O_3 column and the relative amounts of A- and B-type humic acids in each were determined. There were significant differences from one fraction to another with fractions 3 and 4 containing the most A-type acid.

The molecular weights of humic substances in natural water were categorized by adsorption on Sephadex C-11; one main fraction had an estimated mol. wt. $\sim$ 100,000 and another $\sim$ 10,000 (Gjessing, 1965). Both contained organic complexes of Fe^{2+}, and Ca was also present in the lower molecular weight fraction.

Thermogravimetric analysis of humic acid and fulvic acid preparations from podzol was carried out under nitrogen atmosphere (Schnitzer & Hoffman, 1965). In contrast to thermogravimetry under air, the differential thermograms of the two humic compounds did not show maximum indications of definitive decomposition reactions.

Polarographic analyses of carboxyl groups in soil humic compounds were carried out by Schnitzer & Skinner (1966). The humic compounds, after isolation and refluxing with several reagents were analyzed by polarography of the excess reagent. With known aromatic aldehydes and aliphatic and aromatic ketones the results agreed with theoretical calculations for carboxyl groups. Those for quinones, except for phenanthrene quinone, were low.

Paper chromatographic separation was made of humic acid extracts of black earth and podzol previously separated by paper electrophoresis into two fast-travelling zones. Substances separated had the same R_F values as oxidized hydroquinone or pyrogallol. Analysis of degradation products of humic acid and of model substances yielded indoles, phenol and organic acids, mainly pyrocatechol, pyrogallol and other acids. Humic acids extracted with $Ba(OH)_2$ in an acid medium, yielded 2-pyrrole-carboxylic acid and gallic acid spots on paper chromatograms. Brown and gray humic acid fractions have similar chemical structure.

Desalting of humic acids is necessary prior to analysis and may be carried out on Sephadex G-25. A mixture of 1·5 mg humic acid and 120 mg sodium chloride in 4·5 ml water can be separated on a 120 × 2 cm column of Sephadex G-25 using water as eluant (Obenaus & Neumann, 1965).

Study of metal-humic acid complexes by means of paper electrophoresis and emission spectrography was accomplished with Al, Ag, Cu, Fe, Pb and Ti in the pH range 2–8·6 (Muecke & Kleist, 1965). The metals were suggested to occur in stable complexes representing several different structures.

The method of disc electrophoresis has been used to separate colored humic substances, Fe^{3+} organic complexes, substances that reduce nitroblue tetrazolium, and enzymes (esterases) from freshwater sediments (Doyle, 1968). The apparatus used consists of a column in which is packed (1) a 'stacking gel' of 3 % acrylamide monomer in solution (pH 8·3) that has been photopolymerized with a fluorescent lamp and, (2) a separating 'gel' of 7 % acrylamide solution to which has been added ammonium persulfate (pH of gel 9·5) for chemical polymerization. A 40 % sucrose solution is in the tube during packing of the gels but is replaced (after inverting the tube) by the sample slurry, water and glycine buffer (10:20:3) for electrophoresis (fig. 124). The tube is connected to upper and lower buffer reservoirs and a power supply. A 7 mA current from a d.c. power supply capable of putting out 1000 V is used. During electrophoresis, glycine the major anion in true solution, moves through the stacking gel toward the anode. Most of the other organic anions move ahead of the glycine boundary and are stacked in thin bands at a voltage discontinuity in front of the glycine boundary. When the glycine boundary passes into the pH 9·5 separating gel it moves more rapidly than the other organic anions.

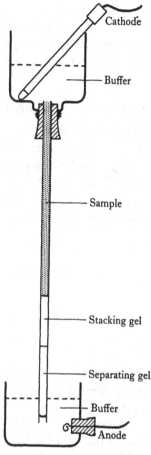

FIGURE 124 Apparatus for separation of organic substances in lake sediments by disc electrophoresis (Doyle, 1968).

These separate into zones according to their electrophoretic mobility in a uniform voltage gradient.

9.4.2 *Hydrocarbons and sterols*

These substances have been found as such only sparingly in terrestrial soils. Hentriacontane ($C_{31}H_{64}$) is one of the few that have been characterized (Schreiner & Shorey, 1910*a*; Schreiner & Lathrop, 1911*a*, *b*). The origin of the hentriacontane is unknown. Chrysene was obtained from soil (Kern, 1947) and according to Cooper & Lindsey (1953) may have come from air contaminated with hydrocarbons. 'Phytosterol' (Schreiner & Shorey, 1909*b*, 1911; Schreiner & Lathrop, 1911*a*, *b*) and 'agrosterol' (Schreiner & Shorey, 1909*b*) a crystalline sterol, obtained with and without saponification, and related to betulin have also been found in soils.

9.4.3 *Fats, waxes, organic acids and alcohols*

One of the higher alcohols in soil extracts is inositol (meso-inositol), one of the polyhydroxycyclohexanes, and which is widely distributed in the organic world (Shorey, 1913; Sallans *et al.* 1937; Yoshida, 1940; Dyer *et al.* 1940; Bower, 1945). The hexaphosphoric acid ester, called phytic acid, is represented as its calcium–magnesium salt (phytin) in plants. It is a component of the vitamin B complex.

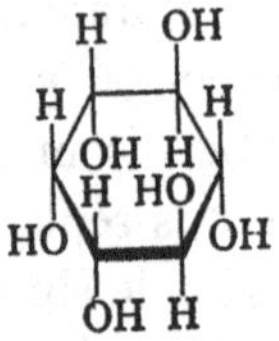

Meso-inositol

Phytin may comprise 33–46% of soil organic-P (Bower, 1945), although the purity of these phytin preparations has been questioned (Smith, 1952).

Glycerol was partially characterized from soil extracts after saponification of the lipid fractions (Schreiner & Shorey, 1910*a*, 1911).

Meinschein & Kenny (1957) examined the constitution of the lipid fractions of soil samples from Texas, Wyoming and Sumatra, representing arid to swampy conditions. As this represents one of the pioneering papers in modern organic geochemistry some of the procedures and results will be briefly discussed. Their samples were extracted with a benzene–methanol mixture (10:1), separated by elution chromatography on silica gel and the weighed benzene eluates were kept for study. These eluates amounted to 4·3–5·5% of the soil extract. The absorption maxima at 2·92 μm are probably due to OH groups; those at 3·4–3·5 μm are CH groups; those at 5·76 μm result from ester carbonyl absorption because after saponification and acidification the $\lambda_{max.}$ of these fractions shifts to 5·84 μm corresponding to the acid carbonyl band. The $\lambda_{max.}$ from 8–9·5 μm represent those found in esters of fatty acids; at 13·9 μm the λ_{max}. are produced by normal alkyl groups containing more than four carbon atoms.

Portions of the benzene eluates were hydrogenated by dissolving them in *n*-heptane and hydrogenating for 12 h using 3,000 p.s.i. hydrogen pressure and with Raney nickel as catalyst. The gaseous reaction products were partly recovered with a liquid nitrogen trap, and the non-volatile products were recovered by removing the *n*-heptane solvent. The hydrogenated non-volatile components were chromatographed on silica gel and found to comprise mainly saturated hydrocarbons, i.e. most of the sample could be eluted with *n*-heptane. The infrared spectra of the *n*-heptane eluates indicate that hydrogenation converted a major part of the benzene eluates to saturated hydrocarbons. Mass spectra were also obtained of the *n*-heptane eluates of the hydrogenated fractions. These spectra were generally similar from one sample to another and the data indicate that odd-numbered normal paraffins are present in greater concentration than even-carbon-numbered homologs. The presence of tetracyclic hydrocarbons in the hydrogenated benzene eluates is indicated by the large values at C 27, 28 and 29; furthermore, the maxima at C 26–28 probably are ions of the tetracyclic hydrocarbon molecules that have lost a methyl group and the maxima at C 16 are due to ring systems. A mass spectrum of cholestane, from cholesterol, was very similar to that of benzene soil extract strongly suggesting a sterol origin of the latter. Other evidence in support of the presence of sterols in the soil benzene eluates is that the original benzene fractions gave positive spectral data for sterols that contain 27, 28 and 29 carbon atoms like the tetracyclic compounds in the soil fractions.

Other hydrocarbons occurring in the hydrogenated benzene eluates as identified in the mass spectra are pentacyclic and hexacyclic hydrocarbons with parent peaks at 30 carbon atoms. Other features of the spectra suggest similar structures for the polycyclic hydrocarbons and the stearanes. The polycyclic hydrocarbons may be the hydrogenolysis products of triterpenes, as it has been shown that the latter are intermediates in the biogenesis of steroids (Dauben & Richards, 1955).

In general, the benzene eluates are composed of waxes that have properties like those of beeswax. Mass spectral analysis of the benzene eluate indicates that waxes with molecular weights of more than 500 are the probable sources of the saturated hydrocarbons in the hydrogenation products of the benzene eluates; these waxes have parent peaks in the mass spectra in the 39–55 carbon atom range. The apparent odd-numbered carbon atom preference starting at C 39 indicates that even-numbered waxes are present in higher concentrations in this fraction than their odd-numbered homologs. The ions in the spectrum are placed as hydrocarbons and even-carbon-number esters of aliphatic acids and alcohols have the same mass as odd-numbered C_nH_{2n-10} hydrocarbons.

Further information about the waxes was obtained from the urea adduct preparations of two of the soil extracts. Urea adducts of the benzene eluates were prepared by adding 5 ml of urea-saturated methanol to 150 mg of eluate fraction in 30 ml of benzene + methanol (4:1) solution. After 24 h the reaction mixture was filtered through sintered glass, and the crystalline product was washed with 5 ml of benzene and decomposed by heating with 10 ml of distilled water. The urea adducts were recovered from the water with benzene. The non-adduct fraction was recovered from the urea-methanol-treated benzene solution after adduction had been completed. Infrared spectra of the urea-adducts and of beeswax are so similar as to indicate similarity in composition. The non-adduct portion of soil sample *b* was found by ultraviolet spectrophotometry to contain, on fractionation into 70 fractions by silica-gel chromatography, to contain possible alkylbenzenes, -naphthalenes and -phenanthrenes, 1, -12-benzperylene, and coronene, totalling less than 5% of the non-adduct fraction. The adduct fraction had no absorption in the u.v. region 230–400 nm.

Hydrogenated fractions of sample 5 were subjected to urea-adduction, chromatographed and mass spectra were obtained. Nearly all the adduct consisted of normal paraffins most of which was of odd-carbon-number atoms. These were derived from even-number aliphatic wax esters, that in turn were formed from even-carbon-number normal aliphatic acids and normal primary aliphatic alcohols. The non-adduct waxes are composed mainly of normal aliphatic acids and cyclic alcohols. The waxes containing the latter were separated from the straight chain waxes by the urea adduction.

Soils have yielded a wide variety of aliphatic organic acids, but only a few, particularly formic, acetic, propanoic and oxalic acid are abundant (Schwartz *et al.* 1954; Jorgensen, 1961). Long-chain fatty acids of low-solubility in water and strong adsorption on clay and humus may persist for long periods in soil, while water soluble acids are much more transitory due to synthesis and destruction by soil micro-organisms (Stevenson, in McLaren & Peterson, 1967). Other aliphatic acids in soil include isocitric, tartaric (Schwartz & Martin, 1955), aconitic? (CH_2:$C(COOH)CH_2COOH$) (Stevenson & Katznelson, 1958), fumaric, glycolic (Louw & Webley, 1959; Sperber, 1957). It has been shown that certain organic acids, such as succinic, fumaric and malic acids are more rapidly used by soil micro-organisms than are other organic acids such as citric, isocitric and *cis*-aconitic (Bose *et al.* 1959).

Water-logged soils such as those of rice paddys (Takijima, 1960) accumulate organic acids in the following general order of decreasing abundance: acetic, butyric, formic, fumaric, propionic, valeric, succinic and lactic. Some of these, particularly butyric, produce the root-rot 'Aki-ochi' condition in rice fields (Mitzui *et al.* 1959). Under anaerobic conditions methane is produced from organic acids in water-logged soils.

Aromatic organic acids identified from soils include benzoic, 3-hydroxy-5-methyl-benzoic (3,5-cresotic) (Shorey, 1914), *p*-hydroxylbenzoic (Walters, 1917) in relatively large amounts; and protocatechuic (Flaig, 1960), *p*-coumaric, vanillic and ferulic (Whitehead, 1964) in much smaller amounts, Lignin is believed to yield, on microbial decay, a number of phenolic acids including ferulic, *p*-hydroxycinnamic, and 4-hydroxy-3-methoxyphenylpyruvic (Ishikawa *et al.* 1963), vanillic and syringic, released by white-rot fungi *Polystictus versicolor* and *Trametes pini* from birch lignin, and vanillic only from spruce lignin (Henderson, 1955, 1957). Phlorizin, a natural constituent of root bark may produce, during microbial decomposition, aromatic acids such as *p*-hydroxyhydrocinnamic and *p*-hydroxybenzoic that are phytotoxic to associated plants (Borner, 1958).

Auxins or plant-growth accelerators that have been extracted from soils are mainly of the nature of indoleactic acid (Parker-Rhodes, 1940). The concentrations of the substances range from 1.7×10^{-7} to 1.5×10^{-8} M in a number of soils (Whitehead, 1963). The possible occurrence of the auxin naphthylacetic acid in peat was noted earlier in the book (p. 47). The level of concentration of auxins in a soil is a balance between their production by soil micro-organisms and plant-root secretion and their destruction and utilization by growing plants and by other micro-organisms.

Organic acids are cited as important chelating agents in soils (Chaberek & Martell, 1959). Oxalic, lactic, citric, 2-ketogluconic acid, and others form chelates with iron, calcium and aluminum; for example:

Calcium citrate (hydrate)   Calcium gluconate

The chelation process has been suggested to explain the increase in solubility of phosphate (Johnston, 1952, 1954, 1959; Struthers & Sieling,

1950) and the prevention of its precipitation by iron and aluminum by addition of organic acids. Addition of organic matter in general made soil phosphates more available judged from ^{32}P-labelled KH_2PO_4 (Mahr & Kaindl, 1954). Micro-organisms activity that resulted in production of organic acids also brought about increase in solubility of phosphates (Sperber, 1958); *Aspergillus niger*, *A. terreus*, and *Sclerotium rolfsii* are among the organisms that were effective (Stevenson, 1967). The chelation of calcium by 2-ketogluconic acid that had been produced by soil bacteria has been demonstrated by chromatographic methods (Duff *et al.* 1963).

Other minerals, particularly silicates, also can enter into chelating processes with organic acids produced by soil fungi. *Aspergillus niger* can bring about the decay of orthoclase, olivine, biotite and muscovite, releasing potassium and other metallic ions (Eno & Reuszer, 1950; Henderson & Duff, 1963; Müller & Forster, 1961, 1964). In addition to *A. niger* and other species of *Aspergillus* as well as *Botrytis*, *Cephalosporium*, *Fusarium*, *Hormodendron*, *Mucor*, *Penicillium*, *Spicaria* and *Trichoderma* produced citric, acetic, formic, fumaric and oxalic acids that dissolved the following silicates: biotite, muscovite, vermiculite, phlogopite, genthite, leucite, nepheline, olivine, wollastonite, apophyllite, harmotome, heulandite and stilbite. Polycarboxylic and hydroxycarboxylic acids have been cited as possibly forming chelates with manganese in soil (Dion *et al.* 1947).

Lichen acids of aromatic nature are believed to be powerful chelating agents in the weathering processes of rocks and minerals, and the term *sequestration* has been applied to their weathering activity (A. Schatz, 1963; A. Schatz *et al.* 1954; V. Schatz *et al.* 1956).

The downward movement of sesquioxides in soils or their removal from soils has been thought the result of chelation with organic constituents and the process is called *cheluviation* (Jackson & Sherman, 1953; Swindale & Jackson, 1956). The process operates to best advantage in podzols. Extracts of pine needles and various leaves and bark yielded carboxylic acids and polyphenols which reduced ferric iron and participated in formation of stable organic complexes with ferrous iron (Bloomfield, 1953–55, 1957).

9.4.4 *Nitrogenous compounds*

The discovery that sizeable portions of the nitrogen in soils may be in the form of clay-fixed ammonium (Rodrigues, 1954; Barshad, 1951; Allison *et al.* 1951; Allison *et al.* 1953) rather than being organically combined was an important recent contribution. Illites and vermiculites are involved in the non-exchangeable fixation of ammonium but the mineralogy has not

TABLE 86   *Distribution of nitrogen in several virgin and*
cultivated soils (Keeney & Bremner, 1964)

Nitrogen	Total soil N (%)			
	Virgin soils		Cultivated soils	
	Range	Av.	Range	Av.
Non-hydrolyzable	18·4–36·7	25·4	19·2–34·3	24·0
Hydrolyzable				
Total	63·3–81·6	74·6	65·7–80·8	76·0
Ammonium	18·6–25·9	22·2	18·7–29·0	24·7
Hexosamine	3·3–6·2	4·9	4·3–7·1	5·4
Amino acid	19·4–34·3	26·5	17·8–31·0	23·4
Unidentified*	17·9–25·1	21·0	19·5–28·9	22·5

* Total hydrolyzable N minus hydrolyzable (ammonium + hexosamine + amino acids)-N.

been thoroughly studied. According to Bremner & Edwards (1965), Bremner (1967), more than 5 % of the total nitrogen in some surface soils and as much as 5 % in some subsurface soils is in the form of fixed ammonium. The distribution of nitrogen in several virgin and cultivated soils is shown in table 86 after Keeney & Bremner (1964). Bound amino acids account for 20–50% of the total nitrogen in most surface soils and hexosamines or amino sugars make up 5–10% (Bremner, 1967). Other organic nitrogenous materials of soil include purine and pyrimidine components of nucleic acids and nucleotides but make up no more than 1 % of the organic nitrogen. Choline, ethanolamine, creatinine and allantoin are present in trace amounts. Thus about half of the total organic nitrogen in soil is of unknown chemical nature.

Amino acids are widespread in soils in bound form for the separation of which acid hydrolysis is necessary. Bremner (1950b, 1952, 1955a), Stevenson (1956), Grov (1963) and Grov & Alvsaker (1963) recorded the amino acids in soils shown in table 87.

The amino compounds of decomposing beechwood sawdust, wheat straw and trefoil inoculated with a high fungal or high bacterial solution were studied after 12 and 20 months (Jacquin, 1961). Amino compounds appeared mainly as amino sugars and there was also a considerable amount of humic acid and other high molecular weight products.

Several of the soil amino acids are of interest in being non-protein in origin: α, ϵ-diaminopimelic acid, ornithine, β-alanine, α-amino-n-butyric acid, γ-amino-n-butyric acid, and 3:4-dihydroxylphenylalanine. Although

TABLE 87   *Amino acids detected by chromatographic analysis of acid hydrolysates of soils (Bremner, 1967)*

Amino acid	
Basic	Isoleucine
Arginine	Phenylalanine
Histidine	3:4-Dihydroxyphenylalanine
Methylhistidine	Tyrosine
Lysine	Serine
Ornithine	Threonine
	Proline
Acidic	Hydroxyproline
Aspartic acid	Methionine
Glutamic acid	Methionine sulfoxide
Cysteic acid	Methionine sulfone
Taurine	Cystine
	α-Amino-n-butyric acid
Neutral	γ-Amino-n-butyric acid
Glycine	α-Amino-n-caprylic acid*
Alanine	β-Alanine
Valine	α,ϵ-Diaminopimelic acid
Leucine	

* Detected in aqueous extract of the H layer of a pine-forest soil profile.

some of these may have formed by decomposition of other amino acids they mostly seem to have been produced by one or another micro-organism; for example α-ϵ-diaminopimelic acid found in some soils is confined almost entirely to bacteria. Although paper chromatography was used widely to identify soil amino acids in the 1950s, in recent years ion-exchange columns employed in automatic amino-acid analyzers have given better and more consistent separations of amino-acid hydrolyzates. The desalting process required in most amino-acid preparations generally results in loss of amino acids. Free amino acids although quantitatively much less than the bound amino acids have been found in certain soil samples (Schreiner & Shorey, 1910*a, b*; Tokuoka & Dyo, 1937; Shorey, 1913; Kivekas, 1939; Paul & Schmidt, 1961). The free amino acids in soils are related to the level of microbial activity.

Although it has been assumed that the bound soil amino acids are largely if not entirely in the form of peptides or proteins it has rarely been possible to isolate either proteins or peptides. Some workers have suggested that the amino acids occur in humic complexes (Swain *et al.* 1959) or in phenolic or quinone complexes, but these possibilities have not yet been investigated sufficiently. Salton (1964) has postulated mucopeptides and techoic acids as hosts for amino acids in soils. Mucopeptides are complexes of amino acids and muramic acid, an amino sugar; alanine and glutamic acid (much of which is in D-form) and lysine or diaminopimelic acid account for most of the amino acids. Techoic acids, all of which contain ester-linked alanine are constituents (up to 50%) of the cell walls of Gram-positive bacteria. Phosphorus is a constituent of techoic acids.

Amino sugars, especially glucosamine and galactosamine, are important hosts for nitrogen in many soils (Bremner, 1950a). Muramic acid may also occur in soils but has not yet been definitely characterized from soil hydrolyzates. According to Stevenson (1957) the proportion of soil nitrogen that occurs as amino sugars increases with depth in some soils and attains 24% of the nitrogen in the B horizon of a planosol. The ratio of glucosamine to galactosamine in soil hydrolyzates varies from 1·6 to 4·1 (Snowden, 1959), and was highest in podzols and lowest in prairie soils; this ratio tends to vary inversely with soil pH and is related to fungal activity.

Derivatives of purine and pyrimidine that have been isolated from soils or soil preparations include: adenine, guanine, cytosine, thymine and uracil all of which are components of nucleic acids (Schreiner & Lathrop, 1912; Bottomley, 1917; Anderson, 1957, 1958, 1961); and xanthine and hypoxanthine which are formed (Schreiner & Lathrop, 1912) by enzymatic deamination of guanine and adenine, respectively. Small amounts of bacterial DNA (deoxyribonucleic acid) was found in Scottish soil (Anderson, 1961) based on the ratio of purines to pyrimidines in soil humic acids. Other soils, however, did not yield either DNA or RNA (Adams *et al.* 1954).

Other nitrogenous substances found in soils are trimethylamine from Georgia marsh soil (Shorey, 1913), ethanolamine and histamine from pine forest soil (Carles & Decau, 1960), choline (Shorey, 1913) from alkali extracts of soils and possibly was in phospholipids, creatinine from soil extracts (Schreiner *et al.* 1911), allantoin from alkali and aqueous soil extracts (Shorey, 1938), cyanuric acid from fulvic acid extract of soil (Shorey & Walters, 1914), α-picoline-γ-carboxylic acid from soils (Schreiner & Shorey, 1909a), urea in forest soils (Fosse, 1916), asparagine and glutamine.

9.4.5 *Carbohydrates*

It goes without saying that organic terrestrial soils are abundantly supplied with carbohydrates, particularly cellulose. In addition, free monosaccharides, hexosans other than cellulose, pentosans, uronic acids, amino sugars and the constituents of probable gums, mucilages, etc. have been obtained from soils.

Free sugars have been obtained from soils by extraction with water and with aqueous ethanol (50–80%) (Alvsaker & Michelsen, 1957; Gupta & Snowden, 1963; Nagar, 1962; Gupta, 1967). For example, a Norwegian pine-forest soil, on extraction with cold water, yielded by paper chromatography glucose, arabinose, fructose, xylose, galactose and ribose. Glucose amounted to two-thirds of the extractable sugars (Alvsaker & Michelsen, 1957). On the other hand four American soils yielded only glucose, while northern Canadian podzol contained galactose, glucose, mannose, arabinose, xylose and fucose (Nagar, 1962).

The common sugars in four soil types were found to be xylose, arabinose, glucose, galactose (Johnston, 1961); the highest amounts were in humic acids and black clay contained less sugars than sandy and silty loam.

The oligosaccharides cellobiose, gentiobiose and cellotriose were obtained from incubated moist soil extracted with 60% ethanol; the oligosaccharide yield increased as a result of the incubation.

The pretreatment of soil samples seems to affect the yield of free sugars; this being markedly reduced by drying and greatly increased by freezing in certain samples (Gupta, 1967).

As far as the adsorption of carbohydrates by clay minerals is concerned, methylated sugars are more strongly adsorbed than others (Greenland, 1956), and polygalacturonic acid was found adsorbed on hydrogen bentonite (Kohl & Taylor, 1961). Rogers (1965) and Swain, Bratt & Kirkwood (1967) found that marine shales exerted a protective effect on thermal degradation of sugars, but very little study of clay–carbohydrate relationships has been made.

Few studies of the actual cellulose content of soils have been made, although numerous empirical analyses are available. The latter involve extraction of non-cellulose components with ether, sulfurous acid and ammonia, under pressure and testing the residue with $Cu(NH_3)_4OH_2$ (Schweitzer's Reagent) or modifications of this method. The results indicate that northern podzol, brown rain forest peat and cultivated peat soils contain from 3 to 10 mg/g of cellulose (Gupta & Snowden, 1964). These values are generally higher, by as much as four times those obtained by acid extraction and chromatography of glucose from similar soil samples. The reasons for this discrepancy have not been satisfactorily explained.

Hemicellulose is a general term for mixtures of hexoses, pentoses and uronic acids in various proportions that are soluble in 17·5% sodium hydroxide solution (the so-called mercerizing strength, Noller, 1951). They are typically in close association with lignin and include D-xylose, L-arabinose, D-glucuronic acid, D- and L-galactose, D-mannose, L-rhamnose and L-fucose (Percival, 1962). The pentosan fraction of the hemicelluloses of sandy loam soils is less than that of podzols and brown earths (Ghildyal & Gupta, 1959; Mehta & Deuel, 1960). They contribute importantly to the humic fraction of soil because of their close association with lignin with which they undergo condensation reactions to form humus compounds.

Uronic acids, chiefly D-glucuronic, D-galacturonic and D-mannuronic acids are rather widely distributed in soils and reportedly vary from 12 to 47·8 mg/g of organic matter in mineral and organic soils (Ivarson & Snowden, 1962; Lynch *et al.* 1957; Dubach & Lynch, 1959; Gupta & Snowden, 1965; Dormaar & Lynch, 1962). These substances are widespread in plants and are probably derived from plant sugars by oxidation of a primary alcoholic group to a carboxyl group. It is believed that about 90% of the uronic acid constituents of organic matter in soil are present in fulvic acid preparations of the soil.

The amino sugar contents of soils were discussed under nitrogenous compounds.

9.4.6 *Organic sulfur compounds*

Although in arid-region soils, gypsum ($CaSO_4.2H_2O$), epsomite ($MgSO_4.7H_2O$) and other inorganic substances are present, most of the sulfur in humid region soils seems to be in organic combination (Freney, 1961; Williams & Steinbergs, 1962). Relatively few studies have been made of the organic sulfur compounds in soil. Shorey (1913) obtained trithiobenzaldehyde ($SCHC_6H_5)_3$ from soil and suggested it formed by reaction of H_2S from bacteria, with benzaldehyde from the decomposition of lignin. Free cystine amounting to 16 μg/kg was extracted with ethanol from a rhizosphere soil (Schmidt, Putnam & Paul, 1960) and methionine sulfoxide, cystine and methionine were found in incubated soils (Paul & Schmidt, 1961). Up to 58% of the total sulfur in Quebec soil was found by treatment with Raney nickel to be directly bonded to carbon (Lowe, 1964; Lowe & DeLong, 1963) and to consist of such compounds as sulfur amino acids but not alkyl sulfones or ester sulfates which do not liberate sulfur with Raney nickel. The rest of the sulfur seems to be present as sulfate esters such as sulfated polysaccharides, phenolic sulfates, choline sulfate, or sulfated lipids that can be reduced to hydrogen sulfide with hydriodic acid and readily hydrolyzed to inorganic sulfate with acid or alkali (Lowe & DeLong, 1961; Taylor & Novelli, 1961; Whitehead, 1963; Haines, 1964; Woolley & Peterson, 1937; Lindberg, 1955; Stevens & Vohra, 1955).

9.4.7 *Free radicals*

The ability of the electron paramagnetic resonance method of spectrometry to detect species which contain unpaired electrons characteristic of free radicals has enabled soil scientists to study the distribution of these paramagnetic organic compounds (Rex, 1960; Steelink & Tollin, 1967). Among

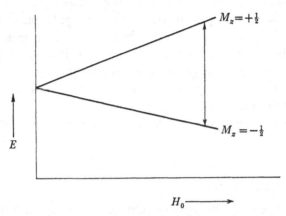

FIGURE 125 Plot of energy (E) versus the applied field (H_0) in electron paramagnetic resonance without application of electromagnetic energy (Steelink & Tollin, 1967).

the huge numbers of chemical reactions continually occurring in soil organisms and soil organic matter there are many opportunities for one-electron reactions to occur, resulting in both the production and the destruction of free radicals. Many of the radicals are of the stable long-lived type.

A detailed discussion of the paramagnetic method of free radical detection is given by Steelink & Tollin (1967). Briefly the method of electron paramagnetic resonance spectrometry involves the interaction between the spin of an unpaired electron and an external magnetic field. If no field is present the energy of an electron is independent of its spin level ($\pm\frac{1}{2}$). If there is a magnetic field of strength H_0 the Zeeman effect gives the following relationship:

$$E = -g\,\beta\,M_2\,H_0, \tag{109}$$

where E is the energy of the unpaired electron, g is the spectroscopic splitting factor which is 2·0023 for free spin, β is the Bohr magneton, and M_2 is the component of spin angular momentum in the direction of the applied field H_0. M_2 may have values of $\frac{1}{2}$ or $-\frac{1}{2}$, and a plot of energy versus

322

the applied field will be as shown in fig. 125, and the value of E between two spin states at a given value of H_0 is:

$$E = hv = g \beta H_0, \tag{110}$$

a sample containing unpaired electrons placed in a magnetic field and irradiated with electromagnetic energy having frequency v that satisfies

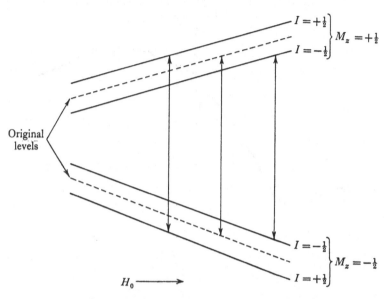

FIGURE 126   Plot of energy versus applied magnetic field where radiation with electromagnetic energy also is involved (Steelink & Tollin, 1967).

equation 110 above, electrons in lower state absorb energy and jump to the upper state, and those in the upper state emit energy and fall to the lower state. A sample that is at thermal equilibrium before irradiation will have a larger number of electrons in the lower than in the upper state, and a net absorption of energy from the radiation field. EPR spectrometry is the measurement of this energy absorption (fig. 126).

The EPR spectra of soil organic matter seems to be almost entirely due to the humic acid + fulvic acid fraction (fig. 127) (Rex, 1960; Steelink & Tollin, 1962). The spin concentrations in soil organic matter and derived humic acid are of the order of 10^{18}/g, considerably more than those in lignins of living plants and lignin preparations (table 88). Some of the non-humic constituents of soil organic matter also contain free radicals: lignin (table 88); tannin has 10^{16}–10^{17} spins/g (Tollin & Steelink, 1966); melanin, 2×10^{18} spins/g (Blois *et al.* 1964); resins, 10^{14}–10^{16} spins/g

(Lagerkrantz & Yhland, 1962); pigments 2×10^{16}–$2 \cdot 8 \times 10^{17}$ spins/g from quinones (Steelink & Tollin, 1967); antibiotics, 10^{17}/g for tetracycline derivatives (Kozlov, 1964), quinone-type vitamins (Weber *et al.* 1965), flavins (Guzzo & Tollin, 1964; Ehrenberg & Ericsson, 1965) and various other soil preparations (Steelink & Tollin, 1967).

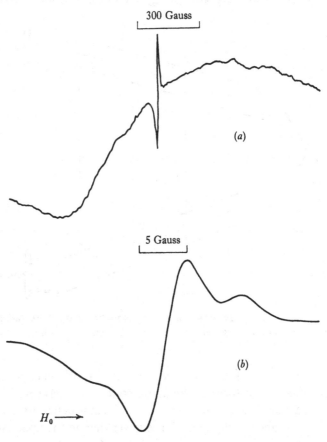

FIGURE 127 (*a*) Electron paramagnetic resonance spectrum of podzol soil; (*b*) EPR spectrum of podzol humic acid (Steelink & Tollin, 1962).

The nature of the organic structure in which the free radicals of humic-acid preparations are concentrated was thought by Steelink *et al.* (1963) to be a polymer containing *ortho-* and *para*-quinhydrones; one property of which is to give rise, on basification, to semi-quinone radical ions on EPR characteristic of humic-acid preparations. However, simple quinones, being diamagnetic in the acid form, do not fit the paramagnetic

TABLE 88 *Free-radical content (spin concentration) in soil organic matter and derived humic acid (Steelink & Tollin, 1967)*

Sample	Spins/g	Mol. wt	Spins/mole
Lignins			
BNL,* spruce	5×10^{16}	$1,000+$	5×10^{19}
Norway spruce sulfuric acid	3×10^{16}	$5,000+$	$1 \cdot 5 \times 10^{20}$
Yellow pine, Kraft	3×10^{17}	$7,000+$	$2 \cdot 1 \times 10^{21}$
BNL* treated under Kraft conditions	4×10^{17}	—	—
Calcium lignin sulfonic acid	3×10^{17}	$10,000+$	3×10^{21}
Western hemlock decayed-wood meal	9×10^{16}	—	—
Saguaro wood meal	5×10^{16}	—	—
Humic acids			
Fulvic acid from Wisconsin podzol	3×10^{17}	—	—
Humic acid from Wisconsin podzol	$0 \cdot 8 \times 10^{18}$	—	—
Humic acid from muck soil	$0 \cdot 3 \times 10^{18}$	—	—
Humic acid from California podzol	2×10^{18}	$20,000+$	4×10^{22}
Humic acid from English podzol	$1 \cdot 4 \times 10^{18}$	$20,000+$	3×10^{22}
Humic acid from Arizona brown-forest soil	$0 \cdot 8 \times 10^{18}$	—	—

* Brauns' native lignin (BNL).

TABLE 89 *Free-radical content (spin concentration) of humic acids from soil and lignite (Steelink, 1966)*

Humic acid	Acid spins/g*	Sodium salt spins/g*	Increase factor
English podzol	$0 \cdot 2$	10	50
Swiss muck	$0 \cdot 6$	24	40
Wisconsin podzol	$0 \cdot 7$	18	26
Wisconsin podzol, fulvic fraction	$0 \cdot 4$	$0 \cdot 7$	2
NALCO Leonardite humic acid	$0 \cdot 4$	$6 \cdot 6$	16
North Dakota Leonardite humic acid	$0 \cdot 4$	30	75
New Mexico lignite humic acid	$2 \cdot 8$	37	12
Artificial catecholamine humic acid	$2 \cdot 3$	54	23

* Spin content $\times 10^{18}$.

properties of free humic acid. Other possible sources are (Steelink & Tollin, 1967):

Semi-quinone polymers, containing electron donors and acceptors.
Adsorbent complexes, retaining mineral ash residue with free radicals.
Hydroxyquinones, known to increase stable radical content on basification.
Polynuclear hydrocarbons, known to yield EPR signals if carbon content is very high.

Trapped radicals in a polymeric matrix, or at sides of imperfections in regular crystals of donor–acceptor systems (Colburn *et al.* 1963; Law & Ebert, 1965; Ray *et al.* 1965; Eastman *et al.* 1962) a unique feature is that while free radicals trapped in molecular sieve are released on solution of the polymer, humates can be repeatedly dissolved and reprecipitated without noticeable loss of free radicals.

According to Steelink & Tollin humic-acid models such as those proposed by Flaig (1960) and Gillet (1956) although divergent, would accommodate the observed paramagnetism

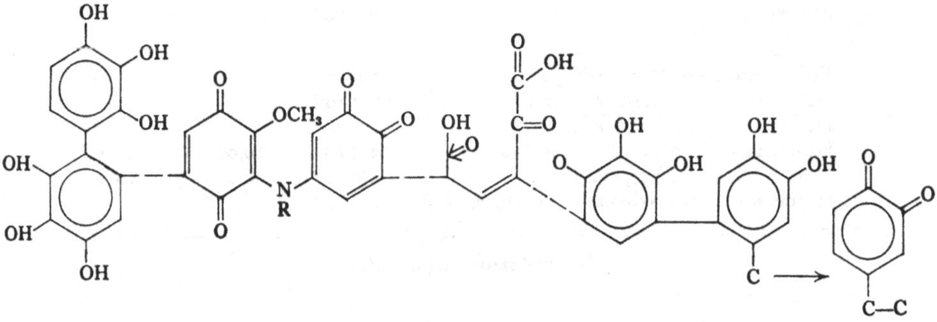

Flaig's humic acid model

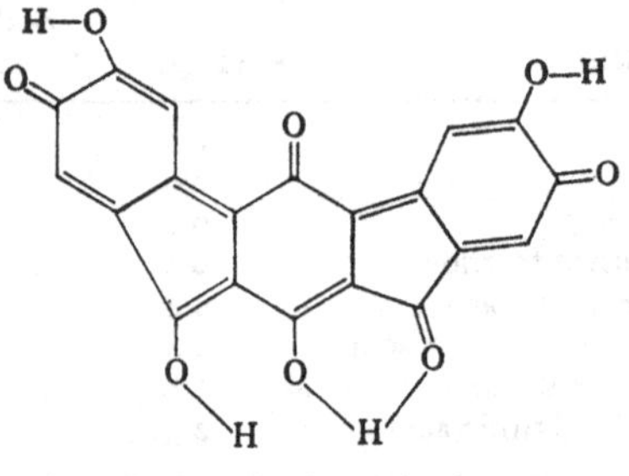

Gillet's humic acid model

According to Friedlander *et al.* (1963) the free radical contents of soil humic acid might inhibit one-electron metabolic processes, catalyze polymerization reactions, or bring about the reduction of nitrogen oxides (Steelink & Tollin, 1967). Presence of free radicals also is believed to enhance the chelating ability of the substrate (Hemmerich *et al.* 1963; Beinert & Hemmerich, 1965).

9.4.8 *Soil enzymes*

The nature, distribution, and reactions of enzymes on organic matter in sediments and soil is of special interest to the organic geochemist because of their possible effect on residual organic matter in sedimentary rocks. A review of knowledge of soil enzymes is that of Skujins (1967).

Dehydrogenases give evidence of the biological activities of microbial populations in the soil, and the enzyme itself is not determined. The test for this enzyme in soil consists of measuring the reduction of 2,3,5-triphenyltetrazolium chloride (TTC) to triphenylformazan (Lenhard, 1957, 1962) and is measured as a function of activity of soil micro-organisms. High dehydrogenase activity is shown by fertile cultivated soils, while saline and alkaline soils shown almost none.

Catalase activity in soils is measured by the release of oxygen after hydrogen peroxide is added. The residual H_2O_2 is titrated with potassium permanganate (Kappen, 1913; Johnson & Temple, 1964). High organic content of soils is typified by high catalase activity, with highest concentration in litter- and humus-accumulating surface and A-zones and with marked decreases below. Catalase activity seems to be higher in alkaline than in acid soils and occasionally in the autumn rather than in other seasons (Runov & Terekhov, 1960). There is some evidence that presence of vegetation is more important than microbial numbers on catalase activity (Zemlyanukhin, 1959). Furthermore, peroxide decomposition of catalytic type in soil may be largely non-enzymatic. Manganese and iron compounds (Skujins, 1967) are suggested as causing as much as 40% of the thermostable (non-biological) so-called catalase activity in soil.

The activities of peroxidase and *polyphenol oxidases*, such as catechol oxidase and tyrosine, are evidently greatest in carbonate-rich soils containing large numbers of micro-organisms (Skujins, 1967), and are also controlled by the season and the type of vegetation. An extracellular *p*-diphenol appearing during soil humification has been shown (cited by Skujins, 1967). Pyrocatechol in the presence of oxygenated water has been used to determine these enzymes (Galstyan, 1965; Kozlov, 1964).

Urate oxidases (uricases) appear to be extracellular enzymes released by micro-organisms from urate-rich soils (Durand, 1961). The enzyme includes two uricolytic components that can be extracted with 0·1 M phosphate at pH 7 and pH 8·4, respectively. Both the urate oxidase and associated uric acid may be adsorbed on clay minerals (Durand, 1961).

A number of *transferase* enzyme activities have been noted in soils. Synthesis of oligosaccharides may occur slowly through transgluocosidation of starch and maltose by soil enzymes (Hoffman, 1963). Not only fructose and glucose form from sucrose by invertase activity but oligosaccharides may also be produced, possibly by transfructosidase activity (Kiss & Peterfi, 1959). Toluene treatment is commonly used to halt enzyme activity in soils but in such a soil alanine forms by transamination from a pyruvate in the presence of several free amino acids, thus indicating *transaminase* activity in the soil.

Phosphatases are widely distributed and important in soils (Rogers, 1942). These enzymes evidently are most active in soils low in biologically available phosphate and high in organic matter (Halstead, 1964; Geller & Dobrotovorskaya, 1961). The enzymes are produced by numerous soil bacteria and by fungi. There is some evidence that alkaline and acid phosphates can be separated (Halstead, 1964; Dubovenko, 1964). Their activity seems to be inhibited by clays in proportion to their base-exchange capacity (Mortland & Gieseking, 1952). They are determined by various methods one of which is to use phenyl phosphate as substrate and to measure the amount of phenol released (Kramer, 1957).

Pyro- and *polymetaphosphatase* hydrolytic activity in soils is evidently quite stable since toluene treatment and heating of soil at 105 °C for 80 h reduced its activity by only 16 and 34% respectively (Rotini, 1933; Rotini & Carloni, 1953). *Phytase* activity has been noted in soils (Jackman & Black, 1952) and is directly related to microbial activity. *Nucleotidases* were studied by Rogers (1942) and high rates of inorganic phosphate at pH 7 and 60 °C in toluene-treated soils were observed.

Acetylesterase activity in soils was found by Haig (1955; cited in Skujins, 1967) to be due to an extracellular enzyme that brought about catalytic decomposition of phenylacetate. A particular clay fraction of the soil seemed to characterize the acetylesterase activity. *Lipiases* were suggested by Pokorna (1964) to be present in several peats and muddy soils.

Amylase activity is present in soils and reaches maxima at pH 5·5–6·0. It increases with organic content and with increasing cation-exchange capacity. Soils apparently contain more β-amylase than α-amylase (Skujins, 1967; Hofmann & Hoffman, 1955). The analytical method for amylase consists of adding 2% buffered starch solution to the dry soil, incubating with toluene and determining the released reducing sugar by the Lehmann–Maquenne thiosulfate method. *Invertase* is widely distributed in soils and is assayed by adding 10% sucrose solution in pH 5·5 buffer to soil containing toluene; after incubation, the amount of reducing sugar is determined by the Lehmann–Maquenne titration method (Hoffmann & Seegerer, 1951) or a colorimetric method (Gettkandt, 1956). A close correlation with microbial numbers, plant root contributions, and metabolic activities is shown by invertase activity; it is highest in neutral calcareous soils. Strong absorption by soil particles and stabilization by clays are characteristic. Typical invertase inhibitors are mercuric chloride, aniline, *p*-toluidine and formaldehyde. Soil is protective of the enzyme *α-glucosidase* (*maltase*) which is only partially inhibited by $AgNO_3$, $HgCl_2$, dihydrostreptomycin and other biological inhibitors even at high concentrations (Kiss & Peterfi, 1960). Hofmann & Hoffman (1953, 1954) detected α-glucosidase in soil using β-phenylglucoside as substrate and assaying the reducing sugar by the Lehmann–Maquenne method. *β-glucosidase* (cellobiase) occurs in soils (Hofmann & Hoffman, 1953) and can be detected using β-phenylglucoside as substrate (Galstyan, 1965). The enzyme seems to have a zero-order reaction rate in soils as no assimilation of its hydrolytic products by micro-organisms has been detected. Both *α-galactosidase* (melibiase) and *β-galactosidase* (lactase) were found in soils by means of phenylgalactosidase substrates (Hofmann & Hoffman, 1953, 1954). Extracellular *cellulase* in soil is suggested by differences in total cellulase activity in response to pH between soils treated and not treated with toluene (Sörensen, 1957). The enzyme appears to be less common than *lichenase* in soils (Kiss & Peterfi, 1961). *Xylanase* is increased in its excretion from micro-organisms by the amount of xylan present. The enzyme is determined by incubating the soil with a xylan solution and phosphate at pH 6·2–6·5, and testing the released reducing sugar with Somogyi's reagent (Sörensen, 1957). *Inulase*, also released in soil by micro-organisms, has been reported by Kiss & Peterfi (1961).

Proteinases have been known since the work of Fermi (1910), who with subsequent workers found enzymes in soil that hydrolyzed gelatin, ovalbumin, peptone, and casein. Pepsin-, cathepsin- and trypsin-activity fractions have been isolated (Nosek & Ambroz, 1957; Vlasyuk *et al.* 1957; McLaren *et al.* 1957; Antoniani *et al.* 1954). Proteinase activity is higher in grasslands and humic soils than in other soils and decreases with depth. *Asparaginase* and *amidase* (deaminase) have been recorded from soils (Drobnik, 1956, cited in Skujins, 1967; Mouraret, 1965; Subrahmanyan, 1927), the latter enzyme was particularly effective on glycine. *Urease* activity in soils is directly related to the number of microbes and organic-matter content (Chin & Kroontje, 1963; Balicka & Sochacka, 1959). Soil contains free urease that can possibly be distinguished from that associated with micro-organisms (Briggs & Segal, 1963). Its maximum activity is at pH 6·5–7·0 and in non-calcareous and non-alkaline soils (Galstyan, 1958). Toluene treatment of soils tends to increase urease activity, perhaps by its release from micro-organisms. Urease activity not previously known to occur in blue-green algae has been reported in *Plectonema calothricoides* and other species of Cyanophyta (Berns *et al.* 1966). A *lyase* in soil has been suggested (Drobnik, 1956, cited in Skujins, 1967) that decarboxylates aspartate to form alanine.

Organic geochemistry of non-marine humus

The enzymes in soil should not be considered apart from the organisms that produced them, but, although specific sources are known in a number of instances, there are many cases in which the specific source-organisms have not been identified for individual enzymes. Enzymatic reactions in soil are due both to microbial and other extracellular enzymes and to typical intracellular enzymes (Skujins, 1967). Which type is the more important although subject to controversy on the part of investigators is evidently a local matter.

Three sources of free soil enzymes are: (1) growing as well as dying micro-organisms that release enzymes in the soil; (2) soil animals; (3) plant roots and plant residues. The first of the three sources is thought to be the most important in the majority of cases.

A remarkable feature about certain enzymes in soil is their resistance to heat and other inactivating agents. Upon being released into the soil the mineral particles appear to exert a protective influence on the enzymatic proteins. Nevertheless, from a geochemical point of view enzymes in soils and sediments are quite short-lived, judged from the fact that analyses of sediments for proteins are typically negative. It appears, therefore, that enzyme activity occurs primarily in the upper part of aquatic sediments if it occurs at all. In aerated soils this activity may extend to greater depths. Even in soils it is generally not possible to know to what extent enzyme activity is of extracellular or other origin.

9.5 RECENT GEOCHEMICAL STUDIES OF SOIL AND AQUATIC ORGANIC MATTER

The organic content of spring waters of similar weakly mineralized composition was determined to be variable as to humic substances (Nevraev et al. 1964). Two groups of springs were recognized, those with low (0·7–2·5 mg/l) and high (4·2–4·7 mg/l) humic content. The amino-acid contents of the spring waters ranged from 1·0 to 4·5 mg/l.

The waters of Dnieper River contain biochemically stable humus of lignin–protein complexes that predominate among the organic substances of the river (Maistrenko, 1963). The C:N ratio of the water varies from 16 to 34; the C:P ratio = 350 and the permanganate O:C = 1·3. The maximum amount of humus in the river is in the spring and the minimum is in the winter.

The toxic effects of humic soils on zooplankton in several Russian lakes was investigated (Czeczuga, 1957). In lakes Blado, Bladko and Mtynek as well as smaller ponds in the Narocy group of lakes, all of which are fairly high in humic-acid contents, the humic compounds were found to be toxic to *Daphnia pulex*, *Leptodora kindtii* and *Simocephalus serrulatus*.

The waters of Suna River, U.S.S.R. have small mineralization of 10–20 mg/l; that of the lakes in its system is 5–200 mg/l because of the heterogeneity of the soils surrounding them (Gritsevskaya, 1958). The humic organic matter varies with the season of the year as to composition and concentration. In the lakes of the upper Suna, that are fed by surface drainage the humic matter is biochemically stable but is easily oxidized.

In the separation of ilmenite from sands containing humic substances by mineral oil flotation (Babenko, 1963), only a portion of the organic matter is retained in the clay-slimes. Addition of sodium carbonate to maintain an alkaline medium improves the separation greatly.

The detrimental effects of presence of humus in sand on strength of concrete are discussed by Karttunin & Sneck (1958). Although the presence of humis is a problem, there does not seem to be quantitative proportionality in the reduced strength of concrete with content of humus.

The phosphorus associated with humic acid from brown soils, when separated electrophoretically, was found to have low mobility and move as a single boundary (Dormaar, 1963). That from black or gray soils, however, had higher mobility and had two or more components.

Gingko biloba L., the maidenhair, the most ancient of living trees, belongs to a group of plants that began more than 200 million years ago. *G. biloba* itself has existed perhaps since early Cretaceous (Wealden) time (Seward, 1959). According to Major (1967) the longevity of the species is due to its resistance to insects and fungi. The *Gingko* leaves contain alcohol-extractable acids, perhaps of the nature of malic and oxalic acid that are inhibitory to the growth of the corn borer larva. The leaves also contain 2-hexenal, an aldehyde, reported to be an insect repellent. The roots also are toxic to insects to some extent. Furthermore, 2-hexenal has antifungal activity which may in part, but not entirely, account for the resistance of *Ginkgo* leaves to fungi. 2-Hexenal appears to have been derived from linolenic acid which is converted by chemical rather than enzymatic action. An oil of non-aromatic hydrocarbon nature and possibly containing a carbonyl group was steam-distilled from *Ginkgo* leaves (10 mg/100 g). In 100 p.p.m. concentrations the oil inhibited growth of *Monilinia fructicola*. A cuticle-wax in *Ginkgo* leaves can also inhibit germ-tube growth and prevent spore germination of some fungi. Several hydroxy-lactones were extracted from dry *Ginkgo* leaves with acetone; two of these had the composition $C_{20}H_{24}O_{10}.1H_2O$ and $C_{20}H_{24}O_{10}.2H_2O$. However, none of the lactones showed activity against *Monilinia fructicola*.

This study brings to mind other 'living fossils' that might be investigated for their pest-resistant substances.

10 Organic geochemistry of non-marine coal

In this brief discussion of a major subject emphasis will be placed on the essential geological and geochemical features of non-marine coal deposits, supplemented by some recent developments in the organic geochemistry of coal.

10.1 MODES OF ACCUMULATION OF COAL-FORMING VEGETABLE MATTER

Coal-forming material may accumulate in seawater, brackish, or freshwater but the last two environments are believed to account for the great majority of coal and lignite deposits. The environments of coal formation are extensively discussed in textbooks on coal, Moore (1940), Francis (1961) and van Krevelen (1961).

Accumulations of vegetable matter in the marine environment may consist of land-derived material that becomes waterlogged and sinks to the bottom beyond the reach of mineral detritus. Such plant material may consist of tree trunks and logs or of floating masses of living freshwater aquatic plants. The material may also comprise masses of fucoid algae. Examples of the first group of plant accumulations formerly were very abundant in the lower flood plain of the Mississippi River where large masses of rafted material were stranded at points along the river or were carried out to sea. Examples of the second (marine) source of vegetable matter are the kelp beds off the coast of California and elsewhere. In both instances it seems doubtful that widespread thick accumulations of coal would be likely to form, so the deposits would tend to be spottily distributed and of unpredictable composition.

Some types of saltwater lagoons such as those along the Texas Gulf Coast have abundant growths of saltwater pondweeds (*Thalassia*, etc.), but it is unlikely that thick accumulations of this material would be expected because of the temporary nature of most such lagoons.

The brackish water environments of estuaries and some types of lagoons and saltwater marshes are better suited to accumulation of coal-forming material than the strictly marine setting. Marsh plants of both freshwater and marine types may form sizeable accumulations in the upper parts of estuaries, such as in Chesapeake Bay, Virginia and Maryland. Large salt marshes typified by the Everglades of southern Florida are favorable places for deposition of herbaceous vegetation of various kinds or of mangrove. In addition the vegetative debris carried by rivers during high-water stages is added to the organic accumulation in estuaries.

The majority of widespread coal beds seem to have formed in a fresh-water environment. Many different types of freshwater and associated terrestrial habitats are potential coal-forming environments.

As discussed in Chapter 3, organic matter may form as lowmoor deposits in open eutrophic lakes to considerable depths but this material is finely divided copropel or gyttja derived from phytoplankton, typically with admixed silt and clay, and would yield a poor grade of coal. A less open, marshy habitat is characterized by numerous pondweeds and the remains of these mixed with algal material and with forest and savannah debris. Such environments as these are typified by pH values ranging from slightly alkaline to slightly acidic and by weakly positive to moderately negative Eh values. Coals that probably formed from these sources have varying amounts of spores, pollen and cuticular plant material that probably have under-gone selective enrichment in the original accumulation because of their re-sistance to decay. Where a forest encroaches upon the marsh and eutrophic lake a terrestrial facies may develop bordering the various aquatic facies.

If cyclical climatic changes occur these facies may migrate back and forth and alternate with each other both laterally and vertically. Such alterations are likely to go unnoticed in most coal deposits except through detailed studies of the individual beds (Ting, 1967; Koppe, 1966).

The terrestrial accumulation of coal-forming vegetable matter takes place with greater difficulty because of the prevailingly positive redox conditions in terrestrial habitats. Highmoor peats formed of either *Sphagnum*-type moss or heather accumulations may develop on moist uplands, such as those of western Wales, north-central England, and parts of Ireland. Forest-litter such as that of the Olympic Mountain rainforest, Oregon and Washington is another example of terrestrial accumulation.

10.2 THE DEVELOPMENT OF THICK LIGNITES

The Mesozoic and Cenozoic deposits of Germany, China, Victoria, Australia and the United States contain thick deposits of lignite or brown coal. That in the Latrobe Valley, southern Victoria, Australia attains 1,200 ft in three superimposed layers (Edwards, 1945). The presence of remains of both conifers and angiosperms and of fungal hyphae and sclerotia indicates that important plant sources of the lignites grew above water level in the swamps. The presence of successive stump horizons and of interlayering of petrographic types of lignite show that cyclical conditions were involved in its formation. Possible steps in the accumulation of thick lignites are as follows (Thomson, 1950): (1) sedge and forest peat accumulated on a clay soil during stillstand of the basin, and under relatively dry-forest conditions; (2) dark peat which later became lignite formed from portions of the foregoing peat in moister areas as well as from plants growing in a progressively more moist climate; partial destruction of the forest by fire or intensive fungal action may occur; (3) rapid subsidence of the basin is accompanied by deposition of mud and sand on the peat and development of lakes; (4) aquatic vegetation grows in water, forming light-colored peat layers; (5) the lake peat is succeeded in normal marsh and bog development by sedge and forest peat.

The cyclical conditions responsible for the formation of lignites may last from several hundred thousand to many million years (Francis, 1961). The Rhenish lignites were deposited over an estimated 50 million years from the Eocene to the Pliocene Epochs and during the time interval the climate changed from subtropical represented by palm, *Liquidambar*, *Mohria* and *Lycopodium* to temperate, represented by pine, fir, oak, alder, birch and chestnut.

10.3 FORMATION OF DRIFT COALS

The Gondwana (Permo-Carboniferous) and Eocene coals of India are examples of the contrast between non-marine and marine coals, respectively that may be of drift origin (Fox, 1931). The Gondwana coals were deposited in a continuous river and lake sequence of sediments representing the following, each of which was a large lake basin:

(1) Raniganj Series: Upper Carboniferous.
(2) Jharia Series: Upper Carboniferous; contains coal beds that pass laterally into carbonaceous shales.
(3) Bihar Series:
(4) Barakar Series: Lower Permian conglomeratic sandstones, shales and clays, carbonaceous shales and coal beds in successive cyclical units.

The relationships of the coal strata suggest material transported by rivers into a large lake. Some of the evidence in support of this concept is: (1) the deposition of the coal-bearing strata was more continuous than would be expected of typical Carboniferous cyclothemic coal swamps; (2) the coal beds commonly pass laterally into carbonaceous shales and then into sandstones and siltstones which suggest sorting due to water transport;

(3) the coal contains woody matter of presumed terrestrial origin but rooted vegetation is lacking indicating a transported rather than an *in situ* origin.

Velikowsky (1956) citing data of D. G. Whitley made several provocative suggestions regarding the possible catastrophic origin of some types of drift coals. The south coast of the New Siberian Islands has deposits up to 300 ft thick in which driftwood tree trunks are interbedded with sandstone. Bone of extinct arctic mammals including mammoth, rhinoceros and bison occur in the sandstone. The mass indicates that some catastrophic events such as hurricanes, 'tidal waves' or landslides were responsible for the accumulation of material. A modern example of the phenomenon is that at Lituya Bay, south-eastern Alaska where in 1958 an earthquake precipitated an avalanche of glacial ice into the bay, creating a giant wave that caused the destruction of a forest bordering the bay. The debris from the forest was deposited at the mouth of the bay as a huge mass of potentially coal-forming material.

10.4 CYCLICAL DEPOSITION OF CARBONIFEROUS COAL STRATA

The cyclical relationships of the Carboniferous coal beds to their associated strata has long been recognized. It remained for the development of the cyclothem concept in Illinois, however, for reasonable alternative explanations of the phenomena to be offered (Weller, 1930, 1931, 1956; Wanless, 1947, 1950; Wanless & Shepard, 1936).

Dawson in Nova Scotia (1854), Newberry in Ohio (1874) and Geikie in England (1882) made early references to the orderly arrangement of Upper Carboniferous coal-bearing sequences. Udden (1912) had noted the repetitive nature of the coal-bed sequences near Peoria, Illinois, and Weller (1930, 1931) described a cycle ideally comprising ten sedimentary units in descending order:

(10) Gray shale containing siderite (clay ironstone) concretions.

(9) Marine fossiliferous limestone and calcareous shale.

(8) Black fissile shale containing large black limestone and marcasite concretions.

(7) Argillaceous marine limestone, commonly absent in individual cyclothems.

(6) Gray silty shale containing marine fossils or land plants or both.

(5) Coal.

(4) Underclay with traces of roots.

(3) Freshwater nodular limestone, with sparse invertebrate fauna.

(2) Shale and sandy shale.

(1) Non-marine sandstone typically with a basal unconformity.

The term *cyclothem* (fig. 128) was applied to the sequence (Wanless & Weller, 1932), and according to Weller (1960) about 100 successive cyclothems have been recognized in the Pennsylvanian System in the United

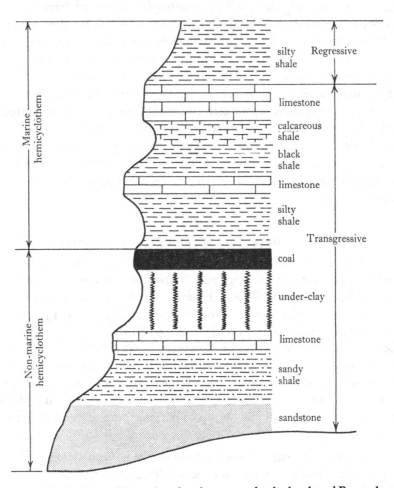

FIGURE 128 Stratigraphic section showing a completely developed Pennsylvanian cyclothem of the kind that occurs in Illinois and neighboring states (After Weller, 1957, *Geol. Soc. Amer.*, Mem. 67, **2**, 331, fig. 2).

States. As originally defined a cyclothem included the deposits from the base of one non-marine sandstone to the base of the next overlying such sandstone. As one proceeds eastward from the eastern Interior Coal Basin (Illinois–Indiana) into the Appalachian Coal Basin (Pennsylvania–West Virginia) of the United States the Pennsylvanian cyclothems progressively lose the marine hemicyclothem (fig. 129) and the non-marine hemicyclothem assumes dominance; the reverse is true as one goes westward to the western Interior Coal Basin of Missouri and Kansas (fig. 129). There is furthermore a tendency (Weller, 1960) for the older Pennsylvanian cyclothems in Illinois to be comparatively simple and incomplete while the later ones show greater differentiation of the marine members. Weller has pointed out that the geographic and stratigraphic complexity of the Penn-

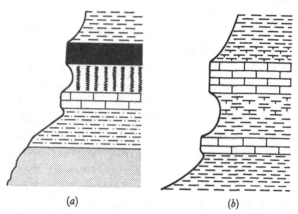

(a)   (b)

FIGURE 129 (a) Stratigraphic section showing a mainly non-marine Pennsylvanian cyclothem of the type that is common in the northern Appalachian region, United States. (b) Similar section showing a mainly marine cyclothem of a type that is common in eastern Kansas (Weller, 1960, *Stratigraphic Principles and Practice*, Harper & Row, 373).

sylvanian cyclothems of the United States makes a simple explanation of them difficult. He interprets them as the result of alternating upward and downward movements of the depositional basin which are balanced delicately with corresponding movements of source areas of sediment. The mechanism by which these presumably vertical oscillations took place may have been related to thermal expansion and contraction in the underlying earth's crust. Build-up and release of radiogenic heat in the crust beneath the area might have produced the necessary uplift by melting part of the crust and providing a 10% volume increase in the melted material.

The diastrophic theory of cyclothem origin does not account for several of their characteristics in the central U.S. (Weller, 1956) i.e. (1) the occurrence of more cyclothems in the deeper parts than in the marginal parts of the Illinois Basin; (2) the change in type of cyclothem with the passage of time in the Pennsylvanian of American coal basins; (3) the apparent cycle of cyclothems ('megacyclothems') of the upper part of the Pennsylvanian of Illinois and in Kansas.

Other diastrophic theories of origin of cyclothems have been proposed by Stout (1931), Hudson (1924) and Rutten (1952).

According to Wanless & Shepard (1936), the rise and fall of sea level necessary to produce the cyclothems, i.e. pulsations of the order of 450 ft, could be accounted for by late Paleozoic glaciations and interglaciations in the southern Hemisphere. Each glaciation represented the stage in which widespread alluviation and local lake development was followed by vegetative growth, chemical weathering and formation of underclay and paludification ('swamping'). The succeeding interglaciations were represented in the coal basins by marine flooding of the coal swamps, reversal of clastic sedimentation by removal of vegetation and later by formation of marine limestones; return of the succeeding glaciation might in some cyclothems be represented by additional clastic deposits. Although the Pennsylvanian and Permian Periods are known to have been marked by glaciations in the southern Hemisphere, only about 11 tillite zones at the most have been recognized there (Weller, 1956). The main glaciations and interglaciations possibly are represented by the megacyclothems of the midcontinent United States while the cyclothems may represent stades (glacial substages) and interstades that are not easily distinguishable in the coarse tills of the southern glacial deposits. The glaciation theory, perhaps in modified form to account for abundant fine clastic material in the marine hemicyclothem seems to the present writer to be the most reasonable one so far proposed.

Other concepts of cyclothem origin include a Plant Control Theory (Robertson, 1948) based on the periodic building and breaching of lagoonal bars and natural levees bordering peat swamps; a Precipitation Control Theory (Brough, 1928) which involved climatic changes in a subsiding depositional basin; a Compaction Control Theory (van der Heide, 1950) based on different rates of compaction of clastic sediments and peat. Thaidens & Haites (1944) did not believe the Dutch coal measures are cyclical, but thought that peat growth was continuous through the productive Carboniferous but was at times interrupted by variations in erosion and deposition.

Certain types of cyclothems involving coal beds have been reported from the Mississippian (DeWitt, 1968), Permian (Elias, 1937), Cretaceous (Young, 1957) and Miocene (Crispin *et al.* 1955) but whether these are basically different from other onlap–offlap sedimentary cycles is not clear. That of the Miocene of Bataan Island, Philippines, at least, represents alternating marine and non-marine units separated by coal which suggests some unusual control on sedimentary processes in that area.

10.5 CHEMICAL CONSTITUENTS OF COAL, LIGNITE AND PEAT

In normal or humic coals, derivatives of woody tissues of plants predominate. The predominant constituents of abnormal or sapropelic coals are material other than woody tissues and include waxes or spore exines in some brown coals or cannel coals and oily algal remains in boghead coals.

Ultimate analyses of several German peats show the following variations (from Francis, 1961; Hausding, 1921):

C	$51 \cdot 13\%$	$\xrightarrow[\text{maturity}]{\text{increasing}}$	$58 \cdot 48\%$
H	$6 \cdot 05\%$		$5 \cdot 64\%$
N	$1 \cdot 83\%$		$2 \cdot 34\%$
O	$40 \cdot 99\%$		$33 \cdot 54\%$

Rational analyses of Scottish moss peats have the following ranges (from Francis, 1961 after Fraser, 1943): cellulose and related carbohydrates, $20 \cdot 98\%$; hemicellulose and related carbohydrates, $9-20\%$; lignin, $22-53\%$; protein, $1 \cdot 5-4 \cdot 2\%$; alcohol soluble fats and waxes, $4-10\%$; water soluble material, $2 \cdot 4-4 \cdot 15\%$.

As peat alters to lignite carbon and hydrogen increase at the expense of oxygen, nitrogen and sulfur which are released as volatiles. The ultimate analyses of Victoria, Australia lignite range as follows (Francis, 1961): C, $66 \cdot 1-71 \cdot 8\%$; H, $4 \cdot 4-5 \cdot 8\%$; O,N,S, $23 \cdot 6-32 \cdot 5\%$. The carbohydrates and lignin of peat condense to a nearly insoluble complex (ulmin) as in Victorian lignite (Francis, 1961): resins, waxes, and hydrocarbons, $3 \cdot 9-7 \cdot 1\%$; cuticular plant remains, $0 \cdot 65-10 \cdot 4\%$; opaque matter, o-traces; fusain, (charcoal-like), o-traces, ulmin, $85 \cdot 5-95 \cdot 45\%$ (by difference).

As to the means by which lignin is formed by plants it is thought that coniferyl alcohol produced by glucosidases from plant carbohydrates is enzymatically dehydrogenated through a series of steps which involves *shikimic acid* and is transformed into a polycondensate, having the properties of lignins (fig. 130) (Freudenberg, 1933, 1950).

The coalification process of accumulated peaty material has remained as one of the incompletely understood geochemical phenomena. Van Krevelen (1961) has pointed out that the moisture content of peat and low rank coals remains high and their hydrophilic character does not alter until bituminous rank is reached. The change to hydrophobic character is thought to be due to the elimination of OH, COOH and other polar groups during progressive coalification.

The transition from lignites to bituminous coal and anthracite results in

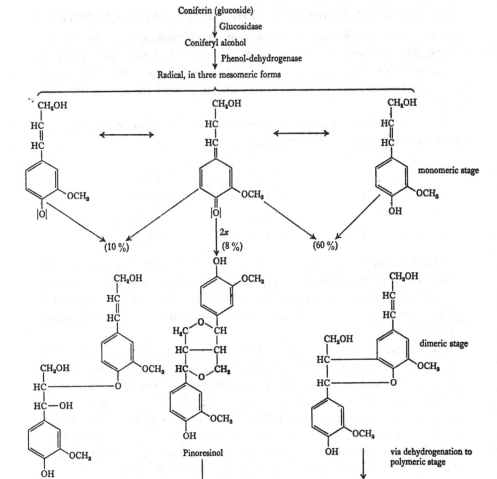

FIGURE 130   Biosynthesis of lignin as proposed by K. Freudenberg
(Van Krevelen, 1961, *Coal*, Elsevier, Amsterdam).

further increase in carbon and decreases in H, O and N, as shown in the
following ultimate analyses compiled from several sources:

	C	H	O	N
Low rank bituminous coal	79	5·45	14	1·55
Medium rank bituminous coal	85–88	5–5·2	7–5	1–1·75
High rank bituminous coal	91·7	4·5	2·2	1·6
Semi anthracite coal	92·3	4·2	2·0	1·5
Anthracite	93–97	3·7–0·6	1·9–1·8	1·4–0·6

Rational analyses of coal have the following general ranges: resins and hydrocarbons, 1–7·8%; inerts (structured plant remains), 1·3–15·3%; ulmins, 79·6–95·7%.

As to the chemical plant constituents that serve as progenitors of coal, carbohydrates are the most important and lignin ranks second. Of the carbohydrates, the structural polysaccharide *cellulose* is considered to be the principal source of coaly material, with other polysaccharides such as starch, pectin, alginic acid, chitin, and pentosans being subordinant in importance.

Lignin is a matrix or cementing substance that firmly binds the cellulose fibers in vegetable tissue. The origin and chemical structure of this dark amorphous material has not been fully worked out but it is believed that phenylpropane is a basic structural unit in lignin (Klason, 1908; Leger & Hibbert, 1938; Baker *et al.* 1948; Freudenberg, 1933; Freudenberg *et al.* 1955; Erdtman & Petterson 1950). Phenylpropane is related to coniferyl alcohol, a dimer of which, dihydroconiferyl alcohol is found in old spruce wood (Fraser, 1955).

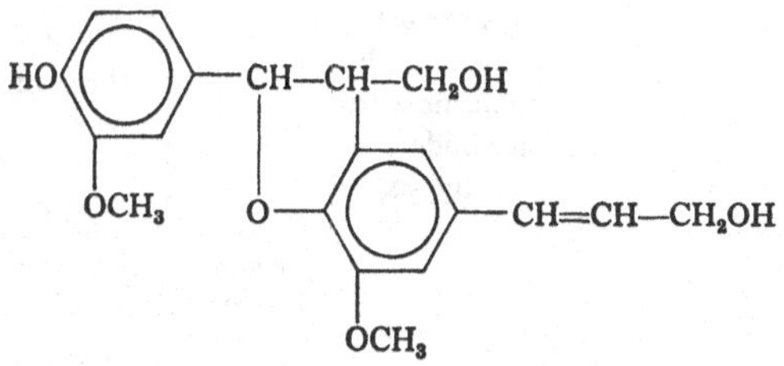

Dihydroconiferyl alcohol

The composition of lignin appears to vary according to the plant species (van Krevelen, 1961); for example, rather than coniferyl alcohol, the lignin of deciduous trees is said to be derived partly from sinapin alcohol which has one more methoxyl (—OCH₃) group than coniferyl alcohol.

The Maillard Reaction (1911) by which brown humic substances can be formed by condensation reactions of sugars and amino acids has influenced ideas as to coalification processes. Enders (1943), however, considered that methylglyoxal was an important intermediate product in degradation of carbohydrates (fig. 131) under conditions (presumably anaerobic) less favorable for the metabolism of micro-organisms. Under normal aerobic conditions carbon dioxide should be the normal product of carbohydrate degradation, while under anaerobic conditions some methane might be produced by certain organisms (Liebmann, 1950):

$$C_6H_{12}O_6 \xrightarrow{\text{\textit{Methanosarcina}}} 3CO_2 + 3CH_4 \tag{111}$$

The coalification theory of Enders seems to be a reasonable one on the basis of present knowledge and is intermediate between the more extreme

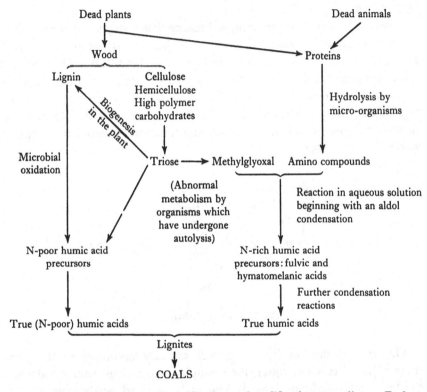

FIGURE 131 Mechanism of humification and coalification according to Enders (Van Krevelen, 1961, *Coal*, Elsevier, Amsterdam).

views of the lignin theory of Fischer & Schrader (1922) and Waksman (1938) and the carbohydrate theory of Maillard (1911), Marcusson (1925–7) and Hilpert & Littman (1934).

Further work by Flaig & Schulze (1952) has shown that phenols and probably lignin also may be altered to humic substances by first the formation of oxyquinone followed by rupturing of the rings to produce reactive keto acids which in turn react to form condensed ring systems.

342

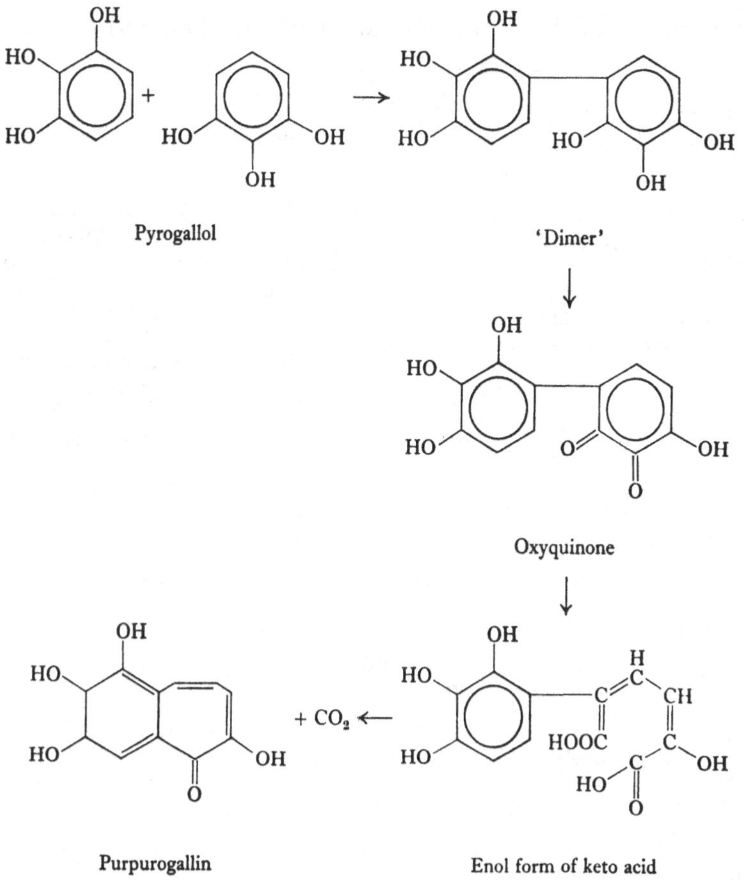

Pyrogallol   'Dimer'

Oxyquinone   (112)

+ CO₂ ←

Purpurogallin   Enol form of keto acid

Methylglyoxal, conversely, may condense readily to yield aromatic quinones:

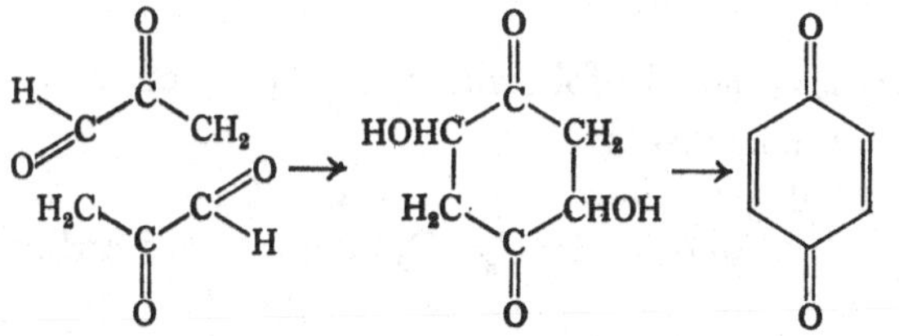

(113)

From this information it appears that either cellulose or lignin or both can easily participate in the coal-forming process.

10.6 RECENT STUDIES IN GEOCHEMISTRY OF COAL AND COAL-FORMING MATERIALS

The degradation of lignin by whiterot fungus *Polystictus versicolor* resulted in a number of phenolic compounds as shown in fig. 132 (Flaig, 1966).

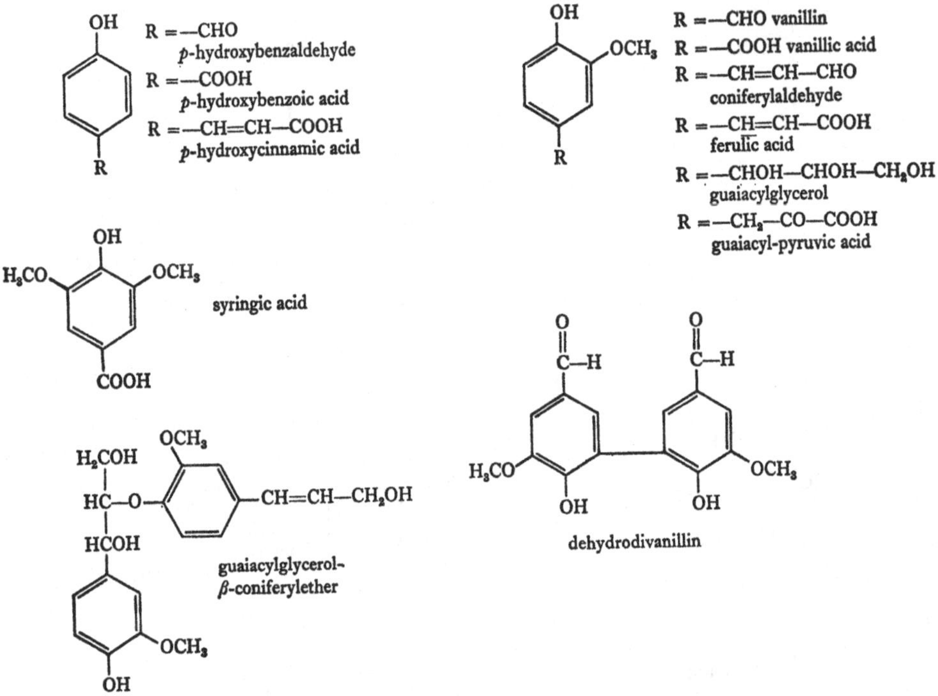

FIGURE 132 Degradation of lignin by whiterot fungus (Flaig, 1966, *Coal Science*, Advances in Chemistry Series no. 55, pp. 58–68, fig. 1, p. 60).

According to that author the following rather involved steps in the decomposition of lignin take place:

(1) Enzymatic degradation of lignin to C_6—C_3 or C_6—C_1 compounds; the latter may arise also from the C_6—C_3 by loss of two carbon atoms in the side chain.

(2) The foregoing compounds may partly repolymerize under oxidizing conditions with loss of carboxyl groups as CO_2 to polymers having a different composition than lignin. There is also some formation, separately, of aliphatic compounds from the lignin.

(3) Demethylation of their corresponding methyl esters leads to the formation of *o*-diphenols or vicinal triphenols; these in turn can form

TABLE 90   *Average organic affinity of some metals as determined by sink-float methods (Zubovic et al. 1961)*

Element	Per cent organic affinity	Per cent organic association
Germanium	87	100
Beryllium	82	75–100
Gallium	79	75–100
Titanium	78	75–100
Boron	77	75–100
Vanadium	76	100
Nickel	59	0–75
Chromium	55	0–100
Cobalt	53	25–50
Yttrium	53	n.d.*
Molybdenum	40	50–75
Copper	34	25–50
Tin	27	0
Lanthanum	3	n.d.*
Zinc	0	50

* n.d.—not determined.

o-benzoquinone and p-benzoquinones, thus adding to the copolymer formation and cross-linking of step 2.

(4) The resulting polymers contain an increasing proportion of aromatic structure, and increasing degradation of the side chain.

The free radical content of humic acids from soils and lignite as examined by EPR spectrometry (Steelink, 1966) and found to be about 10^{18} spins per gram. When the samples of humic acid were converted to their sodium salts the spins were increased by factors of 2–75 (table 89). The radical anion of humic acid can apparently be stabilized by adsorption in a basic solution; according to Weber *et al.* (1965) basification of a naphtho-quinone-naphtho-hydroquinone system results in a spin-content increase. The increase in spin content on basification of humic acid is postulated as being caused by a quinhydrone moiety in the humic acid macromolecule.

The relationship between chelate stability and organic affinity of minor metallic elements in coal was investigated by Zubovic (1966a) on sink-float separates of coal. The average organic affinity of some metals as determined by sink-float methods is shown in table 90 (Zubovic *et al.* 1961). In the decay of plant materials, chlorophylls, amino acids and lignin derivatives are stable, complex-forming ligands that would produce stable chelates. Chlorophylls and amino acids, having nitrogen as the donor element, would be good chelating agents for trivalent vanadium, nickel, copper and perhaps iron. Lignin, in which oxygen is the donor element, would complex with beryllium, germanium, gallium, titanium, cobalt, aluminum and silicon. One of the important possible chelating agents derived from lignin is the 4-hydroxy-3-methoxyphenyl group (guaiacyl group). The chelate

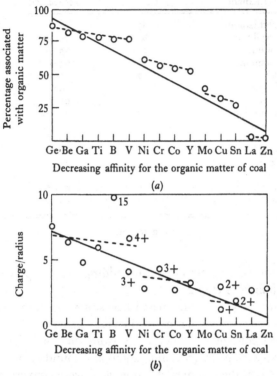

FIGURE 133 Relationship between minor metallic elements in coal and their ionic potential (Zubovic, 1966*b*, *Coal Science*, Advances in Chemistry Series, No. 55, pp. 221–31, fig. 1, p. 223, Table 1, p. 222).

stability of bivalent metals is Be > Cu > Ni > Co > Zn > Fe, which except for copper is the same as that of the organic affinity of the metals in coal. Gallium, yttrium and lanthanum have the same order of organic affinity and chelate stability. The copper may have been in a reduced state in the coal swamp or may have been precipitated as a sulfide. A relationship between organic affinity of the elements and their ionic potential can be demonstrated (fig. 133). In a further study of the minor element distribution in coals of the Interior Province, United States (Zubovic, 1966*b*) observed the following. Of 13 minor elements in 54 coal samples, beryllium, boron, germanium and gallium are concentrated near the source areas of a coal basin. These are metals that form small, highly charged ions, having high-ionic potential. The other elements show variable distribution patterns that may be caused by geological transport factors. Anomalous increases in minor elements occur in weathered coal samples.

The presence of humic acids in brown coals and their absence in higher-rank coals has been used to differentiate the two types. High-rank coals, however, will form humic acids while weathering (Kukharenko, 1959). Treatment of the brown coals with H_2O_2 yields a considerable amount of humic acids, while the younger high-rank coals yield very little humic acid by oxidation with H_2O_2.

A geologically interesting consideration of the effects of weathering on bituminous coals in different stages of metamorphism shows that differences occur between high- and low-rank coals (Kukharenko & Ryzhova, 1960). The first stage of weathering includes addition of oxygen in the form of OH and COOH, followed by formation of hydrolytic bonds at the periphery of the macromol and between its structural links. The second stage, with increased intensity of weathering, is marked by complete hydrolysis of the coal macromol bonds, and the evolution of humic acids, individual organic acids and gaseous substances. In the third stage oxidizing hydrolytic destruction of the humic acids proceeds to form simple, H_2O-soluble, individual organic acids and gases. In low rank bituminous coals, breaking of C—C bonds proceeds along the periphery of the macromols with rupture of side chains and aromatic rings and formation of humic acids. In higher-rank coals the hydrolytic action proceeds along the periphery of the condensed rings of the macromol with breaking of the aromatic links. At the same level of weathering, as measured by the amount of oxygen taken up, the higher the rank of the coal, the lower will be the yield of humic acid, the lower the coagulation threshold, and the higher the optical density of the alkali solutions.

It has been pointed out that the humic acids of weathered coals are products of oxidative destruction, whereas those of peats and brown coals are thought to be largely products of biochemical decomposition of plant residues (Kukharenko & Ekaterinina, 1960). The former have no methoxyl groups and have lower content of hydrogen, higher content of carboxyl groups, lower percentage of phenolic OH, higher optical density and a lower coagulation threshold of their alkaline solutions than the second type of humic acid.

Portions of humic-acid preparations from coals can be separated by ion-exchange and paper chromatography (Hartley & Lawson, 1962). Ether-soluble fractions of subhumic acids, extracted from humic acids with H_2O_2, were separated by use of columns of anion exchange resin and a specially designed technique of paper chromatography. Small portions of less than 0·1 % of the carbon of the original humic acids consist of glycolic, succinic and active tartaric acids. Mellitic acid was also detected. Succinic acid and the malonic acids also found in ether-soluble subhumic acid may have arisen directly from aliphatic groupings in the original coal structure.

The humic acids of oxidized coals and asphaltenes were found to differ from each other and from those of unoxidized materials (Larina & Kasatochkin, 1965). The humic acids of oxidized asphaltenes had higher infrared absorbance than those of oxidized coal or unoxidized asphaltene. Infrared spectra showed that the oxidation proceeded at the side chains attached to the aromatic nucleus. The oxidized-asphaltene humic acids were divided by electrophoresis into a stationary fraction and one that moved toward the anode. The structure of the humic acids is believed to be monomeric, with condensed aromatic nucleus and with side chains of oxygen-containing mainly carboxylic and phenolic hydroxyl groups. The principal difference between the two kinds of amino acid preparations lies in the side chains attached to the aromatic nucleus.

In twenty-two samples of Czechoslovakian lignite the quantity of humic acids decreased with increasing degree of coalification (Hubacek & Lustigova, 1962). In a related study Diaconescu & Doma (1960) found that younger coals contain more humic acids than older coals and had better ion-exchange capacity than the older coals.

The fixation of radioactive elements on humic acids from peat and lignite involves bonding of uranium on humic acid in a ratio of 1 atom of uranium to 2 monomers of humic acid (Jorscik *et al.* 1963).

Leonardite, a naturally occurring partly oxidized lignite with high humic acid content, can be used as an iron-ore pellet binder when treated with alkaline solutions, such as NaOH, to increase its contents of water soluble humic compounds (Fine & Wahl, 1964). The green strength of leonardite-bound iron ore pellets is less than that obtainable with bentonite but can be improved by adding a little $Ca(OH)_2$ to the leonardite-iron ore concentrate mix.

The factors influencing the accumulation of trace elements in coals were stated to be chemical and physical sorption (Bouska & Zdamek, 1963). Humic acids and other humus materials are most prominent in absorption of trace elements because of their high sorption activity. Time is the next most important factor. The source of trace elements in Cenomanian coals from Bohemia and Moravia was suggested to be granitic rocks that had been decomposed by intensive kaolinization.

The relationship between metamorphism, coal components, humic acid and other properties of coals was studied by Bogolyubova (1959). Where the early-diagenetic oxidative 'gelling' of the coal reaches an advanced stage before deep burial there is an increase in humic acid content and in optical density and a lowering of the coagulation limit. Coals that are comprised of non-transparent and semi-transparent components have a high content of humic acid (up to 70%), high optical density and low coagulation limit. Fusains, on the other hand, with thin-walled cells have a low humic-acid content (2–3%).

Infrared analyses of decarboxylated North Dakota lignite (leonardite variety) and humic acids (Falk & Smith, 1963) indicated that little or no H_3O^+ or carboxylate ions were present but that non-ionized carboxyl groups occurred as shown by bands at 2,600, 1,712 and 1,225 cm^{-1}; more in humic acid than in coal.

10.7 CARBON RATIO THEORY. HILT'S RULE AND SCHURMANN'S RULE

A prevailing concept of the changes in grade of coal from peat to lignite to bituminous and to anthracite is that this takes place by progressive increase in sedimentary overburden and in geothermal gradient.

The carbon ratio theory (Rogers, 1860, 1863; White, 1915, 1933; and Thom, 1934) attempts to show a relationship between the grade of coal and amount and nature of associated oil and gas deposits. According to the theory oil and gas occur more or less unaffected by metamorphic processes in areas where lignite occurs in associated strata, and as the effects of earth temperatures and pressures on the sedimentary rocks of a region increase the percentage of fixed carbon increases in the coals and the oils become lower in specific gravity (higher in API gravity). The carbon ratios of an area are calculated by dividing the per cent fixed (uncombined) carbon in the coal by the weight of coal on a dry weight, ash-free basis. The resulting values are plotted on maps and contoured by *isocarb* lines. The carbon ratios of the coals have a generalized relationship to petroleum deposits shown in table 91 after Fuller (1920), but many exceptions occur.

A relationship between petroleum occurrence and carbon ratios can be shown in the Pennsylvanian rocks of the Appalachian and Midcontinent regions of the U.S.A. and in the Lower Cretaceous regions of Saxony (Teichmuller, 1958). In the latter area, the grade of metamorphism of the coal can be conveniently related to its reflectance of light which increases with increasing grade.

In several coal producing areas it has been established that the dryer the coal the less will be the volatile content of the coal. This was observed by Hilt in 1873 in coalfields of south Wales, Pas de Calais and the Ruhr.

TABLE 91 *Relationship between carbon ratios of coal and*
presence of petroleum deposits (Fuller, 1920)

Carbon ratios (surface)	Production
Over 70	No oil or gas with rare exceptions*
65–70	Usually only 'shows' or small pockets No commercial production*
60–65	Commercial pools rare but oil exceptionally high grade when found Gas wells common but usually isolated rather than in pools
55–60	Principal fields of light oils and gas of the Appalachian fields
50–55	Principal fields of medium oils of Ohio–Indiana and mid-Continental fields
Under 50	Fields of heavy coastal plain oils and of unconsolidated Tertiary or other formations

* Lower Devonian gas fields of Appalachians are exceptions.

Hilt's Rule is applicable in various other parts of the world but there may be divergences due to variations in the type of coal at different levels or in effects of differential earth movements.

According to Schurmann's Rule (van Krevelen, 1961), the water content of lignites decreases with increasing depth in mines or boreholes. The water and oxygen contents of lignites generally correlate.

10.8 PETROGRAPHY OF COAL

The petrographic classification of coal components (macerals) has proved to be useful in recognition of the origin of the coal. The terms *anthraxylon* and *attritus* are applied to the bright components consisting of pieces of wood turned to coal, and to the duller components consisting of a miscellaneous assortment of debris, respectively. These substances form the bright and dull bands of typical bituminous coal but also are present in coal of all ranks.

A terminology of macroscopic ingredients of coal was proposed by Stopes (1919) in Europe and is still widely in use there:

Vitrain. Consists of uniform brilliant black bands consisting of the following macerals: vitrinite with or without structure; collinite if structureless; tellinite if structured.

Fusain. These are charcoal-like layers, the macerals of which consist of fusinite having opaque cell walls, and either empty cavities or mineral-filled cells.

Clarain. Represents bright coal, laminated, and with bands of brilliant and duller material; macerals consist of vitrinite, resinite (resin bodies), exinite (cuticles and spores), and smaller quantities of micrinite (granular opaque matter) and fusinite.

Durain. This consists of dull and non-reflecting coal in which laminations are poor or absent; the macerals comprise mainly micrinite, exinite and resinite with a little vitrinite.

Boghead. A sapropelite in contrast to humic coal constituents listed above; dull brown in color conchoidal fracture; alginite a major component.

Cannel. This sapropelite is dull black in color, burns with a long steady flame; composed of micrinite in which large quantities of microspores occur.

In American coal–maceral nomenclature the following are used (Thiessen, 1920):

Anthraxylon. This consists of vitrain.

Translucent attritus. A category referring to humic or to cell-wall material, 14 μm wide and containing spores, pollen, cuticle, fungal sclerotia and hyphae, and resinous bodies.

Opaque detritus. Comprises fusain, particles 30 nm wide, amorphous matter and discrete granules 0·5–1·5 μm wide.

Fusain. Includes mineral charcoal, black and opaque to dark red. A set of petrographic terms has been established (Van Krevelen, 1961) for coal microlithotypes based upon microscopic studies of polished surfaces or thin sections.

Vitrite: (micro-vitrain), vitrinite content 95%.

Clarite: (micro-clarain), vitrinite–exinite mixture having a small amount of resinite, and less than 5% inertinite (fusinite + semi-fusinite + sclerotinite).

Fusite: (micro-fusain), more than 95% fusinite, semi-fusinite and sclerotinite.

Durite: (micro-durain), mixed exinite and micrinite and 5% vitrinite; called pseudo-cannel if exinite is nearly absent. Transitions between these varieties are called durovitrite, vitrodurite, duroclarite, etc. The European *v.* American nomenclature of coal constituents is summarized in table 92.

Petrographic continuity of coal-seam macerals was used by Koppe (1966) for correlation and interpretation of coal-swamp environments. As

TABLE 92 *European and American nomenclature of coal constituents (Van Krevelen, 1961, Coal, Elsevier, Amsterdam)*

European nomenclature			Correlation	U.S.A. nomenclature		
Maceral grouping	Macerals			Components		Constituents (component grouping)
Vitrinite	Telinite		⟷	Megascopic anthraxylon ($>500\,\mu m$)		Anthraxylon
			⤩	Attrital anthraxylon ($500-15\,\mu m$)		
	Collinite		⤨	Sub-anthraxylon ($15-3\,\mu m$)		Translucent attritus
				Humic matter ($<3\,\mu m$)		
				Light brown matter		
Exinite	Resinite		⟷	Red resins		
				Yellow resins		
	Cerinite		⟷	Amorphous wax		
	Sporinite		⟶	Spore coats		
	Cutinite		⟷	Cuticles		
	Suberinite		⟷	Suberin		
	Alginite		⟷	Algal bodies		
Inertinite	Massive micrinite		⟷	Dark brown matter		Opaque attritus
	Granular micrinite			Amorphous opaque matter		
				Granular opaque matter		
	Sclerotinite		⟷	Fusinized fungal matter		Petrologic fusain
	Semi-fusinite		⟷	Dark semi-fusain		
	Fusinite		⟷	Attrital fusain ($<500\,\mu m$)		
				Megascopic fusain ($>500\,\mu m$)		

TABLE 93 *Correlation between coal layers and maceral types in Freeport Coal, Pennsylvania (Koppe, 1966, Coal Science, Advances in Chemistry Series, No. 55, 71–2)*

Bed		Site				
		1	2	3	4	5
Upper	Vitrinite	78·1	77·0	§	73·7	79·0†
	Exinite	2·8	2·7		5·2	3·4
	Micrinite	8·5	6·8		12·1	6·8
	Fusinite*	9·6	13·8		9·0	10·8
Lower	Vitrinite	89·8	90·7	89·6	89·6	92·4‡
	Exinite	2·5	1·3	1·2	3·1	2·0
	Micrinite	3·7	2·3	2·1	2·0	2·3
	Fusinite*	4·1	5·8	7·1	5·3	3·3

* Includes semifusinite.
† Probable correlative of upper bed in the Houtzdale quadrangle.
‡ Average of eight columns.   § Absent.

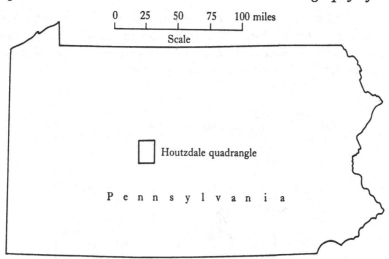

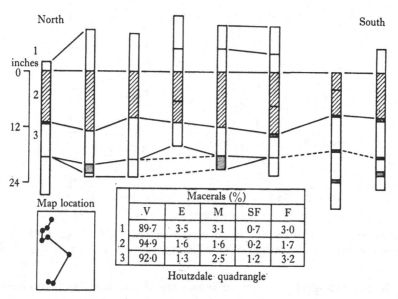

	Macerals (%)				
	V	E	M	SF	F
1	89·7	3·5	3·1	0·7	3·0
2	94·9	1·6	1·6	0·2	1·7
3	92·0	1·3	2·5	1·2	3·2

Houtzdale quadrangle

FIGURE 134 Petrographic continuity of coal-seam macerals in western Pennsylvania (Koppe, 1966. *Coal Science*, Advances in Chemistry Series, No. 55, pp. 69–79, figs. 2, 3, p. 72, Table 1, pp. 71–2).

shown in fig. 134 and table 93, relatively good and uniform correlation of layers and of maceral types occur in the Lower Freeport coal, northwestern part of the Houtzdale quadrangle, Clearfield County, Pennsylvania, which lies in the Clearfield Syncline. On the east and south the

Organic geochemistry of non-marine coal

correlations become uncertain and the maceral types are more variable in thickness and distribution. The explanation for these changes may be localized erosion of part of the coal sequence or discontinuity of the coal-swamp basins, or a combination of these factors.

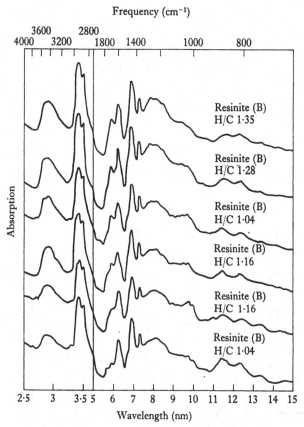

FIGURE 135   Infrared spectra of resinites from coal (Murchison, 1966. *Coal Science*, Advances in Chemistry Series, No. 55, pp. 307–31, figs. 2, 8, p. 322).

The infrared spectra of resinites from coals of various ranks (Murchison, 1966) are in two contrasting groups that are tied only by a transitional spectrum of thermally metamorphosed resinite. The spectra of resinites from bituminous coal have strong non-aromatic CH absorption, weak carbonyl absorbance and weak aromatic bands. Resinites from lower-rank coals have intense aliphatic- or alicyclic-group absorption, a strong carbonyl band, and no $1{,}600$ cm^{-1} absorption (figs. 135, 136). Transitional spectra of heat-treated resinite show a reduction in aliphatic groups and an increase in structureless absorption.

354

Further work on the structure of exinite, vitrinite and micrinite was done by Tschamler & DeRuiter (1966). They found that exinite has the largest aliphatic position and the highest relative amount of CH_2 groups, whereas micrinite has the smallest aliphatic position and the lowest relative amount of CH_2 groups; exinite has the smallest aromatic clusters and micrinite the

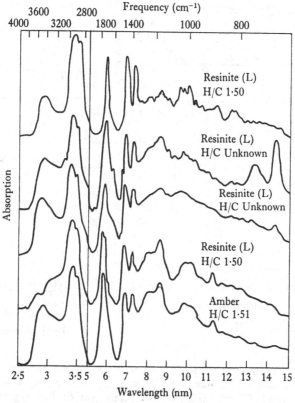

FIGURE 136   Infrared spectra of resinites from coal (Murchison, 1966 *loc. cit.*)

largest. Non-aromatic rings can be demonstrated in the macerals of which part should be hydroaromatic or alicyclic.

The electron spin resonance of coal macerals shows marked differences in north American and British bituminous coals of different ranks (Austen *et al.* 1966). In the case of vitrinites the unpaired spin concentration increase with increasing rank of coal and rises sharply above 90% C. Values for spore-rich exinites lie below those for vitrinites and those for fusinites are much higher and the half-peak width of the signal is much narrower than the other macerals (fig. 137). When the other macerals are pyrolyzed to

550 °C under nitrogen the spin signals increase whereas those of fusinite change very little. The data are used in support of the concept that the woody tissues of fusinites were exposed to moderately high temperatures (i.e. forest fires).

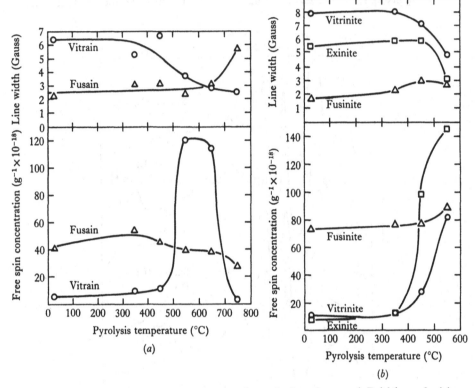

FIGURE 137 Electron spin resonance of north American and British coals (*a*) Fusain and Vitrain. (*b*) Vitrinite, Exinite and Fusinite (Austen *et al.* 1966. *Coal Science*, Advances in Chemistry Series, No. 55, pp. 344–62, figs. 4, 5, pp. 353, 355.)

10.9 STRATIGRAPHIC DISTRIBUTION OF COAL, LIGNITE AND PEAT

Moore (1940) discusses the individual coal deposits of the world as they were known up to about 30 years ago. Since then a large amount of work has been done by the U.S. Geological Survey on American coal deposits, and their publications should be consulted for details.

In table 94 a listing is given of those coal deposits that have some feature of particular geochemical interest.

In a study of coal samples of Permian age from Antarctica, Shopf & Long (1966) found most of the coal to be high rank and non-caking. The

TABLE 94　*Coal deposits having features of geochemical interest*
(data from Moore, 1940 and other sources)

Location	Age	Type of coal	Geological or geochemical features
Nova Scotia, Joggins Area	Pennsylvanian	Bituminous, high ash, 3–5 ft thick	Numerous buried Carboniferous trees
Nova Scotia, Stellarton area	Pennsylvanian	Bituminous, up to 45 ft thick, mining to depths of 3750 ft	Oil shale associated with coal beds, formerly mined for extraction of oil
Yukon Territory, Yukon River, Cliff Creek and Coal Creek areas	Cretaceous and Tertiary	Bituminous coking coal (Cretaceous) and lignite (Tertiary)	Tertiary lignites have high resin content
Parry Islands, Arctic Archipelago	Lower Carboniferous	Bituminous coal up to 50 ft thick	Cannel coal and oil shale are associated with the coal bed
Newfoundland, Humber River Valley	Pennsylvanian	Bituminous and semi-bituminous	Oil shale and albertite-like material lie beneath the Pennsylvanian coal beds, and are pre-Penn. in age
Pennsylvania, Bernice Field, Sullivan County	Pennsylvanian	Semi-anthracite	Marks transition from anthracite fields on the east to the bituminous fields on the west and helps substantiate carbon ratio theory
Pennsylvania, Broadtop Field, Huntingdon and Bedford Counties	Pennsylvanian (Allegheny Series)	Semi-bituminous	Is transitional between anthracite and bituminous fields and helps document carbon-ratio theory
Pennsylvania, Kittanning Coal	Pennsylvanian (Allegheny Series)	Bituminous	Cannel-coal present in large amounts in Clearfield and Beaver Counties
Pennsylvania, Pittsburgh coal	Pennsylvanian (Monongahela Series)	Bituminous	Famous coking coal in Connellsville District, and gas-making coal in Westmoreland County (now largely exhausted)
Pennsylvania, Mercer coals	Pennsylvanian (Pottsville Series)	Bituminous	Shows westward facies from *in situ* origin in Clearfield County to transported origin in western Pennsylvania approaching old shoreline
Rhode Island	Pennsylvanian	Graphite	Formed by metamorphism of coal, up to 30% ash; helps to support carbon-ratio theory
Ohio, Upper Mercer Coal	Pennsylvanian (Pottsville Series)	Bituminous	'Bedford Cannel' coal in Coshocton County represents upper 5 ft of 9 ft Mercer Bed
West Virginia Pittsburgh, Upper Freeport and Upper Kittanning	Pennsylvanian (Monongahela and Allegheny Series)	Semi-bituminous	Important smokeless coals
West Virginia, Cedar Grove and Stockton Coals	Pennsylvanian (Pottsville Series)	Bituminous	Coal beds contain cannel layers

TABLE 94 (*cont.*)

Location	Age	Type of coal	Geological or geochemical features
Virginia, Triassic Basin	Triassic	Bituminous to semi-anthracite	Coal locally changed to natural coke near basic igneous intrusions
Virginia, Frederick and Pulaski Counties	Mississippian (Pocono)	Semi-anthracite to anthracite	Oldest coal mined in the U.S.A.
Kentucky, south-south-eastern fields	Pennsylvanian (Pottsville Series)	Bituminous	Cannel coal in several of the coal seams
Missouri, east of Mendota Field	Pennsylvanian (Des Moines Series)	Bituminous	Isolated pockets of coal up to 80 ft thick, containing cannel layers; possibly related to sink hole topography
Oklahoma, Cherokee Field	Pennsylvanian	Bituminous	Considerable natural gas associated with 10 coal beds
Texas, band extending from Sabine River to the Rio Grande	Eocene	Varies from lignite to sub-bituminous	Lignite consists of tree trunks on the south and of spores, seeds and related vegetal matter on north; changes to bituminous coal approaching Sierra Madre fold-belt in Mexico
New Mexico, Cerillos and other fields	Cretaceous	Sub-bituminous	Coal locally metamorphosed by igneous intrusions
Colorado, Walsenberg District	Upper Cretaceous (Mesa Verde)	Bituminous	Coal intruded by igneous rocks to form spherical 'nigger-head' masses of natural coke, having concentric structure
Colorado, Trinidad Fields	Upper Cretaceous (Laramie)	Bituminous	Natural coke or carbonite formed by sills of igneous rock
Colorado, Pilot Knot area	Upper Cretaceous (Mesa Verde)	Sub-bituminous	Anthracite and natural coke produced near igneous intrusions in Anthracite Range
Colorado, Unita Basin, Crested Butte area	Upper Cretaceous (Mesa Verde)	Sub-bituminous and bituminous	Anthracite formed locally by igneous activity
Wyoming, Powder River Basin	Paleocene (Fort Union)	Lignite	Beds 40–90 ft thick at Gillette, are among thickest in U.S.A.
Washington, Pierce County	Eocene	Bituminous, semi-bituminous	Altered to anthracite near igneous intrusions
Mexico, Santa Clara Field	Triassic	Anthracite and semi-anthracite	Affected by igneous intrusions
Panama, interior district	Tertiary	Lignite	Sulphur content of lignite reported to increase toward volcano Chiriqui
Columbia, Cacagual District	Cretaceous	Bituminous	A laminated cannel-like coal that burns with a glow rather than a long flame like true cannel

TABLE 94 (*cont.*)

Location	Age	Type of coal	Geological or geochemical features
Peru, Cajabma District	Cretaceous and/or Tertiary	Anthracite	Unusually large reserves of anthracite for rocks this young
Argentina, Mendosa Province	Permo-Carboniferous	Albertite	A hydrocarbon derived from petroleum occurring in beds that resemble coal
Wales, south Wales Field	Coal Measures	Ranges from bituminous to anthracite	Problematical variation in grade in coal from anthracite on north and west to bituminous on south and east, perhaps due to intensity of folding-pressure
Ireland, Tyrone Field	Coal Measures	Bituminous	Cannel coal in Gortnashea seam 22 in thick
Ireland, Ballycastle Field	Lower Carboniferous	Bituminous	Coal affected by dolerite intrusions
England, several fields	Coal Measures	Bituminous	'Black band' layers of siderite, a primary deposit in some coal beds
England, Coalbrookdale and Dudley Fields	Coal Measures	Bituminous	Coal beds intruded by igneous rocks
Scotland, Torband Hill	Coal Measures	Boghead Coal	Torbanite, a deposit rich in *Botryococcus*, waxy algae
France, Commentary Basin	Coal Measures (Stephanian)	Bituminous	Large boulders, fossil trees at odd angles, coal pebbles in conglomerates, indicate terrestrial stream or glacial origin
Switzerland	Carboniferous, Jurassic, Eocene, Pleistocene	Lignite and bituminous	Locally changed by folding into anthracite and graphite
Belgium, Namur Field	Lower Coal Measures (Namurian)	Bituminous	Contains some cannel; about half coals are of coking type; a coal mine 3,900 ft deep occurs in this basin
Faeroe Islands and Iceland	Tertiary	Lignite	Locally changed to anthracite by igneous intrusions and lava flows
Germany, Lower Silesia	Lower and Upper Carboniferous	Bituminous	Igneous intrusions produce local deposits of natural coke; gas abundant in mines; cannel coal occurs
Germany, Bavaria	Oligocene	Lignite	The deposit is of drift origin
Bosnia, near Sarajevo	Pliocene	Lignite	Deposits attain thicknesses of 10–20 m
Greece, near Keshan	Pennsylvanian?	Bituminous	A hard cannel coal
Russia, Moscow Basin	Lower Carboniferous	Bituminous	Contains boghead and cannel facies

TABLE 94 (*cont.*)

Location	Age	Type of coal	Geological or geochemical features
Norway, Andö Island	Jurassic	Bituminous?	High-ash cannel coal occurs
Manchuria, Chien-chin-chai section	Tertiary	Sub-bituminous to bituminous	Main coal 130–200 ft thick, with about 100 thin partings
Japan	Mesozoic	Bituminous	Altered to anthracite and natural coke near igneous intrusions
India, Raniganj, Giridih, and Jhenia Fields	Gondwana (Permo-Carboniferous)	Bituminous	Locally altered by igneous intrusions
Borneo, Brooketon, Sarawak	Tertiary	Lignite and bituminous	Some coal beds contain a large amount of resin
Madagascar, Ianapera Area	Carboniferous-Triassic (Karroo)	Cannel	Several seams reaching 8 ft thick
Australia, New South Wales	Permo-Carboniferous	Oil shale	Kerosene shale, resembles torbanite, contains waxy algae *Elaeophyton* (= *Botryococcus*), in leases up to 4 ft thick and 1 mile in extent
Australia, Victoria, Morwell area	Miocene	Lignite	Beds totalling 780 ft thick in 1010 ft of strata, in beds 266, 227 and 166 ft
Australia, Tasmania		Oil shale	Tasmanite, yields 40–50 gallons of oil/ton
Australia, Tasmania	Triassic	Bituminous?	Altered by igneous intrusions

coal is also characterized by unusually high-oxygen content. The authors suggest that the oxygen was added during widespread igneothermal metamorphism as well as perhaps a type of regional metamorphism by exposure to water or water vapor. Field data suggest that the coal had reached 'high volatile A bituminous rank' before metamorphism but infrared and electron spin resonance suggest that the coal had not gone beyond the lignitous or sub-bituminous rank before igneous metamorphism occurred (Given, in Schopf & Long, 1966). In another study of Antarctic coals Schapiro & Gray (1966) found that the high oxygen values of Antarctic coals are higher than those of commercial anthracites in the U.S.A. The alteration features (higher reflectance, hardness, lower electrical resistivity) near igneous sills can be used to judge proximity to the altering igneous mass. The drift origin of the Gondwana coals is supported in a recent study by Babu & Dutcher (1966). They found that the inert content of two Gondwana seams from Chirmiri, Central India is high and the macerals are of varying types which suggest a heterogeneous source and changing environments of origin and deposition. Electron microprobe analyses of the ash showed an association of iron and sulfur as sulfide that was not detectable by other means.

10.10   CHEMICAL STRUCTURE OF COAL

Of the four main components of coal (vitrinite, exinite, fusinite and micrinite) only the chemical structure of vitrinite has been studied extensively. The approach to study of coal structure has been by solvent extraction and comparison of the properties of the extracts. Some of the methods that have been used to study the fractional extracts of coal are infrared spectrophotometry, X-ray scattering technique, nuclear magnetic resonance, electron spin resonance, and mass spectrometry.

When coal is extracted with successive solvents up to 100 °C, benzene, acetone, acetophenone, an acetophenone–dimethylformaldehyde mixture, dimethylformaldehyde and ethylene diamine (the extract totalling 50–60% of the coal), the infrared spectra of all the extracts are closely similar, suggesting the same basic pattern of structure (Rybicka, 1959) other chemical reactions and X-ray scattering show similar results (Given, in Francis, 1961).

X-ray studies of vitrinites of less than 90% C show that the aromatic systems are small, consisting mainly of one to three fused rings. Each molecule in the coal contains many such systems of fused rings. Coal is believed to be comparable to a synthetic high polymer, to the extent that a given sample may contain many different molecules. The molecules may be quite different in molecular weight and several kinds of geometrical and positional isomers may be present. One of the molecules in the coal may provide, however, a satisfactory statistical sample of the entire coal.

Resins, hydrocarbons, alkyl and other derivatives of naphthalene, although they are important constituents of coal do not apparently contribute to the structure of vitrinite and form only about 1% of purified vitrinite. It appears that as much as 60–85% of the carbon in vitrinites is in the form of aromatic nuclei (Francis, 1961). The nuclei, based on the work of several different investigators consists of, on the average, up to three fused rings, represented principally by benzene, naphthalene, diphenyl and phenanthrene or of larger nuclei that contain these systems. One of the four nuclei naphthalene, diphenyl, phenanthrene and triphenylene predominates in individual vitrinites and the indene system also contributes in an important way.

The remainder of the carbon in vitrinite not in aromatic form, and amounting to 10–35% of the total carbon is believed to be arranged in hydroaromatic molecules. Most of the hydrogen in vitrinite is thought to be attached to carbon in aliphatic combination, only a small part of which, however, is present in methyl groups. The oxygen in vitrinites, and evidently in coal in general, is in the form of phenolic hydroxyl groups. Some of it is present in strongly conjugated carbonyl groups that are attached by chelating to the hydroxyls. The rest of the oxygen is thought to link aromatic nuclei in ethers or to occur in fused furan or pyran heterocyclic rings (Given, in Francis, 1961). About 80% of the vitrinite so organized chemically in aromatic and hydroaromatic rings is in molecules that range from about 300–400 up to several thousands molecular weight.

According to Dormans *et al.* (1961) and Kröger & Bürger (1959) exinites contain more hydrogen, less oxygen, and are less aromatic than vitrinites. Given *et al.* (1960) found the hydroxyl contents of spore-rich exinites to be less than in the associated vitrinites, and they probably contain a higher proportion of hydroaromatic structures. Although they probably have higher molecular weights than vitrinites (they are less soluble in organic solvents) they possess other properties in common with vitrinites and probably do not differ greatly from them in chemical structure (Given *et al.* 1960). Fusinites are marked by high-carbon content and may be primarily amorphous carbon rather than having a truly organic chemical structure (Brown, 1959). Micrinites appear to have some definite chemical structure despite their high-carbon content (Brown, 1959). They contain less hydrogen and about the same proportion of oxygen (Kröger & Bürger, 1959), have larger aromatic nuclei and seem to have fewer non-aromatics (Van Krevelen & Schuyer, 1957); their hydroxyl contents are intermediate between vitrinites and exinites.

11 Geologic history of non-marine organic matter

II.I INTRODUCTION

The solutions to problems of the first appearance of the organic compounds unique to living organisms and the possible evolutionary development of these compounds can only be inferred at the present time.

II.2 POSSIBLE EARLY DEVELOPMENT OF LIFE

Many recent papers have speculated on the origin of life through some sort of polymerization process involving abiogenically formed organic substances including lipids, amino acids, purines and other substances in early aquatic environments on the earth or even in preterrestrial bodies (Horowitz, & Miller 1962; Barghoorn, 1957; Fox, 1960; Urey, 1952). The synthesis of such substances under geologically plausible conditions from simple components such as methane, ammonia, hydrogen cyanide and water has been demonstrated (Miller, 1953, 1957; Oró & Kimball, 1961, 1962; Ponnamperuma et al. 1963; Ponnamperuma et al. 1963 a). The synthesis of biopolymers such as proteins, polysaccharides, chlorophyll, and nucleic acids is more difficult to explain other than by living organisms, but some progress has been made in their synthesis (Oró & Guidry, 1960; Oró, 1963; Ponnamperuma et al. 1963 b).

The organization of synthetic organic compounds into a living cell has not been accomplished, although nearly all the necessary components seem to have been made in the laboratory, including chromosomal nucleic acids. The so-called life force for the viable cell has remained elusive.

The writer suggests one possible source of the energy required for the living cell which dates from the early history of the solar system. The early nebulous stages of the solar system ($\sim 5 \times 10^9$ years ago) are thought to have been marked by activity of short-lived radioactive nuclides (mean lifetime 10^6–10^8 years) (Reynolds, 1963). An early heat source for the preterrestrial planetesimals of the solar system and/or earth may have been supplied by these nuclides.

One of the stable isotopes that has been found in several meteorites in excess of its predicted solar abundance is xenon-129. The excess is thought to have originated from iodine-129, an extinct radionuclide of so-called short life ($\tau = 2.4 \times 10^7$ years). The belief that the excess ^{129}Xe in Abee meteorite originated from [129]I was strengthened by irradiation experiments on [127]I from Abee meteorite (Jeffrey & Reynolds, 1961). The [127]I yielded [128]Xe, and [128]Xe and [129]Xe produced in a pile proved to have similar release patterns which indicates that [129]Xe is intimately connected with [127]I as presently located. Studies of other stone meteorites, Richardton and Bruderheim, show presence of lightly bound [127]I not correlated with [129]Xe which was interpreted by Reynolds (1963) to mean that there were originally two components of [129]Xe but the one associated with lightly bound iodine has been lost by diffusion. The K–Ar age of Richardton is 4.6×10^9 years. Calculations can be made (summarized in Reynolds, 1967) which suggest that the known amounts of excess [129]Xe in several chondritic meteorites formed from [129]I over a period of about 50×10^6 years, i.e. a remarkably brief time in the early solar system. The data strongly suggest that solid bodies were in existence in the primeval solar nebula while the early sun was forming.

Iodine is a constituent of thyroxine (α-amino-β [3,5-diiodo-4-hydroxyphenoxy] phenyl) propionic acid) and iodogorgoic acid (3,5-diiodotyrosine) both of which are protein amino acids; thus, they are among the amino acids derived by hydrolysis from present day living cell material. Furthermore iodine is alone among the 'short-lived' radionuclides in being a constituent of a biologically essential group of organic compounds. Aluminum 26 may also have contributed to the radiation energy involved in synthesis of primitive organic substances as it is thought to have been abundant in the primitive solar system, and is much more strongly radioactive than [129]I (V. R. Murthy, oral communication). Other extinct radioactive nuclides are [107]Pd (mean lifetime 9.8×10^6 years) which decays to [107]Ag, [205]Pb (mean lifetime 3.5×10^7 years) which decays to [205]Tl, and [244]Pu which decays by fission to heavy isotopes of xenon (mean lifetime 1.2×10^8 years).

Possible steps in the genesis and subsequent development of biopolymers involving [129]I are as follows:

(1) The early solar nebula contained [129]I, which has been produced in the galactic synthesis at a constant production rate Kr and a mean lifetime τ was synthesized over a duration of time T. Its abundance at the time of isolation of the solar system, sometime around $10\text{–}20 \times 10^9$ years ago was (Murthy, 1963):

$$[129]I = Kr \cdot \tau \, (l - e^{-T/\tau}). \tag{114}$$

The formation of radiogenic [129]Xe from [129]I is believed to proceed according to the following equation:

$$\Delta t = \tau_{129} \ln \left[\left(\frac{[127]I}{[129]Xe} \right) \left(\frac{K_{129}}{K_{127}} \right) \left(\frac{\tau_{129}}{T} \right) \right]. \tag{115}$$

TABLE 95 ^{129}I–^{129}Xe *decay intervals as determined from meteorites*

Sample	^{127}I (p.p.b.)	^{129}Xe (ml STP × 10^{-11}/g)
Abee	145	430
Indarch	270	250
Indarch	270	210
St Marks	70	170
Bruderheim	16	3·6
Richardton	28	10
Murray	230	22
Sardis troilite	3600	3·3

Using values $K_{129}/K_{127} = 1$, $\tau_{129} = 2\cdot4 \times 10^7$ years, and $T = 1 \times 10^{10}$ years based on a model of continuous nucleosynthesis of solar system material (Burbridge *et al.* 1957), the decay intervals calculated for some meteorites is shown in table 95.

(2) The trapping of excess ^{129}I-derived ^{129}Xe in stone meteorites at the time of their formation infers that the meteorites were cold enough to retain xenon. If some carbon, nitrogen, hydrogen and oxygen, and perhaps sulfur and phosphorus, all or in part, were also present with ^{129}I in meteoritic masses the possibility may have existed for the synthesis of hydrocarbons, amino compounds and carbohydrates and for the ^{129}I-induced radiation synthesis of polymers from the organic compounds. The synthesis of a number of nitrogenous as well as other organic compounds in meteorites in the early history of the solar system has been suggested (Studier *et al.* 1965; Hayatsu 1961; Fowler, 1962) on the basis of experimental evidence. The production of peptide and glycose bonds and perhaps the appearance of the 'life force' would have been accomplished at this stage. The time available for such biopolymer synthesis was apparently ample, 60–250 × 10^6 years judging from meteorite evidence (table 95). If the Sun grew within this time period or during the same general time interval the biopolymers would have been protected from the full force of its u.v. radiation and heat. Meteorite masses large enough to protect the contained matter from solar radiation and far enough away from the central Sun to avoid high temperatures would have provided the most favorable sites for development of biopolymers. Just which C, H, O, N and perhaps P and S constituents were available may not have been critical to the development of primitive prokaryotic cells. If something like ^{129}I radioactivity was responsible for the life force in the earliest 'organisms', the power of reproduction may not have been a requisite as long as the

radioactivity was available. At some time before the completion of ^{129}I–^{129}Xe isotopic mixing, however, the power of reproduction would necessarily have developed. If the growth of the Sun was as rapid as postulated by some (e.g. Hayatsu, 1961) the interstitial habitation in meteorites of the pro-karyotic organisms would have been necessary for their protection from

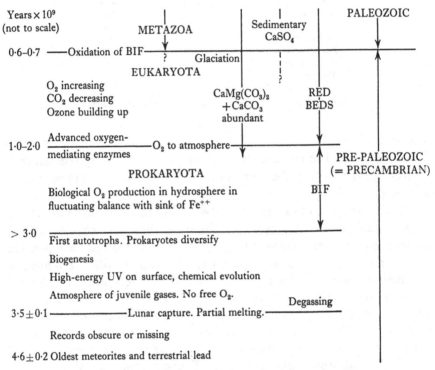

FIGURE 138 Suggested critical events in interacting biological, atmospheric and lithological processes on the primitive earth (Cloud, 1968). BIF, Banded Iron Formations.

extreme heat or cold and u.v. radiation. It is suggested that some of the organized bodies and organic materials found in meteorites may be fossil representatives of this early 'life'. Perhaps the most favorable position for development of the organic material was at mid-distances in the solar system such as the Earth–Mars–Asteroid–Jupiter zone.

(3) During the growth of the planets by accretion of solar system mater-ial, the early life forms would have participated in planetary growth. Presumably most of the life forms would have been destroyed by impact or by planetary remelting, but if only a tiny fraction of the cells survived, perhaps in the presence of steam a mechanism would have existed for the middle planets to become seeded with viable cells.

366

In a discussion of the development of life conditions on the primitive Earth (Cloud, 1968) suggested that the Moon was captured approximately $3 \cdot 5 \times 10^9$ years ago, or approximately 1×10^9 years after formation of the solid Earth. The presence of stromatolitic (algal-like) structures showing evidence of diurnal tidal variations $2 \cdot 5 - 3 \times 10^9$ years old are cited as in support of an orbiting moon that long ago. The pre-Huronian thermal-metamorphic episode that was widespread over the earth may also have been related to lunar capture. A major degassing of the Earth, that con-tributed to its early atmospheric envelope and permitted further develop-ment of primitive life, may also have occurred as a result of lunar capture. Some of these early events as summarized by Cloud are shown in fig. 138.

11.3 DEVELOPMENT OF TERRESTRIAL ORGANIC MATTER

A tabulation of the known ranges in geologic time, modified from Eglinton & Calvin (1967), or organic compounds is shown in fig. 139. Saturated fatty acids, saturated and aromatic hydrocarbons, amino acids, porphyrins and carbohydrates have been recorded from rocks $1 - 2 \times 10^9$ years old. Of these only carbohydrates have been recognized in biopolymeric form (Swain & Bratt, 1969), but the other substances probably also were repre-sented by or in polymeric compounds. Cellular organic structures that are found, presumably indigenously, in Precambrian rocks in Minnesota, South Africa and elsewhere must have included at least primitive forms of proteins as well as of polysaccharides. The primitive prokaryotic (un-nucleated) organisms may not have had other than simple glycans, simple albuminoid proteins, and protective oils. According to Rutten (1968) the level of atmospheric oxygen in the period $> 2 \cdot 7 \times 10^9$ years to $\sim 1 \cdot 45 \times 10^9$ years ago was only $0 \cdot 01$ PAL (present atmospheric level), but was sufficient to allow respiration to occur, i.e. the so-called Pasteur point had been passed. Prior to this period ($> 2 \cdot 7 \times 10^9$ years) the available oxygen was principally that formed by ultraviolet dissociation of atmospheric water, the so-called Urey mechanism, and was of the order of $0 \cdot 001$ PAL. At this level of concentration only fermentation by micro-organisms was possible.

Eukaryotic (nucleated) organisms seem to have appeared sometime after $1 \cdot 45 \times 10^9$ years in PAL of oxygen $\sim 0 \cdot 01$. Presumably most life prior to this time inhabited only aquatic environments where dissolved nutrients were available and where protective effects of the aqueous medium against heat and light could be utilized. Rutten suggests that by Ordovician time $0 \cdot 4 - 0 \cdot 5 \times 10^9$ years ago the oxygen level had reached $0 \cdot 1$ PAL. Metazoan invertebrates were well-established in the oceans by that time, but there is very little information on non-marine life of that time.

Geologic history of non-marine organic matter

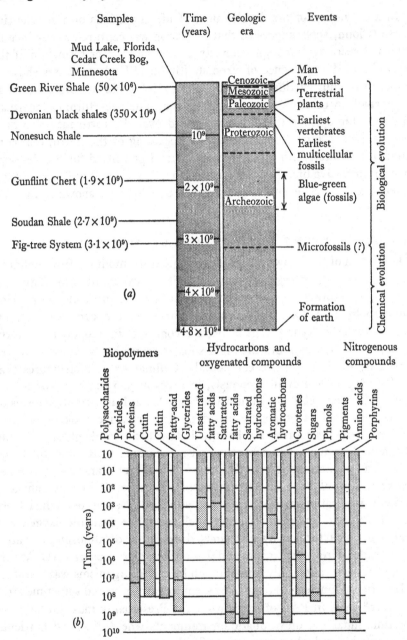

FIGURE 139 (a) Geologic time scale showing some sedimentary rocks studied for organic matter, and organic evolutionary scale. (b) Known distribution of some organic compounds in geologic time (Eglinton & Calvin, 1967, *Chemical Fossils*, Copyright by *Scientific American* Inc. All rights reserved.) (coarse stippling, uncertain occurrence).

In a study of the distribution of organic substances in marine and non-marine Paleozoic rocks of the Appalachian geosyncline the present writer (Swain, 1966) found evidence that organic productivity was relatively low on the lands surrounding the Appalachian seaway until the late Silurian or Devonian, as reflected in both the marine deposits receiving organic debris from the lands, and in non-marine deposits themselves. There is some evidence, however, that shore vegetation may have been an important if not dominant contributor to the organic residues of some of the deposits rather than strictly terrestrial sources. Hydrocarbons and carbohydrates in the marine and non-marine Ordovician and Silurian rocks of the area average less than 50 p.p.m. while that in the marine Devonian is more than 50 p.p.m. The non-marine Devonian and Mississippian rocks, mainly red beds, contain less than 50 p.p.m. The presence of primitive forms of terrestrial vegetation by the late Silurian or early Devonian seems fairly definite from knowledge of Devonian land plants as well as from the bio-geochemical evidence. What evidence there is for the kind of terrestrial vegetation in pre-Upper Silurian is based mostly on presence of *Sporongites*-like megaspores in the early Silurian Tuscarora Sandstone (unpublished data of the writer) of the Mount Union area, Pennsylvania, as well as of published and unpublished records of similar megaspores in marine pre-Devonian rocks. The available biogeochemical evidence from such terrestrial pre-Devonian deposits as the late Ordovician Juniata red beds of Pennsylvania (table 94) strongly suggests that small but significant amounts of lower plants occurred on the land surfaces in Ordovician time and perhaps earlier. These primitive plants may have included fungi, bacteria, algae and perhaps lichens. Organic soils other than those containing mechanical admixtures of organic matter may not have existed until soil microfaunas developed. There is no definite evidence as to when these appeared but presumably the atmospheric oxygen content would have needed to approach present levels for such faunas to have evolved.

Organic soils plausibly may have begun to develop by Devonian time with the inauguration of rooted larger terrestrial plants. These would have presumably required some organic nutrients from the soil and in turn would have supplied organic debris to the soil. These occurrences were very probably restricted to lowlands judging from the scarcity of Devonian plant remains other than in dark colored swampy deposits. Devonian and younger red beds, however, contain traces of sugars and amino acids indicating that organisms were present to a limited extent.

The restriction of larger terrestrial plants to lowland areas apparently continued through the Carboniferous Era. In large part the climate in which the larger plants grew was tropical to subtropical and with poor development of seasons, judging from the absence of growth rings in these plants. According to Seward (1959), although many of the Carboniferous plants grew in or near water, some of the trees probably were confined to higher and drier ground. The lack of flowering plants may indicate that differentiation of organic pigments on the lands had not reached the development that characterized late times. The colorful reconstructions of Paleozoic marine invertebrates do not have support as yet in biogeochemical studies. Triassic echinoderms, however, have been found to contain complex organic pigments. Extensive discussions of the Devonian Carboniferous and Permian floras are presented in textbooks of paleobotany and need not be repeated here. Lowland soils were presumably well-developed by the late Carboniferous.

The Triassic is marked by an important increase in the number of upland plants related on one hand to modern conifers, horsetails, cycads and ferns and on the other to Paleozoic cordaites and lycopods. Upland soils had perhaps begun to develop by the Triassic but the abundance of red beds in the non-marine Triassic as in earlier non-marine deposits suggests that organic soils in areas of moderate to steep relief were poorly developed. Reconstructions of Triassic and Rhaetic (late-Triassic to early Jurassic) landscapes show little ground cover. Organic-rich lacustrine deposits also appeared particularly in the Rhaetic but have not been studied biogeochemically.

The Jurassic Period represents a further development of upland plants and lacustrine organic deposits. Organic soils containing cycadean stems and other fossil plants are represented in the Portlandian (Upper Jurassic) of England. The non-marine Jurassic deposits are typified as much by partly varicolored mudstones as by highly oxidized red beds which probably reflects the increase in organic content of the soils. Amino acids, carbohydrates, hydrocarbons and heteroaromatic substances (Swain, 1963) from Jurassic non-marine deposits having little or no visible particulate organic matter suggests that metabolic products of soil micro-organisms were accumulating in the Jurassic deposits.

Non-marine aquatic invertebrates, particularly pelecypods, gastropods and ostracods became abundant in the late Jurassic Purbeckian facies and continued in the early Cretaceous Wealden facies. The possible effect of these filter-feeding and browsing organisms on the accumulating organic matter has not been ascertained but in some deposits, judging from modern examples it could have been appreciable.

The appearance of widespread angiosperms including hardwood forests in mid-Cretaceous time must have been preceded by the development of extensive organic soils and their accompanying micro-organisms and enzyme systems. Prairie and upland soils presumably had well-developed zonation profiles by the late Cretaceous.

By middle Neogene time all the organic constituents of the modern non-marine environments were in existence. The grasses and related plants appeared at least by the Oligocene Epoch, and from then on there is no reason to expect that non-marine biogeochemistry was much different from that of the present.

The evolutionary development of non-marine organic matter, just as in other organic realms, is the result of the gradual appearance of primary pond-, rock- and soil-inhibiting organisms and the consequent development of other organisms dependent on the primary forms. The contribution of non-marine organic geochemistry to geological history of these kinds of deposits depends on the extent to which the organic residues can be related to their source organisms and diagenetic history.

Appendix A Geologic time scale

Era-Erathem	Period-system	Epoch-series	Approx. age of beginning (yr)
Cenozoic	Neogene	Holocene	1×10^5
		Pleistocene	2×10^6
		Pliocene	7×10^6
		Miocene	26×10^6
	Paleogene	Oligocene	38×10^6
		Eocene	54×10^6
		Paleocene	65×10^6

Era-Erathem	Period-system	Epochs	Ages-stages	
Mesozoic	Cretaceous	Upper Gulf	Maestrichtian	—
			Campanian	—
			Santonian	—
			Coniacian	—
			Turonian	—
			Cenomanian	—
		Lower-Comanchian	Albian	—
			Aptian	—
			Neocomian	136×10^6
	Jurassic	Upper	Portlandian	—
			Kimmeridgian	—
			Oxfordian	—
		Middle	Callovian	—
			Bathonian	—
			Bajocian	—
		Lower	Toarcian	—
			Pliensbachian	—
			Sinemurian	—
			Hettangian	190×10^6
	Triassic	Upper	Rhaetian	—
			Norian	—
			Carnian	—
		Middle	Ladinian	—
			Anisian	—
		Lower	Sythian	225×10^6

Era-Erathem	Period-system	Epochs	Ages-stages	Approx. age of beginning (yr)
Paleozoic	Permian		Tatarian (Lwr. pt.) (U.S.S.R.) = Ochoan (Up. pt.) (U.S.A.)	—
			Kazanian (U.S.S.R) = Ochoan (Lwr. pt.) and Guadalupian (Up. pt.) (U.S.A.)	—
			Artinskian (U.S.S.R.) = Leonardian (U.S.A.)	—
			Sakmarian (U.S.S.R.) = Wolfcampian (U.S.A.)	280×10^6
	Pennsylvanian (= Upper Carboniferous)		Stephanian (Eu.) = Missourian + Virgilian (U.S.A.)	—
			Westphalian (Eu.) = U. Morrowan + Atokan + Demoinsian (U.S.A.)	—
			Namurian (Eu.) = L. Morrowan (U.S.A.)	325×10^6
	Mississippian (= Lower Carboniferous)		Chesterian (U.S.A.) = U. Visean (Eu.)	—
			Meramecian (U.S.A.) = L. Visean (Eu.)	—
			Osagean (U.S.A.) = U. Tournasian (Eu.)	—
			Kinderhookian (U.S.A.) = L. Tournasian (Eu.)	345×10^6
	Devonian	Upper	Fammenian (Eu.) = Conewangoan + Cassadagan (U.S.A.)	—
		Middle	Frasnian (Eu.) = Chemungian + Fingerlakian (Eu.)	—
			Givetian (Eu.) = Taghanian + Tioughniogan + U. Cazenovian.(U.S.A.)	—
			Eifelian (Eu) = L. Cazenovian (U.S.A.)	—

Appendix A

Era-Erathem	Period-system	Epochs	Ages-stages	Approx. age of beginning (yr)
Paleozoic	Devonian	Lower	Coblenzian (Eu.) = Onesquethawian + Deerparkian (U.S.A.)	—
			Gedinnian (Eu.) = Helderbergian (U.S.A.)	395×10^6
	Silurian	Ludlovian (Eu.) = Cayugan (U.S.A.)	—	—
		Wenlockian (Eu.) = U. Niagarian (U.S.A.)	—	—
		Llandoverian (Eu.) = L. Niagaran + Medinan (U.S.A.)	—	430×10^6
	Ordovician	Cincinnatian	Gamachian + Richmondian (U.S.A.) = Ashgillian (Eu.)	—
			Maysvillian + Edenian (U.S.A.) = U. Caradocian (Eu.)	—
		Champlianian	Mohawkian (U.S.A.) = L. Caradocian (Eu.)	—
			Chazyan (U.S.A.) = Llandeilian + Llanvirnian (Eu.)	—
		Canadian	Arenigian (Eu.) = Tremadocian (Eu.)	500×10^6
	Cambrian	Upper (Croixan)	Trempealeauan (U.S.A.)	—
			Franconian (U.S.A.)	—
			Dresbachian (U.S.A.)	—
		Middle (Albertan)	—	—
		Lower (Waucoban)	—	570×10^6
Precambrian	Upper Keeweenawan (N.A.)	—	Grenville Orogeny	$1 \cdot 05 \times 10^9$
	Lower Keeweenawan (N.A.)	—	Penokean Orogeny	$1 \cdot 7 \times 10^9$
	Huronian (N.A.)		Algoman Orogeny	$2 \cdot 4 \times 10^9$
	Timiskamian (N.A.)		Saganagan Orogeny	$2 \cdot 6 \times 10^9$
	Keewatinian (N.A.)	—		$> 3 \cdot 2 \times 10^9$
	Swazilandian (S.Afr.)	—		$> 3 \cdot 2 \times 10^9$

Appendix B Solvent systems and R_F values for separation of some organic compounds by paper chromatography

TABLE B.I *Solvent systems for separation of amino acids by paper chromatography, and R_F values (Block et. al. 1955).*

Compound	Solvent 1	Solvent 2	Solvent 3
α-Alanine	0·37	0·66	—
Allothreonine	—	0·50	—
α-Amino-n-butyric acid	0·45	0·71	—
β-Amino-n-butyric acid	0·50	—	0·64
α-Aminoisobutyric acid	0·48	0·74	—
γ-Aminobutyric acid	—	0·77	—
α-Aminocapyrilic acid	0·55	—	—
ϵ-Aminohexanoic acid	—	0·86	—
α-Amino-ϵ-hydroxycaproic acid	—	0·75	—
α-Aminooctanoic acid	—	0·89	—
γ-Aminopentanoic acid	—	0·82	—
α-Aminophenylacetic acid	—	0·80	—
Arginine	—	0·89	0·43
Asparagine	0·19	0·40	—
Aspartic acid	—	0·19	0·29
Carnosine	—	0·82	—
Citrulline	0·25	0·63	—
Cystathionine	—	0·31	—
Cysteic acid	—	0·08	—
Cysteine	0·07	—	—
Cystine	—	—	0·20
α,γ-Diaminobutyric acid	0·12	—	—
α,ϵ-Diaminopimelic acid	—	0·30	—
Dihydroxyphenylalanine	—	0·24	—
Diiodotyrosine	0·70	0·61	—
β,β-Dimethylcysteine	—	0·59	—
Djenkolic acid	—	0·40	—
Ethanolamine	—	0·89	—
Ethanolamine phosphate	—	0·33	—
Glucosamine	—	0·69	—
Glutamic acid	—	0·31	0·47
Glutamine	—	0·57	—
Glutathione	—	0·12	—
Glycine	—	0·41	0·41
Hippuric acid	0·93	—	—
Histamine	0·22	0·95	—
Histidine	—	0·69	0·37
Homocysteic acid	—	0·12	—
Homoserine	0·30	—	—

TABLE B.1 *(cont.)*

Compound	Solvent		
	1	2	3
Hydroxylysine	—	0·66	—
Hydroxyproline	—	0·63	0·44
Hydroxytryptophan	0·48	—	—
3-Hydroxykynurenine	0·40	—	—
Isoleucine	—	0·84	0·65
Kynurenine	0·45	—	—
Lanthionine	—	0·26	—
Leucine	—	0·84	0·72
Lysine	—	0·81	0·35
Methionine	—	0·81	0·70
Methionine sulfone	—	0·70	—
Methionine sulfoxide	—	0·78	—
α-Methyl-α-amino-*n*-butyric acid	—	0·84	—
Methylhistidine	—	0·87	—
Monoiodotyrosine	—	0·66	—
Norleucine	0·78	0·84	—
Norvaline	0·65	0·80	—
Ornithine	0·15	0·79	—
Phenylalanine	—	0·85	0·73
Proline	—	0·88	0·55
Serine	—	0·36	—
Serine phosphate	—	0·10	—
Taurine	—	0·50	—
Threonine	—	0·86	—
Trimethylalanine	0·66	—	—
Tryptophan	—	0·75	0·67
Tryptamine	0·73	—	—
Tyrosine	—	0·51	0·65
Valine	—	0·78	0·70

Solvent (1): 1-butanol:acetic acid:water (4:1:5); (2) phenol:water = 100:20 v/v Mallinckrodt liquid phenol; (3) pyridine:acetic acid:water (50:35:15). Stain: 25% ninhydrin in acetone.

TABLE B.2   *R*$_F$ *values of some amines in solvent system butanol:acetic acid:water* (4:1:5) *on Whatman no. 1 filter paper* (*Bremner & Kenten* 1951)

Compound	R_F
Methylamine	0·37
Ethylamine	0·45
n-Propylamine	0·58
n-Butylamine	0·70
n-Amylamine	0·77
n-Heptylamine	0·85
1,4-Diaminobutane (putreseine)	0·16
1,5-Diaminopentane (cadaverine)	0·17
Benzylamine	0·68
Ethanolamine	0·33
Dimethylamine	0·43

TABLE B.3   *R*$_F$ *of sugars on Whatman no. 1 paper in two solvent systems* (*Partridge*, 1948; *Bersin & Müller*, 1952; *Block* et al. 1955)

Sugar	Solvent	
	1	2
Monosaccharides		
D-Glucose	0·39	0·18
D-Galactose	0·44	0·16
D-Mannose	0·45	0·20
L-Sorbose	0·42	0·20
D-Fructose	0·51	0·23
D-Xylose	0·44	0·28
D-Arabinose	0·54	0·21
D-Ribose	0·59	0·31
L-Rhamnose	0·59	0·37
D-Deoxyribose	0·73	—
L-Fucose	0·63	0·27
Oligosaccharides		
Lactose	0·38	0·09
Maltose	0·36	0·11
Sucrose	0·39	0·14
Raffinose	0·27	0·05
Uronic acids		
D-Galacturonic acid	0·13	0·14
D-Glucuronic acid	0·12	0·12(0·72)*

* Due to lactone.
Solvent (1): phenol + 1 % NH$_3$ + HCN; (2) *n*-butanol:acetic acid:water (4:1:5 v/v).
Stain: AgNO$_3$ (ammoniacal).

TABLE B.4 *R$_F$ of polyhydroxy alcohols on Whatman no. 1 paper in two solvent systems (Block et al. 1955)*

Compound	Solvent	
	1	2
Inositol	0·06	0·08
Sorbitol	0·18	0·19
Dulcitol	0·16	0·18
Mannitol	0·19	0·19
Ribitol	0·25	0·28
Arabitol	0·22	—
Xylitol	0·20	—
Erythritol	0·35	0·34
Ethylene glycol	0·64	—
Glycerol	0·48	—

Solvents: (1) 1-butanol:acetic acid:water (4:1:5 v/v); (2) ethyl acetate:acetic acid:water (3:1:3 v/v).
Stain: Ammoniacal silver nitrate.

TABLE B.5 *R$_F$ values of some lower fatty acids on Whatman no. 1 filter paper (Block et al. 1955)*

Fatty acid	1-Butanol:1·5 aqueous NH$_3$ (1:1 v/v)
Formic	0·10
Acetic	0·11
Propionic	0·19
n-Butyric	0·29
n-Valeric	0·41
n-Caproic	0·53
n-Heptanoic	0·62
n-Octanoic	0·65
n-Nonanoic	0·67

Stain: Bromophenol blue, 50 mg in 100 ml water, made acid with 200 mg citric acid.

TABLE B.6 R_F *of hydroxamic acids* of some higher fatty acids on acetylated Whatman no. 1 paper (25 % acetyl) (Micheel, 1954)*

Fatty acid	R_F
Valeric acid	0·84
Caproic acid	0·72
Capyrilic acid	0·57
Lauric acid	0·38
Myristic acid	0·34
Palmitic acid	0·30
Oleic acid	0·30
Stearic acid	0·24

* Prepared by mixing aldehyde in absolute ethanol with benzene sulfohydroxamic acid and enough NaOH to make reaction mixture basic.

Stain: alkaline 0·4 % solution of bromothymal blue in methyl cellosolve; detects 30–50 ng of each acid.

Solvent: ethyl acetate:tetrahydrofuran:water (0·6:3·5:4·7).

TABLE B.7 R_F *of some nucleic acid derivatives (purines and pyrimidines) on Whatman no. 1 paper in 1-butanol saturated with* $NH_4OH:H_2O$ *(1:4 v/v) (Tamen, 1953)*

Compound	R_F
Adenine	0·72
Guanine	0·17
Hypoxanthine	0·23
Cytosine	0·50
Uracil	0·55
Thymine	1·02
Thymine deoxyriboside	1·00

Stain: cysteine:sulfuric acid (0·5 g cysteine hydrochloride in 100 ml 3 N-H_2SO_4), pink spots appear.

TABLE B.8 R_F *of some porphyrins on Whatman no. 1 paper in 2,5-dimethylpyridine:*
water (1:1 v/v), NH₃ atmosphere (Kehl, 1951; Block et al. 1955)

Compound	No. of —COOH groups	R_F
Uroporphyrin	8	0·26
Coproporphyrin	4	0·54
Protoporphyrin	2	0·84
Mesoporphyrin	2	0·86
Deuteroporphyrin	2	0·88
Etioporphyrin	o	1·0
Porphyrin esters	o	1·0

Detection: red spots in plain light, pink spots in ultraviolet light.

Appendix C Solvent systems and R_F values used in thin-layer chromatographic analysis of some organic compounds

TABLE C.1 *R_F values and color reactions of tryptophan metabolites (Randerath, 1963)*

| | R_F value in solvent | | Color | |
Compound	1	2	Ultraviolet	Ehrlich
Tryptophan	0·25	0·07	—	Violet
Indole	0·90	0·98	Blue	Violet
Urine indican	0·61	0·14	Brown	Brown
DL-Kynurenine	0·32	0·11	Green-blue	Yellow-brown
3-Hydroxykynurenine	0·16	0·06	Yellow-green	Orange
Kynurenic acid	0·45	0·18	Green	—
Xanthurenic acid	0·45	0·26	Grey	—
3-Hydroxyanthranilic acid	0·31	0·75	Light blue	Yellow
2-Amino-3-hydroxyacetophenone	0·88	0·93	Blue	Yellow-brown

TLC plate: silica Gel G, activated for 30 min at 110 °C.

Solvents: (1) methyl acetate:isopropanol:25 % ammonium hydroxide (s.g. = 0·91) (45:35:20); (2) chloroform:96 % acetic acid (95:5).

Detection: (1) observation in long wave u.v. light; (2) Erlich reagent: 1 % solution of p-dimethylaminobenzaldehyde in 96 % ethanol; after chromatogram has been sprayed it is placed in vessel saturated with hydrochloric acid vapor for 3–5 min.

Appendix C

TABLE C.2 R_F *values of amines (hydrochlorides) on various layers*
(Randerath, 1963)

Compound	Solvent		
	1	2	3
Methylamine	0·11	0·10	0·10
Ethylamine	0·15	0·13	0·20
n-Propylamine	0·30	0·19	0·30
Isoamylamine	0·49	0·39	0·45
Cadaverine	0·06	0·02	0·01
Putrescine	0·04	0·02	0·01
Ethanolamine	0·18	0·11	0·10
Histamine	0·26	0·02	0·03
Tyramine	0·44	0·38	0·55
Phenylethylamine	0·56	0·37	0·55
Benzylamine	0·49	0·36	0·50
Tryptamine	0·54	0·43	0·90

TLC plates: silica gel G, buffered silica gel G.

Solvents: (1) phenol:water (8:3) (silica gel G); (2) butanol:acetic acid:water (4:1:5) (silica gel G); (3) 70% ethanol (buffered silica gel G). The buffered layers are prepared using 0·2 M Sörensen buffer of pH 6·8.

Detection: ninhydrin.

TABLE C.3 *R_F values of steroids on Brockmann basic aluminum oxide of activity III (Randerath, 1963)*

Compound	Solvent					
	1	2	3	4	5	6
Cholesterol	—	—	0·09	0·13	—	—
Cholesteryl acetate	0·33	0·43	0·73	0·84	—	—
Cholesteryl benzoate	0·46	0·61	0·86	—	—	—
Cholesteryl tosylate	0·07	0·11	0·45	—	—	—
Methyl ester of acetoxyetianic acid	—	—	0·18	0·39	0·63	0·88
Testosterone	—	—	—	0·01	0·20	0·39
Testosterone acetate	—	—	—	0·05	0·32	0·59
Diosgenin	—	—	—	—	0·22	0·43
Solasodin	—	—	—	—	0·16	0·41
Dehydroepiandrosterone	—	—	0·01	0·02	0·20	0·42
Dehydroepiandrosterone acetate	—	—	0·04	0·16	0·36	0·65
Progesterone	—	—	—	—	0·32	—
11-Oxoprogesterone	—	—	—	—	—	0·54
11 α-Hydroxyprogesterone	—	—	—	—	—	0·34
17 α-Hydroxyprogesterone	—	—	—	—	—	0·43
17 α-Acetoxyprogesterone	—	—	—	—	0·27	0·57
21-Acetoxyprogesterone	—	—	—	—	—	0·58
17,21-Dihydroxyprogesterone	—	—	—	—	—	0·22
Estrone	—	—	—	—	—	0·32
Estradiol	—	—	—	—	—	0·30
Estradiol methyl ether	—	—	—	—	—	0·53
Pregnenolone	—	—	—	0·03	0·19	0·41
Pregnenolone acetate	—	—	—	0·28	0·57	0·70

Solvents: (1) petroleum ether:10% benzene; (2) petroleum ether:20% benzene; (3) petroleum ether:50% benzene; (4) benzene; (5) benzene:2% ethanol; (6) benzene:5% ethanol.

Detection: (1) antimony trichloride; after spraying, warm at 100 °C for 10 min and view with u.v. or plain light; (2) morin dye, 0·005–0·01% soln. in methanol; u.v. light; (3) conc. sulfuric acid:methanol (1:1), heated 110 °C for 15 min.

Appendix C

TABLE C.4 *R_F values of carotenoids and chlorophylls on impregnated Kieselgur layers (Randerath, 1963)*

Compound	—OH	O	=O	—OR	$\overset{\parallel}{-COR}$	Paraffin impreg.*	Fat impreg.†
		No. of substituents				R_F values	
Capsanthin	2	—	1	—	—	—	0·74
Isozeaxanthin	2	—	—	—	—	—	0·49
Zeaxanthin	2	—	—	—	—	—	0·54
Isozeaxanthin dimethyl ether	—	—	—	2	—	0·60	—
Cryptoxanthin	1	—	—	—	—	0·90	0·07
Rhodoxanthin	—	—	2	—	—	—	0·26
Torularhodin methyl ester	—	—	—	—	1	0·48	—
Echinenone	—	—	1	—	—	0·61	—
β-Apo-8′-carotenoate (C$_{30}$)	—	—	1	—	—	0·83	—
Methyl β-apo-8′-carotenoate (C$_{30}$)	—	—	—	—	1	0·69	—
Methyl β-apo-6′-carotenoate (C$_{35}$)	—	—	—	—	1	0·58	—
Lutein (xanthophyll)	2	—	—	—	—	—	0·56
Lutein epoxide	2	1	—	—	—	—	0·72
Violaxanthin	2	2	—	—	—	—	0·84
Neoxanthin	3	1	—	—	—	—	0·95
β-Carotene	—	—	—	—	—	0·10	0·00
α-Carotene	—	—	—	—	—	0·15	0·00
Lutein dipalmitate (helenien, physalien)	—	—	—	—	2	0·02	—

* Solvent: methanol:acetone (5:2) (saturated with liquid paraffin).

† Solvent: methanol:acetone:water (20:4:3) (saturated with vegetable oil).

Chlorophylls: The following R_F values are obtained on Kieselgur plates impregnated with vegetable oil (see above) and developed with methanol:acetone:water (20:4:3) (saturated with oil); chlorophyll *a*, 0·13; chlorophyll *b*, 0·25; phaeophytin *a*, 0·01; phaeophytin *b*, 0·07.

TABLE C.5   *R$_F$ values and color reactions of phenols, phenol aldehydes and phenol carboxylic acids. Silica gel G layer (Randerath, 1963)*

Compound	R$_F$ value in solvent		Color with tetrazotized benzidine
	4	5	
Phenol	0·76	0·60	Yellow
Benzaldehyde	0·68	0·62	Pale yellow
Benzoic acid	0·74	—	*
Catechol	0·58	0·54	Gray-green
Resorcinol	0·56	0·52	Reddish brown
Hydroquinone	0·54	0·46	Yellow
Benzoquinone	0·65	0·69	Yellow (own color)
Phloroglucinol	0·34	0·32	Violet
Pyrogallol	0·32	0·45	Dark brown
Salicylaldehyde	0·82	0·83	*
Salicylic acid	0·64	0·68	*
m-Hydroxybenzoic acid	0·49	0·51	Light yellow
Ethyl m-hydroxybenzoate	0·76	0·61	Yellow
p-Hydroxybenzaldehyde	0·63	0·56	Brownish
p-Hydroxybenzoic acid	0·55	0·60	Yellow
Gentistic acid	0·30	0·40	Orange
Methyl gentisate	0·71	0·61	Gray
β-Resorcylic acid	0·54	0·52	Rust red
Protocatechualdehyde	0·32	0·45	Gray-brown
Protocatechuic acid	0·32	0·39	Brown
Ethyl protocatechuate	0·50	0·62	Brown
Gallic acid	0·18	0·23	Light brown
Ethyl gallate	0·36	0·38	Gray-brown
Cinnamic acid	0·63	0·65	*
p-Coumaric acid	0·49	0·52	Gray-brown
o-Coumaric acid	0·47	0·62	Reddish brown
Caffeic acid	0·24	0·43	Yellowish brown
Guaiacol	0·83	0·72	Reddish brown
Resorcinol dimethyl ether	0·87	0·87	—
Veratrole	0·80	0·75	—
Anisaldehyde	0·79	0·66	Orange
Anisic acid	0·84	0·69	—
o-Vanillin	0·74	0·59	Brown
Isovanillin	0·56	0·47	Yellowish brown
Vanillin	0·70	0·64	Yellow
Vanillic acid	0·54	0·61	Reddish brown
Veratraldehyde	0·80	0·72	Reddish violet
Veratric acid	0·73	0·70	—
Syringaldehyde	0·60	0·57	Orange
Syringic acid	0·48	0·60	Orange
Ferulic acid	0·50	0·58	Grayish brown
Isoferulic (hesperitic) acid)	0·43	0·66	Carmine
Kojic acid	0·06	0·06	Light red
Umbelliferone	0·55	0·58	—
α-Hydrindone	0·85	0·75	—

* Detected by spraying with 0·1 potassium permanganate solution containing sodium carbonate.

TLC plate: silica gel G dried 105 °C for 30 min. Solvents: see (4) and (5) TABLE C.6.

Detection: tetrazotized benzidine. Solution 1:5 g benzidine dissolved in 14 ml conc. hydrochloric acid and diluted with water to 1 l. Solution 2: aqueous 10 % sodium nitrite solution. Equal volumes of the two solutions are mixed just before use. After spraying the plates dried at 105 °C for a few minutes.

TABLE C.6 R_F *values of monocarboxylic acids (Randerath*, 1963)

| | R_F value in solvent | |
Compound	1	2
Formic acid	0·52	0·64
Acetic acid	0·58	0·66
Lactic acid	0·63	0·51
Pyruvic acid	0·48	0·42

TLC plate: silica gel G.

Solvents: (1) pyridine:petroleum ether (1:2); (2) ethanol:ammonium hydroxide:water (80:4:16).

Detection: spray with solution of 0·04 g bromocresol purple in 50 % ethanol adjusted to pH 10 with sodium hydroxide; or with bromocresol green or bromophenol blue.

TABLE C.7 R_F *values of aliphatic dicarboxylic acids. Silica gel G.*
Without tank saturation

| | R_F values in solvent | | |
Compound	3*	4†	5†
Oxalic acid	0·05	0·00	0·00
Malonic acid	0·14	0·13	0·05
Succinic acid	0·30	0·28	0·23
Glutaric acid	0·39	0·35	0·28
Adipic acid	0·43	0·42	0·34
Pimelic acid	0·53	0·47	0·36
Suberic acid	0·54	0·50	0·40
Azelaic acid	0·56	0·53	0·43
Sebacic acid	0·67	0·55	0·47

* 250 μm layer prepared with the Desaga spreader.

† Hand-made layer prepared by pouring Silica gel G suspension on to glass plates. 4 g adsorbent/4·5 × 17·5 cm plate.

Solvents: (3) 96 % ethanol:water:25 % ammonium hydroxide (100:12:16); (4) benzene:methanol:acetic acid (45:8:4); (5) benzene:dioxane:acetic acid (90:25:4).

Solvents (4) and (5) can also be used for chromatographing phenols and phenolcarboxylic acids.

Detection: See TABLE C.6.

TABLE C.8   *R_F values of miscellaneous carboxylic acids on silica gel G*

Compound	R_F value
Citric acid	0·05
Tartaric acid	0·08
Phthalic acid	0·26
Terephthalic acid	0·73
Benzoic acid	0·76
p-Toluic acid	0·76

Solvent: solvent 3 above (TABLE C.7).
Detection: See TABLE C.6.

References

Abelson, P. H. (1954). Organic constituents of fossils. *A. Rep. Dir. geophys. Lab. Carnegie Instn.* Yearbook, 1953–54, **53**, 97–101.

Abelson, P. H. (1955). Paleobiochemistry. *A. Rep. Dir. geophys. Lab. Carnegie Instn.* Yearbook, 1954–55, **54**, 107–9.

Abelson, P. H. (1957). Some aspects of paleobiochemistry. *Ann. N.Y. Acad. Sci.* **69**, 276–85.

Abelson, P. H. (1959). Geochemistry of organic substances. In *Researches in Geochemistry*, P. H. Abelson (ed.), pp. 79–103. New York: John Wiley and Sons, Inc.

Abelson, P. H. (1963). Geochemistry of amino acids. In *Organic Geochemistry*, I. A. Breger (ed.) pp. 431–55. Oxford: Pergamon Press.

Abelson, P. H. (ed.) (1968). *Researches in Geochemistry*, vol. 2, pp. i–x, 1–511. New York: John Wiley and Sons.

Abelson, P. H. & Hare, P. E. (1968). Recent origin of amino acids in the Gunflint Chert. *Prog. Geol. Soc. Am. 81st Ann. Mtg. Mexico City*, p. 2.

Abelson, P. H., Hoering, T. C. & Parker, P. L. (1964). Fatty acids in sedimentary rocks. In *Adv. in Organic Geochemistry*, U. Columbo and G. D. Hobson (eds.), pp. 169–74. New York: Macmillan.

Abelson, P. H. & Parker, P. L. (1962). Fatty acids in sedimentary rocks. *Carnegie Instn.* Yearbook **61**, 181–4.

Adams, A. P., Bartholomew, W. V. & Clark, F. E. (1954). Measurement of nucleic acid components in soil. *Soil Sci. (Ann. Proc.)* **18**, 40.

Agapov, A. I. (1966). Cobalt complex formation with organic compounds of the soil. I. Potentiometric titration of humic acids of soil and peat. *Agrokhimiya* 1966 (9), 88–94.

Allison, F. E., Doetsch, J. H. & Roller, E. M. (1951). Ammonium fixation and availability in Harpster clay loam. *Soil Sci.* **72**, 187.

Allison, F. E., Kefauver, M. & Roller, E. M. (1953). Ammonium fixation in soils. *Soil Sci. (Soc. Am. Proc.)* **17**, 107.

Alvsaker, E. & Michelson, K. (1957). Carbohydrates in a cold-water extract of a pine forest soil. *Acta chem. scand.* **11**, 1794.

Andersen, S. T. & Gundersen, K. (1955). Ether soluble pigments in interglacial gyttja. *Experientia* **11**, 345–8.

Anderson, G. (1957). Nucleic acid derivatives in soils. *Nature, Lond.* **180**, 287.

Anderson, G. (1958). Identification of derivatives of deoxyribonucleic acid in humic acid. *Soil Sci.* **86**, 169.

Anderson, G. (1961). Estimation of purines and pyrimidines in soil humic acid. *Soil Sci.* **91**, 156.

Antoniani, C., Montanari, T. & Camoriano, A. (1954). Soil enzymology. I. Cathepsin-like activity. A preliminary note. *Annali Fac. Agr. Univ. Milano* **3**, 99.

Aurich, H., Muecke, D. & Obenaus, R. (1963). Biochemistry of humic acids. II. The amino acids of the water soluble and water insoluble humic acids of *Cephalosporium gordoni*. *Acta biol. med. germ.* **11** (3), 311–22.

References

Austen, D. E. G., Ingram, D. J. E., Given, P. H., Binder, C. R. & Hill, L. W. (1966). Electron spin resonance study of pure macerals. In *Coal Science, Am. chem. Soc., Adv. Chem. Ser.* **55**, 344–62.

Babenko, S. A. (1963). Flotation of ilmenite from sands. *Izv. sib. Otdel. Akad. Nauk SSSR, Suom. Tekhn. Nauk* **3**, 104–8.

Babu, S. K. & Dutcher, R. R. (1966). Petrographic investigation of two Gondwana seams from Madhya Pradesh, India. In *Coal Science, Am. chem. Soc., Adv. Chem. Ser.* **55**, 284–306.

Bailey, T. L. (1947). Geologic development of the Ventura Basin, California, and its petroleum deposits. *Oil Gas J.* **45**, 119.

Baker, D. R. (1962). Organic geochemistry of Cherokee Group in south-eastern Kansas and north-eastern Oklahoma. *Bull. Am. Ass. Petrol Geol.* **46**, 1621–42.

Baker, S. B., Evans, J. H. & Hibbert, H. (1948). Lignin and related compounds. LXXXV. Synthesis and properties of dimers related to lignin. *J. Am. chem. Soc.* **70**, 60.

Balicka, N. & Sochacka, Z. (1959). Biological activity in light soils. *Zesz. probl. Postep. Nauki pol.* **21**, 257.

Banse, K. (1957). Results of the hydrographical and biological longitudinal sections through the north Sea during the summer of 1956. II. Distribution of oxygen, phosphate, and suspended matter. *Kieler Meeresforsch.* **1**, 186–201.

Barghoorn, E. S. (1957). Origin of life. In *Treatise on Marine Ecology and Paleoecology. Geol. Soc. Am. mem.* **67**, 2, 76–86.

Barnes, F. F. & Cobb, E. H. (1959). Geology and coal resources of the Hower district, Kenai Coal Field, Alaska. *U.S. Geol. Survey Bull.* (1058-F) 217–60.

Barshad, I. (1951). Cation exchange in soils. I. Ammonium fixation and its relation to potassium fixation and to determination of ammonium exchange capacity. *Soil Sci.* **72**, 361.

Basu, A. N., Mukherjee, D. C. & Mukherjee, S. K. (1961). Interaction between humic acid fraction of soil and trace element cations. *J. Indian Soc. Soil Sci.* **12**, 311–18.

Baturia, V. P. (1937). Genesis of the productive formation of the Apsheron Peninsula and of neighboring regions. *Seventeenth Int. geol. Congr.* Moscow, U.S.S.R. **4**, 279–93.

Baudisch, O. & Von Euler, H. (1934). Über den Gehalt eineger Moor-Erdarten an Carotinoiden. *Ark. Kemi Miner. Geol.* 11 A (21), 1–10.

Bazanova, T. D. (1960). Distribution of organic substances in Neogene deposits in the southwestern part of the Northern Sakalin Island. *Geol. i Geokhim., Uses, Neft. Nauch.-Issled. Geologorazved. Inst. Nauchn.-Tekhn. Obshes-Foo Neft. i. Gaz. Prom., Dokl. i Stat'i* **3**, 231–41.

Bear, F. E. *et al.* (1955), *Chemistry of the Soil*, pp. 1–373. New York: Reinhold Co.

Beatty, R. A. (1941). The pigmentation of cavernicolous animals. II. Carotenoid pigments in the cave environment. *J. exp. Biol.* **18**, 144–52.

Beauchamp, R. S. A. (1964). The Rift Valley lakes of Africa. *Verh. int. Verein. theor. angew. Limnol.* Stuttgart, **15**, 91–9.

Bedford, J. W., Roelofs, E. W. & Zabik, M. J. (1968). The freshwater mussel as a biological monitor of pesticide concentrations in a lotic environment. *Limnol. Oceanogr.* **13**, 118–26.

Beinert, H. & Hemmerich, P. (1965). Evidence for semiquinone-metal interaction in metal flavoproteins. *Biochem. biophys. Res. Commun.* **18** (2), 212–20.

389

References

Bergmann, W. (1963). Geochemistry of lipids. In *Organic Geochemistry*, I. A. Breger (ed.), pp. 503–42. Oxford: Pergamon Press.

Berner, R. A. (1968). Calcium carbonate concretions formed by the decomposition of organic matter. *Science, N.Y.* **159**, 195–7.

Berns, D. S., Holohan, P. & Scott, E. (1966). Urease activity in blue-green algae. *Science, N.Y.* **152**, 1077–8.

Bersin, T. & Muller, A. (1952). Die Ring-Papierchromatographie von Zuckern. *Helv. chim. Acta* **35**, 475–8.

Black, C. A. (1968). *Soil–Plant Relationships*, 2nd ed. pp. i–viii, 1–792. New York: Wiley and Sons.

Block, R. J., Durrum, E. L. & Zweig, G. (1955). *Paper Chromatography and Paper Electrophoresis*, pp. 1–484. New York: Academic Press.

Blois, M. S., Jr., Zahlan, A. B. & Maling, J. (1964). Electron spin resonance studies on melanin. *Biophys. J.* **4**, 471.

Bloomfield, C. (1953). Sesquioxide immobilization and clay movement in podzolized soils. *J. Soil Sci.* **4**, 5, 17.

Bloomfield, C. (1954). Podzolization. III. The mobilization of iron and aluminum by Rimu (*Dacrydium cupressinum*). IV. The mobilization of iron and aluminum by picked and fallen larch needles. V. The mobilization of iron and aluminum by aspen and ash leaves. *J. Soil Sci.* **5**, 39, 46, 50.

Bloomfield, C. (1955). Podzolization. VI. The immobilization of iron and aluminum. *J. Soil Sci.* **6**, 284.

Bloomfield, C. (1957). Possible significance of polyphenols in soil formation. *J. Sci. Fd Agric.* **8**, 389.

Blumentals, A. & Swain, F. M. (1956). Comparison of amino acids obtained by acid hydrolysis of lake sediments, central Minnesota. *Bull. geol. Soc. Am.* **67**, 1673.

Blumer, M. (1950). Porphyrenfarbostoffe und Porphyrin-Metallkomplexe in schweizerischen Bitumina. *Helv. chim. Acta* **33**, 1627–37.

Blumer, M. (1952). Chemical studies of bituminous rocks. *Bull. V. Sch. Pet. geol.* **19**, 17–26.

Bogert, M. T. & Marion, S. J. (1933). A chemical examination of the volatile oils of *Sarothra Gentianoides* L. and the detection therein of normal nonane. *J. Am. chem. Soc.* **55**, 4187–94.

Bogolyubova, L. I. (1959). Characteristics of certain types of coals from the Volchensk and Bogoslovsk deposits. *Genezio Tverd. Goryuch. Iskopaem., Akad. Nauk SSSR, 1st. Goryuch, Iskopaem. 1959*, 137–42.

Bone, W. A., Horton, L. & Ward, S. G. (1930). Researches on the chemistry of coal. VI. Its benzenoid constitution as shown by its excitation with alkaline permanganate. *Proc. Roy. Soc. Lond.* **127**A, 480–510.

Bone, W. A., Parsons, L. G. B., Sapiro, R. H. & Grocock, C. M. (1935). Researches on the Chemistry of Coal. VIII. The development of benzenoid constitution in the lignin–peat–coal series. *Proc. Roy. Soc. Lond.* **148**A, 492–522.

Borger, H. D. (1952). Case history of Quiriquire Field, Venezuela. *Bull. Am. Ass. Petrol. Geol.* **36**, 2291–2330.

Borner, H. (1958). Untersuchungen über den Abban von Phlorizin im Boden. *Naturwiss enschaften* **45**, 138–9.

Bose, B., Stevenson, I. L. & Katenelson, H. (1959). Further observations on the metabolic activity of the soil microflora. *Proc. natn. Inst. Sci. India* B**25** (5), 223–9.

Bottomley, W. B. (1917). Isolation from peat of certain nucleic acid derivatives. *Proc. Roy. Soc. Lond.* B 90, 39.

Bouska, H. V. & Zdanek, J. (1963). Geochemistry and petrography of the Cenomanian coals from Bohemia and Moravia. *Rozpr. české. Akad. Věd Řada mat. Privod. věd.* 73 (8), 3–78.

Bower, C. A. (1945). Separation and identification of phytin and its derivatives from soils. *Soil Sci.* 59, 277–85.

Bradley, W. H. (1966). Tropical lakes, copropel and oil shale. *Bull. geol. Soc. Am.* 77, 1333–8.

Brandt, C. W. (1952). Constitution of phyllocladene and related diterpenes. *N.Z. Jl. Sci. Technol.* 34B, 46–57.

Bray, E. E. & Evans, E. D. (1961). Distribution of *n*-paraffins as a clue to the recognition of source beds. *Geochim. cosmochim. Acta* 22, 2–15.

Breger, I. A. (1963a). Origin and classification of naturally occurring carbonaceous substances. In *Organic Geochemistry*, I. A. Breger (ed.), pp. 50–86. Oxford: Pergamon Press.

Breger, I. A. (ed.) (1963b). In *Organic Geochemistry*, pp. 1–658. Oxford: Pergamon Press.

Bremner, J. M. (1950a). The amino acid composition of the proteinaceous material in soil. *Biochem. J.* 47 (5), 538–42.

Bremner, J. M. (1950b). Amino acids in soil. *Nature, Lond.* 165, 367.

Bremner, J. M. (1952). The nature of soil-nitrogen complexes. *J. Sci. Fd Agric.* 3, 497.

Bremner, J. M. (1955a). Soil humic acids. I. The chemical nature of humic nitrogen *J. agric. Sci., Camb.* 46, 247.

Bremner, J. M. (1955b). Nitrogen distribution and amino acid composition of portions of a humic acid from a chernozem soil. *Z. PflErnähr Düng. Bodenk.* 71, 63–6.

Bremner, J. M. (1967). Nitrogenous compounds. In *Soil Biochemistry*, pp. 19–66. New York: Marcel Dekker.

Bremner, J. M. & Edwards, A. P. (1965). Determination and isotope-ratio analysis of different forms of nitrogen in soils. I. Apparatus and procedure for distillation and determination of ammonium. *Soil Sci. Soc. Am. Proc.* 24 (5), 504–7.

Bremner, J. M. & Kenten, R. H. (1951). Paper chromatography of amines. *Biochem. J.* 49, 651–5.

Briggs, M. H. & Segal, L. (1963). Preparation and properties of a free soil enzyme. *Life Sci. 1963,* 69.

Briggs, M. H. & Spedding, D. J. (1963). Soil enzymes. *Sci. Prog. Lond.* 51, 217.

Brough, J. (1928). On rhythmic deposition in the Yoredale Series. *Proc. Univ. Durham phil. Soc.* 8, 116–26.

Brown, J. K. (1959). Infrared spectra of solvent extracts of coals. *Fuel, Lond.* 38, 55.

Brown, S. R. (1968). Bacterial carotenoids from freshwater sediments. *Limnol. Oceanogr.* 13, 233–41.

Brown, S. R. & Coleman, B. (1963). Oscillaxanthin in lake sediments. *Limnol. Oceanogr.* 8, 352–3.

Burbridge, E. M., Burbridge, G. R., Fowler, W. A. & Hoyle, F. (1957). Synthesis of the elements in stars. *Rev. mod. Phys.* 29, 548.

Burg, S. P. & Burg, E. A. (1966). Auxin-induced ethylene formation: its relation to flowering in the pineapple. *Science, N.Y.* 152, 1269.

References

Burlingame, A. C. & Simonet, B. R. (1968). Isoprenoid fatty acids isolated from the kerogen matrix of the Green River Formation (Eocene). *Science, N.Y.* **160**, 531–3.

Cameron, F. K. & Bell, J. M. (1905). The mineral constituents of the soil solution. *U.S. Dept. Agric. Bur. Soils Bull.* **30**, 7–70.

Carles, J. & Decau, J. (1960). Variations in the amino acids of soil hydrolysates. *Scient. Proc. R. Dubl. Soc.* A**1**, 141.

Carlisle, D. B. (1964). Chitin in a Cambrian Fossil, *Hyolithellus. Biochem. J.* **90**, 1C, 2C.

Carter, C. L. & Heazlewood, W. V. (1949). The essential oil of *Pittosporum Eugenioides. J. Soc. chem. Ind., Lond.* **68**, 34–6.

Chaberek, S. & Martell, A. E. (1959). *Organic Sequestering Agents*, pp. 1–449. New York: Wiley and Sons.

Chin, W. T. & Kroontje, W. (1963). Urea hydrolysis and subsequent loss of ammonia. *Soil Sci. (Soc. Am. Proc.)* **27**, 316.

Clisby, K. H. & Sears, P. B. (1955). Palynology in southern North America, Part III. Microfossil profiles under Mexico City correlated with the sedimentary profiles. *Bull. geol. Soc. Am.* **66**, 511–20.

Cloud, P. E., jr. (1968). Atmospheric and hydrospheric evolution on the primitive earth. *Science, N.Y.* **160**, 729–36.

Colburn, C. B., Ettinger, R. & Johnson, F. A. (1963). Isolation and storage of free radicals on molecular sieves. I. The electron paramagnetic resonance (E.P.R.) spectrum of nitrogen dioxide. *Inorg. chem.* **2**, 1305.

Colombo, U. & Hobson, G. D. (eds.) (1964). *Advances in Organic Geochemistry*, pp. 1–488. New York: Macmillan.

Cooper, J. E. (1962). Fatty acids in Recent and ancient sediments and petroleum reservoir waters. *Nature, N.Y.* **193**, 744–6.

Cooper, R. L. & Lindsay, A. J. (1953). Atmospheric pollution by polycyclic hydrocarbons. *Chemy. Ind.* 1953, 1177–8.

Cramer, M. & Cox, E. H. (1922). Sur la constitution de la glucosene. *Helv. chim. Acta* **5**, 844.

Creitz, G. I. & Richards, F. A. (1955). The estimation and characterization of planktonic populations by pigment analysis. III. A note on the use of 'millipore' membrane filters in the estimation of plankton pigments. *J. mar. Res.* **14** (3), 211–16.

Crispin, O., Ibañez, C. & Weller, J. M. (1955). Geology and coal resources of Batan Island, Albay, Republic of the Philippines. *Philipp. Bur. Min. Spec. Projects Ser.*

Cushing, E. J. & Wright, H. E. (1963). Piston corers for lake sediments. *Limnol. Res. Center, U. Minnesota, Minneapolis. Mimeographed report.*

Czeczuga, B. (1957). The effect of humic compounds on zooplankton. *Akad. med. biol.* **3**, 85–126.

Dall, W. H. & Harris, G. D. (1892). Correlation papers: Neocene. *U.S. Geol. Survey Bull.* (84) 1–349.

Dauben, W. G. & Richards, J. H. (1955). The biogenesis of the triterpene eburicoic acid. *Chem. Ind.* 1955, 94–5.

Dawson, J. W. (1854). On the Coal measures of the South Joggins, Nova Scotia. *Q. Jl geol. Soc. Lond.* **10**, 1–42.

Deevey, E. S. (1939). Studies on Connecticut lake sediments. I. A Post-glacial climatic chronology for southern New England. *Am. J. Sci.* **237**, 691–724.

References

Deevey, E. S. (1940). Limnological studies in Connecticut. V. A contribution to regional limnology. *Am. J. Sci.* **238**, 717–41.

Degens, E. T. (1965). *Geochemistry of sediments: a brief survey*, pp. 1–342. Englewood Cliffs, N. J.: Prentice Hall, Inc.

Degens, E. T. & Bajor, M. (1962). Die Verteilung von Aminosäuren in bituminözen Sedimenten und ihre Bedeutung für die Kohlen- und Erdölgeologie. *Glückauf.* **24**, 1525–34.

Degens, E. T., Prashnowsky, A., Emery, K. O. & Pimenta, J. (1961). Organic materials in recent and ancient sediments. Part II. Amino acids in marine sediments of Santa Barbara Basin, California. *Neues Jb. Geol. Paläont. Mh.* 413–26.

Degering, E. F. (1962). *Organic Chemistry*, 6th ed., pp. 1–422. New York: Barnes and Noble.

DeWitt, W. (1968). Correlation of the Lower Mississippian rocks in western Maryland and adjacent states. *Prog. geol. Soc. Am.*, south-eastern Section, Durham, N.C. 31.

Dhéré, C. (1934). Fluorescence spectral research in oil shale (additional note), *Swiss Bull. Min. Pet.* **14**, 518.

Dhéré, C. & Hradil, G. (1934). Spectroscopic investigation of the fluorescence of oil shales. *Schweiz. miner. petrogr. Mitt.* **14**, 279–95.

Diaconescu, E. & Doma, M. (1960). Relation between the geological age of coal and exchange capacity of ions derived from coal. *Bul. Inst. politeh. Iasi* **6** (1–2), 139–56.

Dion, H. G., Mann, P. J. G. & Heintze, S. G. (1947). The 'easily reducible' manganese of soils. *J. agric. Sci., Camb.* **37** (1), 17–22.

Dormaar, J. F. (1963). Humic acid associated phosphorus in some soils of Alberta. *Can. J. Soil Sci.* **43** (2), 235–41.

Dormaar, J. F. & Lynch, D. L. (1962). Amendments to the determination of 'uronic acids' in soils with carbazole. *Soil Sci. Soc. (Am. Proc.)* **26**, 251.

Dormans, H. M. N., Huntjens, F. J. & van Krevelen, D. W., unpublished observations quoted in Francis, W. (1961). *Coal*, 2nd ed., p. 753. London: Arnold.

Douglas, A. G., Douraghi-Zadeh, K., Eglinton, G., Maxwell, J. R. & Ramsay, J. N. (1969). In *Advances in Organic Geochemistry*, 1966, G. D. Hobson and G. C. Speers (eds.). Oxford: Pergamon Press. pp. 315–34.

Doyle, R. W. (1968). Isolation of enzymes and other organic ions from fresh water sediment. *Limnol. Oceanogr.* **13**, 518–22.

Drake, J. A. & Owen, R. (1950). Report on the experimental removal of phosphorus with lime. *Minn. Health Dep.* (mimeographed report). 1–17.

Dubach, P. & Lynch, D. L. (1959). Comparison of the determination of uronic acids in soil extracts with carbazole and by decarboxylation. *Soil Sci.* **87**, 273.

Dubovenko, E. K. (1964). Phosphatase activity of different soils. *Zhivlennya ta Vdobr.* 1964, 29.

Duff, R. B., Webley, D. M. & Scott, R. O. (1963). Solubilization of minerals and related minerals by 2-ketogluconic acid-producing bacteria. *Soil Sci.* **95**, 105.

Dunning, H. N. (1963). Geochemistry of organic pigments. In *Organic Geochemistry*, I. A. Breger (ed.), pp. 367–430. Oxford: Pergamon Press.

Dunning, H. N., Moore, J. W. & Myers, A. T. (1954). Properties of porphyrins in petroleum. *Ind. Engng. Chem.* **46**, 2000–7.

Dunton, M. L. & Hunt, J. M. (1962). Distribution of low molecular-weight hydrocarbons in Recent and ancient sediments. *Bull. Am. Ass. Petrol. Geol.* **46**, 2246–8.

393

References

Durand, G. (1961). Degradation of purine and pyrimidine bases in soil; aerobic degradation of uric acid. *Compte rend.* **252**, 1687.

Duxbury, A. C. & Yentsch, C. S. (1956). Plankton pigment nomographs. *J. mar. Res.* **15** (1), 92–101.

Dyer, W. J., Wrenshall, C. L. & Smith, G. R. (1940). Isolation of phytin from the soil. *Science, N.Y.* **91**, 319–20.

Eastman, J. W., Androes, G. M. & Calvin, M. (1962). Election-spin resonance absorption and other properties of some solid hydrocarbon-quinone complexes. *J. chem. Phys.* **36**, 1197.

Eckelmann, W. R., Broeker, W. S., Whitlock, D. W. & Allsup, J. R. (1962). Implications of carbon isotopic composition of total organic carbon of some Recent sediments and ancient oils. *Bull. Am. Ass. Petrol. Geol.* **46**, 699–704.

Edwards, A. B. (1945). The composition of Victorian brown coals. *Proc. Australas. Inst. Min. Metall.* n.s. **140**, 206.

Eglinton, G. & Calvin, M. (1967). Chemical Fossils. *Sci. Am.* **216**, 32–43.

Eglinton, G. Douglas, A. G., Maxwell, J. R., Ramsay, J. N. & Stallberg-Stenhagen, S. (1966*a*). Occurrence of isoprenoid fatty acids in the Green River Shale. *Science, N.Y.* **153**, 1133–4.

Eglinton, G., Maxwell, J. R., Murphy, Sister M. T. J., Henderson, W. & Douraghi-Zadeh, K. (1966*b*). Hydrocarbons and fatty acids in algal shales and related materials. *Prog. Geol. Soc. Am.* *79th Ann. Mtg., San Francisco*, 59.

Eglinton, G., Scott, P. M., Belsky, T., Burlingame, A. L., Richter, W. & Calvin, M. (1966*c*). Occurrence of isoprenoid alkanes in a Precambrian sediment. In *Adv. in Organic Geochemistry, 1964*, G. D. Hobson and M. Louis (eds.), pp. 41–74. Oxford: Pergamon Press.

Eglinton, G. & Murphy, Sister M. T. J. (eds.) (1970). *Organic Geochemistry—Methods and Results*, pp. 1–828. Berlin: Springer-Verlag.

Ehrenberg, A. & Eriksson, L. E. G. (1965). E.S.R. spectra of flavine free radicals; assignment of hyperfine splittings to the benzenoid part. *Archs. Biochem. Biophys.* **110**, 628.

Elias, M. K. (1937). Depth of deposition of the Big Blue (Late Paleozoic) sediments in Kansas. *Bull. geol. Soc. Am.* **48**, 403–32.

Emery, K. O. & Rittenberg, S. C. (1952). Early diagenesis of California basin sediments in relation to origin of oil. *Bull. Am. Ass. Petrol. Geol.* **36**, 735–806.

Enders, C. (1943). Origin of humus in nature. *Angew. Chem.* **56**, 281.

Eno, C. F. & Reuszer, H. W. (1950). Availability of potassium in certain minerals to *Aspergillus niger*. *Soil Sci.* (*Soc. Am. Proc.*) **15**, 155.

Erdman, J. G. (1961). Some chemical aspects of petroleum genesis as related to the problem of source bed recognition. *Geochim. cosmochim. Acta* **22**, 16–36.

Erdman, J. G., Marlett, E. M. & Hanson, W. E. (1958*a*). The quantitative determination of low molecular weight aromatic hydrocarbons in aquatic sediments. *Progm. Am. chem. Soc.* *134th An. Mtg, Chicago*, Sept. 7–12.

Erdman, J. G., Marlett, E. M. & Hanson, W. E. (1958*b*). The occurrence and distribution of low molecular weight aromatic hydrocarbons in recent and ancient carbonaceous sediments. *Progm. Am. chem. Soc.* *134th Ann. Mtg., Chicago*, Sept. 7–12.

Erdtman, H. & Pettersson, T. (1950). Studies on the sulfonation of lignin. II. Sulfonation of ethoxylated lignin sulfonic acids of low sulphur content. *Acta chem. scand.* **4** (7), 971–7.

394

Evans, E. D., Kenny, G. S., Meinschein, W. G. & Bray, E. E. (1957). Distribution of N-paraffins and separation of hydrocarbons from Recent marine sediments. *Analyt. Chem.* **29**, 1858–61.

Faber, W. & Krejci-Graf, K. (1936). Zur Frage des geologischen Vorkommes organischer Kalkverbindungen. *Mineral Petrol. Mitt.* **48**, 305–16.

Falk, M. & Smith, D. B. (1963). Structure of carboxyl groups in humic acids. *Nature, Lond.* **200**, 569.

Felts, W. M. (1954). Occurrence of oil and gas and its relation to possible source beds in continental Tertiary of intermountain region. *Bull. Am. Ass. Petrol. Geol.* **38**, 1661–70.

Fermi, C. (1910). The presence of enzymes in soil, water and dust. *Zentbl. Bakt. ParasitKde* Abt. II, **26**, 330.

Fine, M. M. & Wahl, W. C. (1964). Iron ore pellet binders from lignite deposits. *U.S. Bur. Min. Rep. Inv.* (6564) 1–18.

Fischer, F. & Schrader, H. (1922). *Entstehung und chemische Struckten der Kohle.* Brenns. Chem. **3**, 67.

Flaig, W. (1960). Chemistry of humus materials. *Suom. Kemistilehti* **33** A, 229.

Flaig, W. (1964). Effects of micro-organisms on the transformation of lignin into humic substances. *Geochim. cosmochim. Acta* **28**, 1523–35.

Flaig, W. (1966). Chemistry of humic substances in relation to coalification. In *Coal Science, Am. chem. Soc., Adv. Chem. Ser.* **55**, 58–68.

Flaig, W. & Schulze, H. (1952). The mechanism of formation of synthetic humic acids. *Z. PflErnähr. Düng. Bodenk.* **58**, 59.

Fogg, G. E. (1956). The comparative physiology and biochemistry of the blue-green algae. *Bact. Rev.* **20**, 148–65.

Fogg, G. E. & Belcher, J. H. (1961). Pigments from the bottom deposits of an English lake. *New Phytol.* **60**, 129–42.

Folger, D. W., Burckle, L. H. & Heezen, B. C. (1967). Opal phytoliths in a North Atlantic dust fall. *Science, N.Y.* **155**, 1243–4.

Foreman, F. (1955). Palynology in southern North America. Part 2. Study of two cores from lake sediments of the Mexico City Basin. *Bull. geol. Soc. Am.* **66**, 475–509.

Forsman, J. P. (1963). Geochemistry of kerogen. In *Organic Geochemistry*, I. A. Breger (ed.), pp. 148–82. Oxford: Pergamon Press.

Forsman, J. P. & Hunt, J. M. (1958). Insoluble organic matter (kerogen) in sedimentary rocks. *Geochim. cosmochim. Acta* **15**, 170–82.

Forsman, J. P. & Hunt, J. M. (1958). Insoluble organic matter (kerogen) in sedimentary rocks of marine origin. *Habitat of Oil. Tulsa, Am. Ass. Petrol. Geol.* 747–78.

Fosse, R. (1916). Origin and distribution of urea in nature. *Annali. Chim.* (Paris) **6**, 13.

Fowler, W. A. (1962). Nuclear clues to the early history of the solar system. *Science* **135**, 1037–45.

Fox, C. S. (1931). Coal in India. I. The natural history of Indian coal. *Mem. geol. Surv. India* **57**, pp. i–viii, 1–241, 10 pls.

Fox, D. L. (1937). Carotenoids and other lipoid-soluble pigments in the sea and in deep marine mud. *Proc. natn. Acad. Sci. U.S.A.* **23**, 295–301.

Fox, D. L. & Anderson, L. J. (1941). Pigments from marine muds. *Proc. natn. Acad. Sci. U.S.A.* **27**, 333–7.

References

Fox, D. L., Isaacs, J. D. & Corcoran, E. F. (1952). Marine leptopel, its recovery, measurement and distribution. *Sears Found: J. mar. Res.* **11**, 29–46.

Fox, D. L., Updegraff, D. M. & Novelli, D. G. (1944). Carotenoid pigments in the ocean floor. *Archs. Biochem.* **5**, 1–23.

Fox, S. W. (1960). How did life begin? *Science* **132**, 200–8.

Fox, S. W. (1963). Prebiological formation of biochemical substances. In *Organic Geochemistry*, I. A. Breger (ed.) pp. 36–49. Oxford: Pergamon Press.

Francis, W. (1961). *Coal*, 2nd ed., pp. 1–806. London: Edward Arnold Ltd.

Fraser, G. K. (1943). *Peat deposits of Scotland*. D.S.I.R. Wartime Pamphlet (36).

Fraser, G. K. (1955). In *Chemistry of the Soil*, Bear, F. E. *et al.* pp. 149–76. New York: Reinhold.

French, C. S., Smith, J. H. C. & Virgin, H. I. (1957). Fluorescence spectra of protochlorophyll, chlorophylls *c* and *d*, and their pheophytins. In *Research in Photosynthesis*, H. Brown *et al.* (eds.), pp. 17–18. New York: Interscience Pub.

Freney, J. R. (1961). The nature of organic sulfur compounds in soil. *Aust. J. agric. Res.* **12**, 424.

Freudenberg, K. (1933). *Tannin, Cellulose and Lignin*. Berlin.

Freudenberg, K. (1950). Uber die Entstehung des Lignins und den Verholzungssoorgang. *Angew. Chem.* **62**, 30.

Freudenberg, K., Reznik, H., Fuchs, W. & Reichert, M. (1955). Formation of lignin and of wood. *Naturwissenschaften* **42**, 29.

Frey, D. G. (1949). Morphometry and hydrography of some natural lakes of the North Carolina coastal plain. *J. Elisha Mitchell scient. Soc.* **65**, 1–37.

Frey, D. G. (1953). Regional aspects of the late-glacial and post-glacial pollen succession of south-eastern North Carolina. *Ecol. Monogr.* **23**, 289–313.

Friedel, R. A. & Orchin, M. (1952). *Ultraviolet spectra of aromatic compounds*, pp. 1–52, 579 spectra. New York: John Wiley and Sons.

Friedlander, H. Z., Saldick, J. & Frink, C. R. (1963). Electron spin resonance spectra in various clay minerals. *Nature, Lond.* **199**, 62.

Fruton, J. S. & Simonds, S. (1958). *Gen. Biochemistry*, p. 1077. New York: Wiley.

Fuller, M. L. (1920). Carbon ratios in Carboniferous coals of Oklahoma and their relation to petroleum. *Econ. Geol.* **15**, 225–35.

Galstyan, A. S. (1958). Fermentative activity of some Armenian soils. *Dokl. Akad. Nauk armyan. SSR* **26**, 29.

Galstyan, A. S. (1959). Glucose oxidose activity in soil. *Izv. Akad. Nauk armyan SSR Biol. Nauk* **12** (4), 75.

Galstyan, A. S. (1965). O degridrogenazakh pochvy [on soil dehydrogenases]. *Dokl. Akad. Nauk SSSR* **156** (1), 166–7.

Geikie, A. (1882). *Textbook of Geology*, pp. 1–971. London: Macmillan.

Geissman, T. A. (1968). *Principles of Organic Chemistry*, 3rd edn. pp. 1–854. San Francisco: W. H. Freeman.

Geller, I. A. & Dobrotovorskaya, K. M. (1961). Phosphatase activity of soils in beet root-seedling areas. *Trudy Inst. Mikrobiol. Akad Nauk SSSR* **11**, 215.

Gelpi, E., Oró, J., Schneider, H. J. & Bennett, E. O. (1968). Olefins of high molecular weight in two microscopic algae. *Science, N.Y.* **161**, 700–1.

Gettkandt, G. (1956). Colorimetric determination of glucose in soil solutions and its application to Hofmann's enzyme method. *Landw. Forsch.* **9**, 155.

Ghildyal, B. P. & Gupta, U. C. (1959). Biochemical and microbiological changes during the decomposition of *Crotolaria juncea* (Sunn hemp) at different stages of growth in soil. *Pl. Soil* **11**, 312.

Gillbricht, M. (1952). The production of living matter in the Bay of Kiel. *Kieler Meeresforsch.* **8**, 174–91.

Gillet, A. (1956). From cellulose to anthracite. The development of humins ligneous matter. *Brennst. Chem.* **37**, 395.

Gillman, C. (1933). The hydrology of Lake Tanganyika. *Geol. Surv. Dep. Bull.* **5**, 1–27.

Gilson, H. C. (1964). Lake Titicaca. *Verh. Internat. Verein. Limnol.* **15**, 112–27.

Gimpelwich, E. D. (1959). Chemical composition of bitumens in Tertiary deposits of the central and north-eastern Caucasus region. *Trudȳ vses. nauchno-issled. geol.-razv. neft. Inst.* 1959 (17), 54–105.

Given, P. H. (ed.) (1965). *Coal Science, Am. chem. Soc. Adv. Chem. Ser.* **55**, pp. 1–743.

Given, P. H., Peover, M. E. & Wyss, W. F. (1960). Chemical properties of coal macerals. I. Introductory survey and some properties of exinites. *Fuel* **39**, 323.

Gjessing, E. T. (1965). Use of Sephadex gel for estimation of molecular weight of humic substances in natural water. *Nature, Lond.* **208**, 1091–2.

Glebovskaya, E. A. & Volkenschtein, M. V. (1948). Spectra of porphyrins in petroleums and bitumens. *J. gen. Chem. U.S.S.R.* (Eng. trans.) **18**, 1440–51.

Glover, S. R. (1947). Oil and gas exploration in Washington, *Wash. Div. Mines Geol. Inf. Circ.* (15).

Glover, S. R. (1953). Oil and gas exploration in Washington, *Wash. Div. Mines Geol. Inf. Circ.* (15) (suppl).

Goedheer, J. C. (1958). Investigations on bacteriochlorophyll in organic solutions. *Biochim. biophys. Acta* **27**, 478–90.

Goguetidze, K. D. (1937). The petroleum excursion–the Georgian Soviet Socialist Republic. *Seventeenth Int. geol. Congr.* Moscow, U.S.S.R., **4**, 22–5.

Golumbic, C., Hofer, L. J. E., Peebles, W. C. & Orchin, M. (1950). Detection of abietic acid in bituminous coal. *J. Soc. chem. Ind., Lond.* **69**, 100–2.

Gorham, E. (1960). Chlorophyll derivatives in surface muds from the English lakes. *Limnol. Oceanogr.* **5**, 29–33.

Gorham, E. (1961). Chlorophyll derivatives, sulphur, and carbon in sediment cores from two English lakes. *Can. J. Bot.* **39**, 333–8.

Goryunova, S. V. (1954). Characterization of dissolved organic substances in water of Beloe Lake (White Lake). *Trudȳ Inst. Mikrobiol., Akad Nauk SSSR*, **3**, 185–93.

Greenland, D. J. (1956). The absorption of sugars by montmorillonite. *J. Soil Sci.* **7**, 319, 329.

Greenwood, D. J. (1961). The effect of oxygen concentration on the decomposition of organic materials in soil. *Pl. Soil* **14**, 360–76.

Gritsevskaya, G. L. (1958). Hydrochemistry of the Suna River basin. *Trudȳ karel'. Fil. Akad. Nauk SSSR* **18**, 158–93.

Grov, A. (1963). Amino acids in soil. III. Acids in hydrolyzates of water-extracted soil and their distribution in a pine forest soil profile. *Acta chem. scand.* **17**, 2319.

Grov, A. & Alvsakar, E. (1963). Amino acids in soil. I. Water-soluble acids. *Acta chem. scand.* **17**, 2307.

Guild, F. N. (1922). The occurrence of terpin hydrate in nature. *J. Am. chem. Soc.* **44**, 216.

References

Gupta, U. C. (1967). Carbohydrates. In *Soil Biochemistry*, A. D McLaren and G. H. Peterson (eds.), pp. 91–118. New York: Marcel Dekker, Inc.

Gupta, U. C. & Snowden, F. J. (1963). Occurrence of free sugars in soil organic matter. *Soil Sci.* **96**, 217.

Gupta, U. C. & Snowden, F. J. (1964). Isolation and characterization of cellulose from soil organic matter. *Soil Sci.* **97**, 328.

Gupta, U. C. & Snowden, F. J. (1965). Determination of sugars and uronic acids in soils. *Can. J. Soil Sci.* **45**, 237.

Guzzo, A. V. & Tollin, G. (1964). Electron paramagnetic resonance (E.P.R.) studies of riboflavine and its derivatives. III. Isoalloxazine, semi-quinones at neutral and basic pH. *Archs. Biochem. Biophys.* **105**, 380.

Haines, T. H. (1964). A new sulfolipid in microbes. *Diss. Abstr.* **25**, 2203.

Halstead, R. L. (1964). Phosphate activity of soils influenced by lime and other treatments. *Can. J. Soil Sci.* **44**, 137.

Harris, D. G. & Zscheile, F. P. (1943). Effects of solvents upon absorption spectra of chlorophyll *a* and *b;* their ultraviolet absorption spectra in ether solution. *Bot. Gaz.* **104**, 515–27.

Hartley, R. D. & Lawson, G. J. (1962). Chemical constitution of coal. IX. Ion exchange separation of sub-humic acids with hydrogen peroxide. *Fuel, Lond.* **41**, 447–56.

Hartman, R. T. & Brown, D. L. (1966). Methane as a constituent of the internal atmosphere of kascular hydrophytes. *Limnol. Oceanogr.* **11**, 109–12.

Haug, P., Schnoes, H. K. & Burlingame, A. L. (1967). Isoprenoid and dicarboxylic acids isolated from Colorado Green River Shale (Eocene). *Science, N.Y.* **158**, 772–3.

Hausding, A. (1921). *A Handbook on the Winning and Utilization of Peat.* London: D.S.I.R.

Hayashi, T. (1956). Components of soil humic acid. III. Nitrogenous constituents of A-type humic acids. *J. Sci. Soil Manure, Tokyo* **26**, 371–5.

Hayashi, T. & Nagai, T. (1959). The components of soil humic acids. V. The colloid titration of each component. *Nippon Dojo Hiryogaku Zasshi* **28**, 369–73 (1957).

Hayatsu, R. (1961). Stellar evolution in early phases of gravitational attraction. *Publs. astr. Soc. Japan* **13**, 450–2.

Hedberg, H. D. (1945). Review of 'Geologia do Brasil' by de Oliviera and Leonardos. *Bull. Am. Ass. Petrol. Geol.* **29**, 289–95.

Hedberg, H. D. (1968). Significance of high-wax oils with respect to genesis of petroleum. *Bull. Am. Ass. Petrol. Geol.* **52**, 736–50.

Heide, S. van der (1950). Compaction as a possible factor in Upper Carboniferous rhythmic sedimentation. *Rep. 18th Int. geol. Congr.*, Pt. 4, 38–45.

Hemmerich, P., Dervartanian, D. V., Veeger, C. & van Voorst, J. D. W. (1963). Evidence from electron-spin resonance spectra for metal chelation shifting flavin–leucoflavin equilibria toward radical state. *Biochim. biophys. Acta* **77**, 504–6.

Henderson, M. E. K. (1955). Release of aromatic compounds from birch and spruce sawdusts during decomposition by white-rot fungi. *Nature, Lond.* **175**, 634–5.

Henderson, M. E. K. (1957). Metabolism of methoxylated aromatic compounds by soil fungi. *J. gen. Microbiol.* **16**, 686–95.

Henderson, M. E. K. & Duff, R. B. (1963). The release of metallic and silicate ions from minerals, rocks, and soils by fungal activity. *J. Soil Sci.* **14**, 236.

Heslop-Harrison, J. (1968). Pollen wall development. *Science, N.Y.* **161**, 230–7.

Hilpert, R. S. & Littman, E. (1934). The resinification of sugars by acids and its bearing on the determination of lignin. *Ber. dt. chem. Ges.* **67** *B*, 1551.

Hobbie, J. E., Crawford, C. C. & Webb, K. C. (1968). Amino acid flux in an estuary. *Science, N.Y.* **159**, 1463–4.

Hobson, G. D. & Louis, M. (eds.) (1966). *Advances in Organic Geochemistry* (1964), pp. 1–300. New York: Macmillan Co.

Hobson, G. D. & Speers, G. C. (1969). *Advances in Organic Geochemistry* (1966). pp. 1–577. Oxford: Pergamon Press.

Hodgson, G. W., Hitchon, B., Elofson, R. M., Baker, B. L. & Peake, E. (1960). Petroleum pigments from Recent freshwater sediments. *Geochim. cosmochim. Acta* **19**, 272–88.

Hoefs, J. & Schidlowski, M. (1967). Carbon isotope composition of carbonaceous matter from the Precambrian of the Witwatersrand System. *Science, N.Y.* **155**, 1096–7.

Hoering, T. C. (1965). The extractable organic matter in Precambrian rocks and the problem of contamination. *Carnegie Inst.* Year Book **64**, 215.

Hoering, T. C. (1968). The organic geochemistry of Precambrian rocks. In *Researches in Geochemistry*, vol. 2, P. H. Abelson (ed.), pp. 87–111. New York: John Wiley and Sons, Inc.

Hoffman, G. (1963). Synthetic effects of soil enzymes. In *Recent Progress in Microbiology*, vol. 8, N. E. Gibbons (ed.), pp. 230–4, 720. University of Toronto Press.

Hofmann, E. & Hoffman, G. (1953). α- and β-Glycosidases in the soil. *Naturwissenschaften* **40**, 511.

Hofmann, E. & Hoffman, G. (1954). Enzyme system of cultivated soil. V. α- and β-Galactosidase and α-glucosidase. *Biochem. Z.* **325**, 329.

Hofmann, E. & Hoffman, G. (1955). The enzyme system of arable soils. *Z. PflErnähr. Düng. Bodenk.* **70**, 97.

Hofmann, E. & Seegerer, A. (1951). Enzyme systems of our cultivated soil. I. Sucrase. *Biochem. Z.* **322**, 174.

Holton, R. W., Blecker, H. H. & Stevens, T. S. (1968). Fatty acids in blue-green algae: possible relation to phylogenetic position. *Science, N.Y.* **160**, 545–7.

Horowitz, N. H. & Miller, S. L. (1962). Current theories on the origin of life. *Fortschr. Chem. Org. Nuturestoffe* **20**, 423–459.

Hough, L., Jones, J. K. N. & Wadman, W. H. (1952). An investigation of the polysaccharide components of certain fresh water algae. *J. chem. Soc.* 3393–9.

Huang, T. K. (1947). Report on geological investigation of some oilfields in Sinkiang. *Nat. Geol. Survey, China*, Ser. A, (21).

Hubacek, J. & Lustigova, H. (1962). Humic acids in Czech. brown coals and their determination. *Pr. Úst. Výzk. Paliv* **4**, 357–80 (in Czech).

Hudson, R. G. (1924). On the rhythmic succession of the Yoredale Series in Wensleydale. *Proc. Yorks. geo. Soc.* n.s. **20**, 125–35.

Hunt, J. M. & Jamieson, G. W. (1956). Oil and organic matter in source rocks of petroleum. *Bull. Am. Ass. Petrol. Geol.* **40**, 477–88.

Hunt, J. M., Stewart, F. & Dickey, P. A. (1954). Origin of hydrocarbons of Uinta Basin. *Bull. Am. Ass. Petrol. Geol.* **38**, 1671–98.

Hurd, C. D. & Edwards, O. E. (1949). Thermal degradation of sugars. *J. org. Chem.* **14**, 680.

Hutchinson, G. E. (1937). A contribution to the limnology of arid regions. *Trans. Conn. Acad. Arts. Sci.* **33**, 47–132.

References

Hutchinson, G. E., Patrick, R. & Deevey, E. S. Jr. (1956). Sediments of Lake Patzcuaro, Michoacan, Mexico. *Bull. geol. Soc. Am.* **67**, 1491–1504.

Hutchinson, G. E. & Wollack, A. (1940). Studies on Connecticut lake sediments. II. Chemical analyses of a core from Linsley Pond, North Branford. *Am. J. Sci.* **238**, 493–517.

Hvorslev, M. J. & Stetson, H. C. (1946). Free-fall coring tube: a new type of gravity bottom sampler. *Bull. geol. Soc. Am.* **57**, 935–50.

Ilag, L. & Curtis, R. W. (1968). Production of ethylene by fungi. *Science, N.Y.* **159**, 1357–8.

Iloff, P. M. & Mirov, N. T. (1954). Composition of turpentines of pines, XIX. *J. Am. pharm. Ass.* **43**, 373–8.

Ishikawa, H., Schubert, W. J. & Nord, F. F. (1963). Lignins and lignification XXVII. The enzymic degradation of softwood lignin by white-rot fungi. *Archs. Biochem. Biophys.* **100**, 131–9.

Ishiwatari, R. & Hanya, T. (1965). Infrared spectroscopic characteristics of humic substances extracted from some Recent sediments. *Nippon Kagaku Zasshi* **86** (12), 1270–4 (in Japanese).

Ivanov, E. Ya. (1958). Peat shrinkage. *Inzh.-fiz.Zh., Akad. Nauk Belorus SSR.* (11), 113–16.

Ivarson, K. C. & Snowden, F. J. (1962). Carbohydrate material in soil. I. Colorimetric determination of uronic acids, hexoses, and pentoses. *Soil Sci.* **94**, 245.

Jackman, R. H. & Black, C. A. (1952). Phytose activity in soils. Hydrolysis of phytate phosphorus in soils. *Soil Sci.* **73**, 117, 167.

Jackson, M. L. & Sherman, M. L. (1953). Chemical weathering of minerals in soils. *Adv. Agron.* **5**, 219.

Jaquin, F. (1960). Chromatographic study of various types of humic acids. *Compte rend.* **250**, 1892–3.

Jaquin, F. (1961). Evolution of amino acids during the decomposition of the organic matter of soil. *Compte rend.* **251**, 1810–11.

Jeffery, P. M. & Reynolds, J. H. (1961). Origin of excess ^{129}Xe in stone meteorites. *J. Geophys. Res.* **66** (10), 3582–83.

Jeffrey, L. M., Pasby, B. F., Stevenson, B. & Hood, D. W. (1964). Lipids of ocean water. In *Advances in Organic Geochemistry*, U. Columbo and G. D. Hobson (eds.), pp. 175–97. New York: Macmillan Co.

Johnston, H. H. (1961). Common sugars found in soils. *Soil Sci. Soc. Am. Proc.* **25**, 415.

Johnston, H. H. (1964). Relation of brown humus to lignin. *Pl. Soil* **21**, 191–200.

Johnston, H. W. (1952). The solubilization of phosphate. I. The action of various organic compounds on dicalcium and tricalcium phosphate. *N.Z. Jl Sci. Technol.* B**33**, 436–46.

Johnston, H. W. (1954). The solubilization of 'insoluble' phosphate. II. A quantitative and comparative study of the action of selected aliphatic acids on tricalcium phosphate. *N.Z. Jl Sci. Technol.* B**36**, 49.

Johnston, H. W. (1959). Solubilization of 'insoluble' phosphates. V. The action of some organic acids on iron and aluminum phosphates. *N.Z. Jl Sci* **2**, 215–18.

Johnson, J. L. & Temple, K. L. (1964). Some variables affecting the measurement of catalase activity in soil. *Soil Sci. (Soc. Am. Proc.)* **28**, 207.

Jones, J. D. & Vallentyne, J. R. (1960). Biogeochemistry of organic matter. I. Polypeptides and amino acids in fossils and sediments in relation to geothermometry. *Geochim. cosmochim. Acta* **21**, 1–34.

Jones, L. H. P., Milne, A. A. & Sanders, J. V. (1966). Tabashir: an opal of plant origin. *Science, N.Y.* **151**, 464–6.

Jorgensen, J. R. (1961). Studies on the nature and occurrence of organic acids in soils. *Univ. Minn. Doct. Diss. (Univ. Microfilm Ann Arbor, Mich.*, (63–1253).

Jorscik, I., Upor, E., Hohmann, E. & Juhass, S. (1963). Fixation of radioactive elements on coal and the nature of the bond between uranium and humic acid. *Acta chim. hung.* **35** (2), 225–32.

Judson, S. & Murray, R. C. (1956). Modern hydrocarbons in two Wisconsin lakes. *Bull. Am. Ass. Petrol. Geol.* **40**, 747–61.

Kappen, H. (1913). The catalytic power of agricultral soil. *Fuhlings, Landw. Ztg.* **62**, 377.

Karrer, P. & Jucker, E. (1950). *Carotenoids* (translated and revised by Ernest A. Bruade), pp. 1–384. New York: Elsevier Press.

Karrer, P. & Koenig, H. (1940). Carotinoide der Purpurbakterien v. über Rhodoviolascin. *Helv. chim. Acta* **23**, 460–3.

Karttunin, T. & Sneck, T. (1958). The effect of humus in sand on concrete strength. *Nord. Betong* **2**, 389–96.

Kasatochkin, V. I., Kinoneva, M. M., Larina, N. K. & Egorova, O. I. (1964). Spectral and X-ray investigations of soil humus substances. *Fiz., Khim., Biol. i Mineralog. Pochv. SSSR, Akad. Nauk SSSR Dokl. k VIII–mu[Vos'memu] Mezhdunai Kongr. Pochvovedov, Bucharest*, 195–205.

Kates, M. *et al.* (1965). Aliphatic diether analogues of glyceride-derived lipids. *Biochemistry* **4** (8) 1595–9.

Kaurichev, S. S., Fedorov, E. A. & Shnabel, S. A. (1960). Separation of humic acids by the methods of continuous electrophoresis on paper. *Pochvovedenie* 1960 (10), 31–6.

Keeney, D. R. & Bremner, J. M. (1964). Effect of cultivation on the nitrogen distribution in soils. *Soil Sci. (Soc. Am. Proc.)* **28** (5), 653–6.

Kent, P. E. (1954). Oil occurrences in Coal Measures of England. *Bull. Am. Ass. Petrol. Geol.* **38**, 1699–1713.

Kern, W. (1947). Über das Vorkommen von Chrysen in der Erde. *Helv. chim. Acta* **30**, 1595–9.

Khanov, S. (1959). Bituminological studies of the red rocks of the Cheleken Formation. *Izv. Akad. turkmen SSR.* (3), 87–9.

Kindscher, E. (1924). Über ein Vorkommen von Kautschuk in Mitteldeutschen Braukohlenlagern. *Ber.* **57**, 1152–7.

Kiss, I. & Peterfi, I. (1959). Synthetic action of maltase and invertase in the soil. *Studia Univ. Babes-Bolyai* Ser. 2, 1959 (2), 179.

Kiss, I. & Peterfi, I. (1960). Inhibiting the activity of soil maltase. *Pochvovedenie* **8**, 84.

Kiss, I. & Peterfi, I. (1961). Paper-chromatographic method for identifying the phosphomonoesterases of the soil. *Studia Univ. Babes-Bolyai* Ser. 2, 1961 (2), 292.

Kivekas, J. (1939). Studies on the organic nitrogen compounds of soil. I. The water-soluble fraction. *Acta chem. fenn. Contrti Bot. Univ. Babes-Bolyai*, 12B, 1.

References

Klason, P. (1908). Chemical Composition of Deal (Fir Wood). *Ark. Kemi Miner. Geol.* **3**, 17.

Kleerekoper, H. (1957). Une etude limnologique de la chimie des sediments de fond des lacs de l'Ontario meridional Canada. *Uitgeverig Excelsior, s' Gravenhage*, 1–205.

Kleinhempel, D. & Heike, W. (1965). Separation of humic acids by gel diffusion. *Albrecht-Thaer-Arch.* **9** (2), 165–72.

Kleist, H. & Muecke, D. (1966). Stable free radicals in humic acids. *Experientia* **22**, 136–7.

Klimov, B. K. & Kazakov, E. F. (1937). Carotin (provitamin A) in sapropel. *C. r. Acad. Sci. URSS* **16**, 321–3.

Kohl, R. A. & Taylor, S. A. (1961). Hydrogen bonding between the carbonyl group and Wyoming bentonite. *Soil Sci.* **91**, 223.

Kononova, M. M. (1961). *Soil Organic Matter*, pp. 1–450. Oxford: Pergamon Press.

Kononova, M. M. & Tutova, N. A. (1961). Application of paper electrophoresis in the fractionation of soil organic matter and a study of its complex ion compounds. *Pochvovedenie* 1961, 81–8.

Koppe, E. F. (1966). Petrographic continuity of Pennsylvania coal. In *Coal Science, Am. chem. Soc., Adv. Chem. Ser.* **55**, 69–79.

Koslov, K. (1964). Enzymatic activity of the rhizosphere and soils in the East Siberia area. *Folia microbiol., Praha* (Prague), **9** (3), 145.

Koyama, T. (1966). Ratios of organic carbon, nitrogen and hydrogen in recent sediments. In *Coal Science, Am. chem. Soc., Adv. Chem. Ser.* **55**, 43–57.

Kozhov, M. (1963). *Lake Baikal and its life*, pp. xi, 1–344. The Hague: Dr W. Junk Publ., Biol. Mon.

Kozminski, Z. (1938). Amount and distribution of the chlorophyll in some lakes of north-eastern Wisconsin. *Trans. Wis. Acad. Sci. Arts Lett.* **31**, 411–38.

Kraft, F. (1907). Über fraktionierte Destillation der hoheren normal Paraffins aus Braunkohle im Vacuum des Kathodenlichts. *Ber.* **40**, 4479–84.

Kramer, M. (1957). Phosphatase-enzym-Aktivitat als Anzeiger des biologisch nutzbaren Phosphors im Boden. *Ndtúrwissenschaften* **44**, 13.

Kröger, C. & Bürger, H. (1959). Physiochemical properties of hard-coal constituents. X. Autoxidation and chemical constitution. *Brennst.-Chem.* **40**, 76.

Kroplein, H. (1964). Some organic compounds extracted from *Posidonia* shale. In *Adv. in Organic Geochemistry*, U. Columbo and G. D. Hobson (eds.), pp. 165–7. New York: Macmillan Co.

Kugler, H. G. (1939). A visit to Russian oil districts. *J. Instn. Petrol. Technol.* **25**, 83.

Kukharenko, T. A. (1948). Investigations on hymatomelanic acids by the chemosorption method. *Zh. prikl. Khim. Leningr.* **21**, 2.

Kukharenko, T. A. (1959). Changes in structure and properties of humic acids during coalification. *Genezis Tverd. Goryuch. Iskopaem., Akad. Nauk SSSR, Inst. Goryuch. Iskopaem.* 1959, 319–37.

Kukharenko, T. A. & Ekaterinina, L. N. (1960). The humic acids of weathered coals. *Tr. Inst. Goryuch. Iskop. Akad. Nauk SSSR* **14**, 58–72.

Kukharenko, T. A., Ekaterinina, L. N., Kulakov, V. K. & Mel'nikov, V. Z. (1965). Utilization of the Irsha-Borodino and Nazarovo surface-pit brown coals for obtaining humic fertilizers. *Khim. Pererabotka Topliv (Khim. i. Tekhnol) Akad. Nauk SSSR, Inst. Goryuch. Iskop.* 1965, 43–54.

Kukharenko, T. A. & Ryzhova, Z. A. (1960). The weathering of bituminous coals of different stages of metamorphosis. *Tr. Inst. Goryuch. Iskop. Akad. Nauk SSSR* **14**, 44–57.

Kuksar, G., Benko, K., Hernadi, F., Valu, G. & Kiss, S. (1962). Electron microscope studies of Streptomyces spores. *Acta biol. hung.* **12** (4), 231–41.

Kumada, K. & Kuawanura, Y. (1965). Intrinsic viscosity and functional groups of humic acids. *Nippon Dojo-Hiryogaky Zasshi* **36** (12), 367–72.

Laatsch, W. (1944). Dynamik der deutschen Acker und Waldböden, Untersuchungen über die Bildung und Anreicherung von Humusstoffen. *Beitr. Agrarwiss.* 3.

Ladd, J. N. (1964). The metabolism of model compounds related to soil humic acid. I. The decomposition of *N*-(*o*-carboxyphenyl) glycine. *Aust. J. biol. Sci.* **17** (1), 153–69.

Lagercrantz, C. & Yhland, M. (1962). Free radicals in some natural resins. *Acta chem. scand.* **16**, 505.

Langenheim, J. H. (1969). Amber: a botanical inquiry. *Science, N.Y.* **163**, 1157–69.

Larina, N. K., Kasatochkin, V. I. (1965). Comparative investigation of humic acids of oxydized lithoietal coals and asphaltenes. *Obogasheh. i Kompleksn. Ispol' 2 Topliva. Inst. Goryuch. Iskop.* 1965, 20–28.

Larkin, P. A. (1964). Canadian Lakes. *Verh. int. Verein. theor. angew. Limnol* **15**, 76–90.

Latter, P. & Burges, A. (1960). Experimental decomposition of humic acid by fungi. *Trans. Int. Congr. Soil Sci. 7th*, Madison, Wisc. **3**, 643.

Law, J. & Ebert, M. (1965). Electron spin resonance spectra of poly (methyl) methacrylate after exposure to nitric oxide and nitrogen dioxide. *Nature, Lond.* **205**, 1193.

Lawlor, D. L. & Robinson, W. E. (1965). Organic acids from the Green River Form. *Div. Petrol. Chem. Am. Chem. Soc.* Detroit Meeting May 9.

Lederer, E. (1938). Sur les carotenoides des cryptogames. *Bull. Soc. Chim. biol.* **20**, 611–34.

Leger, F. & Hibbert, H. (1938). Lignin and related compounds. XXXIV. Acetovanilline and acetosyringone as degradation products of lignin-sulfonic acids. *J. Am. chem. Soc.* **60**, 565.

Lenhard, G. (1957). Dehydrogenase activity of the soil as a measure of microbial activity in the soil. *Z. PflErnähr. Düng. Bodenk.* **73**, 1.

Lenhard, G. (1962). Bestimmung der verfugbaren Pflanzennahrstoffe durch Ermitteung der Dehydrogenase-Aktivitat des Bodens. *Z. PflErnähr Düng. Bodenk.* **99**, 182–90.

Leo, R. F. & Parker, P. C. (1966). Branched chain fatty acids in sediments. *Science, N.Y.* **152**, 649–50.

Lepper, G. W. & Evans, P. (1938). Burma and Assam. In *Science of Petroleum*, pp. vi, 133–6. Oxford: Oxford University Press.

Lewis, H. P. (1922). The age of coals. *Fuel* **1**, p. 74.

Liebmann, H. (1950). Zur Biologie der Methanbacterien. *Gesundh. Ing.* **71**, 14–22, 56–59.

Lindberg, B. (1955). The chemistry of lichens. VIII. Investigation of a *Dermatocarpon* and some *Roccella* species. *Acta chem. scand.* **9**, 917.

Lindeman, R. L. (1941). The developmental history of Cedar Creek Bog, Minnesota. *Am. Midl. Nat.* **25**, 101–12.

References

Livingston, R. (1949). The photochemistry of chlorophyll. In Frank and Loomis *Photosynthesis in Plants*, pp. 179–96. Ames: Iowa State Press.

Livingstone, D. A. & Boykin, J. C. (1962). Vertical distribution of phosphorus in Linsley Pond mud. *Limnol Oceanogr.* **7**, 57–62.

Loeffler, H. (1959). Beitrage zu kenntnis der iranischen Binnegewasser. I. Der Nirez-See und sein Einzugsgebeit. *Int. Rev. Gesamt. Hydrobiol.* **44**, 227–76.

Louw, H. A. & Webley, D. M. (1959). A study of soil bacteria dissolving certain mineral phosphate fertilizers and related compounds. *J. appl. Bact.* **22** (2), 227–33.

Lowe, L. E. (1964). An approach to the study of the sulfur status of soils and its application to selected Quebec soils. *Can. J. Soil Sci.* **44**, 176.

Lowe, L. E. & De Long, W. A. (1961). Aspects of the sulfur status of three Quebec soils. *Can. J. Soil Sci.* **41**, 141.

Lowe, L. E. & De Long, W. A. (1963). Carbon-bonded sulfur in selected Quebec soils. *Can. J. Soil Sci.* **43**, 151.

Lundquist, G. (1927). 'Bodenablagerungen und Entwicklungstypen der Seen' (Lake sediments and development types). *Die Binnengewasser*, II, pp. 1–124. Stuttgart: E. Schweizerbart'sche Verlagsbuchhandlung.

Lynch, D. L., Hearns, E. E. & Cotnoir, L. J. (1957). Determination of polyuronides in soils with carbazole. *Soil Sci. (Soc. Am. Proc.)* **21**, 160.

McConnel, W. J. (1968). Limnological effects of organic extracts of litter in a southwestern empoundment. *Limnol. Oceanogr.* **13** 343–9.

McLaren, A. D. & Peterson, G. H. (eds.) (1967). *Soil Biochemistry* pp. 1–509. New York: Marcel Dekker, Inc.

McLaren, A. D., Reshetko, L. & Huber, W. (1957). Sterilization of soil by irradiation with an electron beam, and some observations on soil enzyme activity. *Soil Sci.* **83**, 497.

McLaughlin, D. B. (1955). The Temiskaming Series—a pre-Cambrian analogue of the Newark?. *Penna. Acad. Sci. Proc.* **29**, 171–80.

McLean, J., Rettie, G. H. & Spring, F. S. (1958). Triterpenoids from peat. *Chemy Ind.* 1515–16.

Mackareth, F. J. H. (1958). A portable core sampler for lake deposits. *Limnol. Oceanogr.* **3**, 181–91.

Mahr, H. H. & Kaindl, K. (1954). Untersuchungen über die wirkung organischer düngmittel auf die verfügbarkeit der phosphorsäure. *Agrochimica* **3**, 270–8.

Maistrenko, Yu G. (1963). Organic substances in the Dnieper River water and changes they suffered during regulation of the river discharge. *Pratsi Inst. Hidrobiol. Akad. Nauk Ukr. RSR* **39**, 16–27.

Major, R. T. (1967). The Gingko, the most ancient living tree. *Science, N.Y.* **157**, 1270–3.

Malliard, L. C. (1911). Action of amino acids on sugars. Formation of melanoidins in a methodical way. *Compte rend.* **154**, 66.

Manning, W. M. & Strain, H. H. (1943). Chlorophyll *d*, a green pigment of red algae. *J. biol. Chem.* **151**, 1–19.

Marcusson, J. (1925–27). The composition of peat and the lignin theory (38); Lignin and oxycellulose theory (of the origin of coal) (39); *Z. angew. Chem.* **38**, 339; **39**, 898; **40**, 1233.

Mateles, R. I., Baruah, J. N. & Tannenbaum, S. R. (1967). Growth of a thermophilic bacterium on hydrocarbons; a new source of cell protein. *Science, N.Y.* **157**, 1322–3.

Meader, R. W. (1956). *Stratigraphy and limnology of a hard water lake in western Minnesota*. Master of Science Thesis, University of Minnesota (unpublished).

Megard, R. O. (1964). Biostratigraphic history of Dead Man Lake, Chuska Mountains, New Mexico. *Ecology* **45**, 529–46.

Mehta, N. C. & Deuel, H. (1960). The estimation of pentosan in soils. *Z. PflErnähr Düng Bodenk.* **90**, 209.

Meinschein, W. G. (1965). Soudan Formation: organic extracts of early Precambrian rocks. *Science, N.Y.* **150**, 601.

Meinschein, W. G. (1967). Paleobiochemistry: *McGraw-Hill Yearbook of Science and Technology*, 1967, pp. 283–5.

Meinschein, W. G., Barghoorn, E. S. & Schopf, J. W. (1964). Biological remnants in a Precambrian sediment. *Science, N.Y.* **145**, 262–3.

Meinschein, W. G. & Kenny, G. S. (1957). Analyses of chromatographic fraction of organic extract of soils. *Analyt. Chem.* **29**, 1153–61.

Merktiev, Sh. F. & Aliev, Ad. A. (1963). Geochemistry of organic substances in the Tertiary formations of the Caspian–Kuban territory. *Uch. Zap., Azerb. Gos. Univ., Ser. Geol-Geogr. Nauk* 1963 (3), 3–17.

Miller, S. L. (1953). A production of amino-acids under possible primitive earth conditions. *Science* **117**, 528–9.

Miller, S. L. (1957). The mechanism of synthesis of amino acids by electrical discharges. *Biochim. Biophys. Acta* **23**, 480–9.

Mitsui, S., Kumazawa, L. & Mukai, N. (1959). *Nippon Dojo-Hiryogaku Zasshi* **30**, 345; through Soil Plant Food (Japan) **5**, 150 (1959).

Moore, E. S. (1940). *Coal, its Properties, Analysis, Classification, Geology, Extraction, Uses, and Distribution*, pp. 1–473. New York: Wiley.

Moore, J. W. & Dunning, H. N. (1955). Interfacial activities and porphyrin contents of oil-shale extracts. *Ind. Engng. Chem. analyt. Edn* **47**, 1440–4.

Mora, P. T. & Wood, J. W. (1958). Synthetic polysaccharides. I. Polycondensation of glucose. *J. Am. chem. Soc.* **80**, 685.

Morrison, R. T. & Boyd, R. N. (1966). *Organic Chemistry*, 2nd ed., pp. 1–1204. Boston: Allyn and Bacon.

Mortland, M. M. & Gieseking, J. E. (1952). The influence of clay minerals on the enzymic hydrolysis of organic phosphorus compounds. *Soil Sci. (Soc. Am. Proc.)* **16**, 10.

Mouraret, M. (1965). *Contribution à l'étude de l'activité des enzymes du sol: l'asparaginase*. Paris: office de la Recherche Scientifiqe et Technique Outre-Mer.

Moyle, J. B. (1954). Some aspects of the chemistry of Minnesota surface waters as related to game and fish management. *Minn. Dept. Cons., Inv. Rep.* **151**, 1–36.

Muecke, D. & Kleist, H. (1965). Paper electrophoresis of metal–humic acid compounds. *Albrecht-Thaer-Arch.* **9** (4), 327–36.

Mueller, G. (1963). Organic cosmochemistry. In *Organic Geochemistry*, I. A. Breger (ed.), pp. 1–35. Oxford: Pergamon Press.

Müller, G. & Forster, I. (1961). The release of minerals from culture materials by soil fungi. *Zentbl. Bakt. ParasitKde* **114**, 1.

405

References

Müller, G. & Forster, I. (1964). The influence of microscopic soil fungi on liberation of nutrients from primary minerals as a result of biological dissociation. *Zentbl. Bakt. ParasitKde* 118, 589, 594.

Muraveisky, S. & Chertok, I. (1938). Carotenoids in lacustrine silts. *Dokl. Acad. Nauk SSSR* 19, 521–3.

Murchison, D. G. (1966). Properties of coal macerals. In *Coal Science, Am. chem. Soc., Adv. Chem. Ser.* 55, 307–31.

Murthy, V. R. (1964). Stable isotopes of some heavy elements in meteorites. In *Isotopic and Cosmic Chemistry*, A. Craig, S. L. Miller and G. J. Wasserburg (eds.), pp. 488–515. Amsterdam: North Holland Pub. Co.

Nagar, B. R. (1962). Free monosaccharides in soil organic matter. *Nature, Lond.* 194, 896.

Nagata, T. (1957). Relative ease of release of exchangeable cations absorbed on some soils, clays, crude humus and plant roots. *Nippon Dojo-Hiryogaku Zasshi* 28, 259–61.

Nakao, M. & Shibuye, C. (1924). Constituents of sanna oil. *J. Pharm. Soc.* (Japan), 513, 913–21.

Naumann, E. (1930). 'Einführung in die Bodenkunde der Seen' (Introduction to the soil science of lakes). *Die Binnengewasser*, IX, pp. 1–126. Stuttgart: E. Schweizerbart'sche Verlagsbuchhandlung.

Nevraev, G. H., Vadkobskaya, A. D. & Bakhman, V. I. (1964). Organic matter in weakly mineralized springs in the Truskavets resort and malye Skhodnitsy village. *Mater. po Izych. Lechebn. Mineralin Vod i Gryazei i Balneotekln*, Moscow, Sb. 64–8.

Newberry, J. S. (1874). The Carboniferous System. *Rep. geol. Surv. Ohio* 2 (1), 81–180.

Newell, N. D. (1949). Geology of the Lake Titicaca region, Peru and Bolivia. *Geol. Soc. Am. Mem.* 36, 1–111.

Noller, C. R. (1951). *Chemistry of Organic Compounds*, pp. 1–885. Philadelphia: W. B. Saunders.

Nooner, D. W. & Oró, J. (1967). Organic compounds in meteorites—I. Aliphatic hydrocarbons. *Geochim. cosmochim. Acta* 31, 1359–94.

Nosek, J. & Ambroz, Z. (1957). La faune du sol et l'activite microbenne du sol forestier. *Schweiz. Z. Forstw.* 108 (10/11), 559–70.

Obenaus, R. & Neumann, H. J. (1965). Desalting of humic acids with Sephadex G-25. *Naturwissenschaften* 52 (6), 131.

O'calla, P. S. & Lee, E. E. (1956). Synthetic polysaccharides. *Chem. Ind.* 1956, 522.

O'calla, P. S., Lee, E. E. & McGrath, D. (1962). The action of cation-exchange resins on D-glucose. *J. Am. chem. Soc.* 1962, 2730.

Ohle, W. (1965). Nährstoffanreicherung der Gewässer durch Düngemittel und Meliorationen. In Liebmann, H., *Münchuer Beiträge* 12, 54–83.

Okuda, A. & Hori, S. (1954). Chromatographic investigation of amino acids in humic acids and alkaline alcohol lignins. *Mem. Res. Inst. Food Sci.* (Kyoto) (7), 1–5.

Okuda, A. & Hori, S. (1956). Identification of amino acids in humic acid. *J. Sci. Soil Manure, Tokyo* 26, 346–8.

Oliviera de, A. I. & Leonardos, A. H. (1943). *Geologia do Brasil*, 2nd edn., pp. 1–813, pls. 1–37. Serviço de Informação, Ministeria da Agricultura. Rio de Janeiro.

Olson, W. S. (1954). Source-bed problem in Velasquez Field, Columbia. *Bull. Am. Ass. Petrol. Geol.* 38, 1645–52.

Orr, W. L. & Emery, K. O. (1956). Composition of organic matter in marine sediments: preliminary data on hydrocarbon distribution in basins off southern California. *Bull. geol. Soc. Am.* **67**, 1247–58.

Orr, W. L. & Emery, K. O., Grady, J. R. (1958). Preservation of chlorophyll derivatives in sediments off southern California. *Bull. Am. Ass. Petrol. Geol.* **42**, 925–62.

Orr, W. L. & Grady, J. R. (1957). Determination of chlorophyll derivatives in marine sediments. *Deep Sea Res.* **4**, 263–71.

Oró, J. (1963). Studies in experimental cosmo-chemistry. *Ann. N.Y. Acad. Sci.* **108**, 464–81.

Oró, J. & Guidry, C. L. (1960). A novel synthesis of polypeptides. *Nature, Lond.* **186**, 156–57.

Oró, J. & Kimball, G. P. (1961). Synthesis of purines under primitive earth conditions, **94**, 217–27.

Oró, J., Nooner, D. W., Zlatkis, A., Wikstrom, S. A. & Barghoorn, E. S. (1965). Hydrocarbons of biological origin in sediments about two billion years old. *Science, N.Y.* **148**, 77–9.

Osbon, C. C. (1919). Peat in the Dismal Swamp, Virginia and North Carolina. *U.S. geol. Surv. Bull.* **711 C**, 41–59.

Palacas, J. G. (1959). *Geochemistry of carbohydrates*, pp. 1–103. University of Minnesota Doctoral Dissertation.

Palacas, J. G., Swain, F. M. & Smith, F. (1960). Presence of carbohydrates and other organic substances in ancient sedimentary rocks. *Nature, Lond.* **185**, 234.

Pariser, R. (1950). A study of some oxidation-reduction reactions photosynthesized by chlorophyll and related substances, pp. 1–172. University of Minnesota Doctoral Dissertation.

Park, R. & Epstein, S. (1961). Carbon isotope fractionation during photosynthesis. *Geochim. cosmochim. Acta* **21**, 110–26.

Parker, P. L., Van Baalen, C. & Maurer, L. (1967). Fatty acids in eleven species of blue-green algae: geochemical significance. *Science, N.Y.* **155**, 707.

Parker-Rhodes, A. F. (1940). Preliminary experiments in the estimation of traces of heteroauxin in soils. *J. agric. Sci., Camb.* **30**, 654.

Parsons, T. R. & Strickland, J. D. H. (1963). Discussion of spectrophotometric determination of marine-plant pigments, with revised equations for ascertaining chlorophylls and carotenoids, *J. mar. Res.* **21** (3), 155–71.

Partridge, S. M. (1948). Filter paper partition chromatography of sugars. *Biochem. J.* **42**, 238–49.

Patrick R. (1954). The diatom flora of Bethany Bog. *J. Protozool.* **1**, 34–47.

Paul, E. A. & Schmidt, E. L. (1961). Formation of free amino acids in rhizosphere and nonrhizosphere soil. *Soil Sci. (Soc. Am. Proc.)* **25**, 359.

Pauling, L. (1951). The configuration of polypeptide chains in proteins. *Rec. chem. Progr. (Kresge-Hooker Sci. Lib.)* **12**, 155.

Paulsen, G. W. (1962). *Preservation and stratigraphic distribution of pigments in Minnesota lake sediments*. Master of Science Thesis, University of Minnesota (unpublished).

Percival, E. (1962). *Structural Carbohydrate Chemistry*, 2nd edn. London: J. Garnet Miller, Ltd.

Pershina, M. N. & Bykova, N. V. (1959). Composition of humic substances in desert alluvial soils. *Dokl. mosk. sel'-khoz. Akad. K.A. Timiryazeva* (42), 95–100.

References

Petrova, Y. N., Karpova, J. P. & Mandrykina, Y. M. (1956). Hydrocarbons of dispersed organic matters. *Dokl. Akad. Nauk SSSR* **108**, 885–8.

Phinney, H. K. (1946). A peculiar lake sediment of algal origin. *Am. Midl. Nat.* **35**, 453–9.

Phleger, F. B. (1951). Ecology of Foraminifera, north-west Gulf of Mexico. Part 1. Foraminifera distribution. *Mem. geol. Soc. Am.* **46**, 1–88.

Pictet, A. & Costan, P. (1920). Sur la glucosene. *Helv. chim. Acta* **3**, 645.

Pigman, W. (ed.) (1957). *The Carbohydrates*, pp. 1–902. New York: Academic Press.

Pinfold, E. S. (1954). Oil production from Upper Tertiary fresh-water deposits of west Pakistan. *Bull. Am. Ass. Petrol. Geol.* **38**, 1653–60.

Piper, S. H., Chibnall, A. C., & Williams, E. F. (1934). Melting points and long crystal spacings of the higher primary alcohols and W-fatty acids. *Biochem. J.* **28**, 2175–88.

Plunkett, M. A. (1957). The qualitative determination of some organic compounds in marine sediments. *Deep Sea Res.* **5**, 259–62.

Pokorna, V. (1964). Determination of lipolytic activity of high and low moor peats and muds. *Pochvovedenie* 1964 (1), 106.

Pommer, A. M. & Breger, I. A. (1960a). Potentiometric titration and equivalent weight of humic acid. *Geochim. cosmochim. Acta* **20**, 30–44.

Pommer, A. M. & Breger, I. A. (1960b). Equivalent weight of humic acid from peat. *Geochim. cosmochim. Acta* **20**, 45–50.

Ponnamperuma, C. Lemmon, R. M. Mariner R. & Calvin M. (1963a). Formation of adenine by electron radiation of methane, ammonia, and water. *Nat. acad. Sci. Proc.* **49**, 737.

Ponnamperuma, C. Mariner R. & Sagan C. (1963b). Formation of adenosine by ultraviolet irradation of a solution of adenine and ribose. *Nature, Lond.* **199**, 222.

Prashnowsky, A. A., Degens, E. T., Emery, K. O. & Pimenta, J. (1961). Organic materials in Recent and ancient sediments. Part 1. Sugars in marine sediments of Santa Barbara Basin. *Neues Jb. Geol. Paläont*, Mh. 400–13.

Prowse, G. A. & Talling, J. F. (1958). The seasonal growth and succession of plankton algae in the White Nile. *Limnol. Oceonogr.* **3**, 223–38.

Puddington, I. E. (1948). The thermal decomposition of carbohydrates. *Can. J. Res.* **26B**, 415–31.

Rabinowitch, E. (1951). *Photosynthesis and Related Processes*, pp. 600–1208, 2 (1). New York: Interscience Pub.

Ramsay, J. N. (1966). *Organic geochemistry of fatty acids*. Thesis, University of Glasgow.

Randall, R. B., Benger, M. & Grocock, C. M. (1938). The alkaline permanganate oxidation of organic substances selected for their bearing upon the chemical constitution of coal. *Proc. Roy. Soc. Lond.* **165A**, 432–52.

Randerath, K. (1963). *Thin Layer Chromatography* (translated by D. D. Libman), pp. 1–250. New York: Academic Press.

Ray, H., Guzzo, A. V. & Tollin, G. (1965). Nuclear complexes of flavins and phenols. II. Electrical and magnetic properties of crystalline solids. *Biochim. biophys. Acta* **94**, 258–70.

Reed, L. C. (1946). San Pedro oil field, Province of Salta, Northern Argentina. *Bull. Am. Ass. Petrol. Geol.* **30**, 541–605.

Reilly, J. & Emlyn, J. A. (1940). Preliminary note on Irish peat wax (mona wax). *Scient. Proc. R. Dubl. Soc.* **22**, 267–72.

408

References

Reilly, J., Kilbride, D. A. & Wilson, J. P. (1943). The isolation of behenic acid from refined mona wax. *Biochem. J.* **37**, 195–8.

Rex, R. W. (1960). Electron paramagnetic resonance (E.P.R.) studies of stable free radicals in lignins and humic acids. *Nature, Lond.* **188**, 1186.

Reynolds, J. H. (1960). Isotopic composition of xenon from enstatite chondrites. *Z. Naturforsch.* 15a, 1112–14.

Richards, F. A. (1952). The estimation and characterization of plankton populations by pigment analysis. I. The absorption spectra of some pigments occurring in diatoms, dinoflagellates, and brown algae. *J. mar. Res.* **9** (2), 147–72.

Richards, F. A. & Thompson, T. G. (1952). The estimation and characterization of plankton populations by pigment analyses. II. A spectrophotometric method for the estimation of plankton pigments. (Sears Foundation), *J. mar. Res.* **11**, 156–72.

Robertson, T. (1948). Rhythm in sedimentation and its interpretation; with particular reference to the Carboniferous sequence. *Trans. Edinb. geol. Soc.* **14**, 141–75.

Robinson, W. E., Lawlor, D. L., Cummins, J. J. & Fester, J. I. (1963). Oxidation of Colorado Oil Shale. *U.S. Bur. Mines Rept. Inv.* (6166), 1–33.

Rock, S. M. (1951). Qualitative analysis from mass spectra. *Analyt. Chem.* **23**, 261–8.

Roelofs, E. W. (1944). Water soils in relation to lake productivity. *Bull. Mich. agric. Coll. Exp. stn.* **190**, 1–31.

Rodrigues, G. (1954). Fixed ammonia in tropical soils. *J. Soil Sci.* **5**, 264.

Rogers, H. D. (1860). On the distribution and probable origin of rock oil of western Pennsylvania, New York, and Ohio. *Phil. Soc. Glasgow, Proc.* **4**, 355–9.

Rogers, H. D. (1866). On petroleum. *Phil. Soc. Glasgow, Proc.* **6**, 48–60.

Rogers, H. T. (1942). Dephosphorylation of organic phosphorus compounds by soil catalysts. *Soil Sci.* **54**, 439.

Rogers, M. A. (1965). Carbohydrates in aquatic plants and associated sediments from two Minnesota lakes. *Geochim. cosmochim. Acta* **29**, 183–200.

Rotini, O. T. (1933). Soil phosphatases. *Atti della Soc. Italiana per il Progresso delle Scienze, XXI Riunione, Roma, 1932, Frat. Fresi, Pavia*, **2**, 1933.

Rotini, O. T. & Carloni, L. (1953). Transformation of the metaphosphates into orthophosphates, promoted by the agrarian soil. *Annali Sper. agr.* **7**, 1789.

Rubey, W. W. (1951). Geologic history of sea water: an attempt to state the problem. *Bull. geol. Soc. Am.* **62**, 1111–48.

Ruhemann, S. & Rand, H. (1932). Über die Harze der Braunkohle. I. Die Sterine des Harzbitumens. *Brennst. Chem.* **13**, 341–5.

Runov, E. V. & Terekhov, O. S. (1960). Catalase activity in several forest soils. *Pochvovedenie* 1960 (9), **75**.

Russell, I. C. (1885). Geological history of Lake Lahonton a quaternary lake of north-western Nevada. *U.S. Geol. Surv. Monogr.* **11**, 1–288.

Rutten, M. G. (1952). Rhythm in sedimentation and erosion, *3rd Congr. Avanc. Étud. Stratigr. carb.* 529–37.

Ruttner, F. (1953). *Fundamentals of limnology* (translated by D. G. Frey and F. E. Fry), pp. 1–242. Toronto: University of Toronto Press.

Rybicka, S. M. (1959). The solvent extraction of a low-rank vitrain. *Fuel* **38**, 45.

Ryzicka, L. & Hosking, J. R. (1929). Über die agaten-disäure, die krystalliziete Hartsäure, $C_{20}H_{30}O_4$, des Kaurikopals, des Hart und des Weidimanidakopals. *Am. chem. J.* **469**, 147–92.

References

Rzoska, J., Brook, A. J. & Prowse, G. A. (1955). Seasonal plankton development in the White and Blue Nile near Khartoum. *Verh. int. Verein. theor. angew. Limnol.* **12**, 327–34.

Sakabe, T. & Sassa, R. (1952). An aromatic hydrocarbon isolated from coal bitumen. *Bull. chem. Soc. Japan* **25**, 353–5.

Sallans, H. R., Snell, J. M., MacKinney, H. W. & McKibbon, R. R. (1937). Water-soluble acid substances in the raw humus of podzol soils. *Can. J. Res.* **15B** 315–20.

Salton, M. R. J. (1964). *The bacterial cell wall*, pp. i–xii, 1–293. New York: Elsevier Press.

Sanchez, R. A., Ferris, J. P. & Orgel, L. E. (1966). Cyanoacetylene in prebiotic synthesis. *Science, N.Y.* **154**, 784–5.

Sanger, J. E. (1968). *A quantitative study of leaf pigments from initiation in buds to decomposition in soils.* University of Minnesota Doctoral Thesis, 1–252.

Savinov, B. G., Mikhaĭlovnina, A. A. & Shapiro, S. A. (1950). Carotene in medicinal muds of U.S.S.R. *Dokl. Acad. Nauk SSSR* **72**, 1087–90 (in Russian). (Also *C. r. Und. Sci. URSS* **72**, 1087–9.)

Schall, C. (1892). Notizen vermischen Inhalts. *Ber.* **25**, 1489–91.

Schapiro, N. & Gray, R. J. (1966). Physical variations in highly metamorphosed Antarctic coals. In *Coal Science, Am. chem. Soc., Adv. Chem. Ser.* **55**, 196–210.

Scharpenseel, H. W. & Krausse, R. (1962). Investigations on amino acids in various organic sediments, especially gray and brown humic acid fractions of different types of soil (including C^{14}-labelled humic acids). *Z. PflErnähr. Düng. Bodenk.* **96**, 11.

Schatz, A. (1963). Soil micro-organisms and soil chelation. The pedogenic action of lichens and lichen acids. *J. agric. Fd Chem.* **11**, 112.

Schatz, A., Cheronis, A. D., Schatz, V. & Trelawny, G. S. (1954). Chelation (sequestration) as a biological weathering factor in pedogenesis. *Proc. Pa. Acad. Sci.* **28**, 44.

Schatz, V., Schatz, A., Trelawny, G. S. & Barth, K. (1956). Significance of lichens as pedogenic (soil forming) agents. *Proc. Penn. Acad. Sci.* **30**, 62.

Schlidowski, M. (1965). Probable life forms from Precambrian of the Witwatersrand System (South Africa). *Nature, Lond.* **205**, 895–6.

Schmidt, E. L., Putnam, H. D. & Paul, E. A. (1960). Behavior of free amino acids in soil. *Soil Sci. (Soc. Am. Proc.)* **24** (2), 107–9.

Schnitzer, M. & Hoffman, I. (1965). Thermogravimetry of soil humic compounds. *Geochim. cosmochim. Acta* **29** (8), 859–70.

Schnitzer, M. & Skinner, S. I. M. (1966). Polarographic method for the determination of carboxyl groups in soil humic compounds. *Soil Sci.* **101** (2), 120–4.

Schopf, J. W. & Barghoorn, E. S. (1967). Alga-like fossils from the early Precambrian of South Africa. *Science, N.Y.* **156**, 508–11.

Schopf, J. W., Kvenvolden, K. A. & Barghoorn, E. S. (1968). Amino acids in Precambrian sediments: an assay. *Proc. natn. Acad. Sci. U.S.A.* **59**, 639–46.

Schopf, J. M. & Long, W. E. (1966). Coal metamorphism and igneous associations in Antarctica. In *Coal Sciences, Am. chem. Soc., Adv. Chem. Ser.* **55**, 156–95.

Schreiner, O. & Lathrop, E. C. (1911a). Examination of soils for organic constituents, especially dihydroxystearic acid. *Bull. Div. Soils U.S. Dep. Agric.* **47**, 7–33.

Schreiner, O. & Lathrop, E. C. (1911b). Dihydroxystearic acid in good and poor soils. *J. Am. chem. Soc.* **33**, 1412–17.

Schreiner, O. & Lathrop, E. C. (1912). The chemistry of steam heated soils. *J. Am. chem. Soc.* **34**, 1242.

Schreiner, O. & Shorey, E. C. (1908). The isolation of dihydroxystearic acid from soils. *J. Am. chem. Soc.* **30**, 1599–607.

Schreiner, O. & Shorey, E. C. (1909*a*). The isolation of harmful organic substances from soils. *Bull. Div. Soils U.S. Dep. Agric.* **53.**

Schreiner, O. & Shorey, E. C. (1909*b*). The presence of cholesterol substances in soils; agrosterol. *J. Am. chem. Soc.* **31**, 116–18.

Schreiner, O. & Shorey, E. C. (1910*a*). Chemical nature of soil organic matter. *Bull. Div. Soils U.S. Dep. Agric.* **74.**

Schreiner, O. & Shorey, E. C. (1910*b*). The presence of arginine and histidine in soils. *J. biol. Chem.* **8**, 381.

Schreiner, O. & Shorey, E. C. (1910*c*). Some acid constituents of soil humus. *J. Am. chem. Soc.* **32**, 1674–80.

Schreiner, O. & Shorey, E. C. (1911). Paraffin hydrocarbons in soil. *J. Am. chem. Soc.* **33**, 81–3.

Schreiner, O., Shorey, E. C., Sullivan, M. X. & Skinner, J. J. (1911). A Beneficial Organic Constituent of Soils: Creatinine. *Bull. Div. Soils U.S. Dep. Agric.* **83.**

Schwartz, S. M. & Martin, W. P. (1955). Influence of soil organic acids on soluble phosphorus in Miami and Wooster silt loam soils. *Soil Sci.* (*Soc. Am. Proc.*) **19**, 185–8.

Schwartz, S. M., Varner, J. E. & Martin, W. P. (1954). Separation of organic acids from several dormant and incubated Ohio soils. *Soil Sci.* (*Soc. Am. Proc.*) **18**, 174–7.

Seward, A. C. (1959). *Plant Life through the Ages*, pp. 1–603. New York: Hafner.

Shaler, N. S. (1890). Fresh water morasses of the United States. *U.S. geol. Surv. A. Rep.* **10**, 261–339.

Shapiro, J. (1957). Chemical and biological studies on the yellow organic acids of lake water. *Limnol. Oceanogr.* **11**, 161–79.

Shorey, E. C. (1913). Some Organic Soil Constituents. *Bull. Div. Soils U.S. Dep. Agric.* **88**, 5–41.

Shorey, E. C. (1914). Some benzene derivatives in soils. *J. agric. Res.* **1**, 357–63.

Shorey, E. C. (1938). The presence of allantoin in soils. *Soil Sci.* **45**, 177.

Shorey, E. C. & Walters, E. H, (1914). Ammonia-soluble inorganic soil colloids. *J. agric. Res.* **3**, 175.

Siegel, S. M., Roberts, K., Nathan, H. & Daly, O. (1967). Living relative of the microfossil *Kakabekia. Science, N.Y.* **156**, 1231–4.

Silverman, S. R. & Epstein, S. (1959). Carbon isotopic compositions of petroleums and other sedimentary organic materials. *Bull. Am. Ass. Petrol. Geol.* **42**, 988–1012.

Skujins, J. J. (1967). Enzymes in soil. In *Soil Biochemistry*, A. D. McLaren and G. H. Peterson (eds.), pp. 371–414. New York: Marcel Dekker Inc.

Smith, D. H. & Clark, F. E. (1952). Chromatographic seperations of inositol phosphorus compounds. *Soil Sci.* (*Soc. Am. Proc.*) **16**, 170.

Smith, P. V., Jun. (1952). The occurrence of hydrocarbons in recent sediments from the Gulf of Mexico. *Science, N.Y.* **116**, 437.

Smith, P. V., Jun. (1954). Studies on origin of petroleum: occurrence of hydrocarbons in recent sediment. *Bull. Am. Ass. Petrol. Geol.* **38**, 377–404.

References

Snow, L. M. & Fred, E. B. (1926). Some characteristics of the bacteria of Lake Mendota. *Trans. Wis. Acad. Sci. Arts Lett.* **22**, 143–54.

Snowden, F. J. (1959). Investigations on the amounts of hexosamines found in various soils and methods for their determination. *Soil Sci.* **88**, 138.

Soltys, A. (1929). Über das Iosen, ein neuen Kohlenwasserstoff aus steirschen Braunkohlen. *Mh. Chem.* **53–4**, 175–86.

Sondheimer, E., Dence, W. A., Mattick, L. R. & Silverman, S. R. (1966). Composition of combustible concretions of the Alewife, *Alosa pseudoharengus*. *Science, N.Y.* **157**, 221–3.

Sörenson, H. (1957). Microbial decomposition of xylan. *Acta agric. scand. Suppl.* **1**, 1, 4–82.

Spackman, W., Dobson, C. P. & Riegel, W. (1966). Phytogenic organic sediments and sedimentary environments in the Everglades-mangrove complex. *Palaeontographia Abt. B* **117**, 135–52.

Sperber, J. I. (1957). Solution of mineral phosphates by soil bacteria. *Nature, Lond.* **180**, 994–5.

Sperber, J. I. (1958). Solution of apatite by soil microorganisms producing organic acids. *Aust. J. agric. Res.* **9**, 782–7.

Stadnikov, G. & Weizmann, A. (1929). Ein Beitrag zur Kenntnis der Umwandlung der Fettsäuren in Laufe der geologischen Zeitperioden. I. *Brennst.-Chem.* **10**, 61–3; II. *Ibid.*, 81–2; III. *Ibid.*, 401–3.

Stankovic, S. (1960). *The Balkan Lake Ohrid and its Living World*. The Hague, Dr W. Junk, Biol. Mon. *IX*, 1–357.

Steelink, C. (1966). Electron paramagnetic resonance studies of humic acid and related model compounds. In *Coal Science, Am. chem. Soc., Adv. Chem. Ser.* **55**, 80–90.

Steelink, C., Reid, T. & Tollin, G. (1963). Structure of humic acid. IV. Electron paramagnetic resonance studies. *Biochim. biophys. Acta* **66**, 444.

Steelink, C. & Tollin, G. (1962). Stable free radicals in soil humic acid. *Biochim. biophys. Acta* **59**, 25.

Steelink, C. & Tollin, G. (1967). Free radicals in soil. In *Soil Biochemistry*, A. D. McLaren and G. H. Peterson (eds.) pp. 148–69. New York: Marcel Dekker, Inc.

Stern, A. & Wenderlein, H. (1935). Über die Lichtabsorption der Porphyrine II. *Z. phys. Chem.* A **174**, 81–103.

Stevens, C. M. & Vohra, P. (1955). Occurrence of choline sulfate in *Penicillium chrysogenum*. *J. Am. chem. Soc.* **77**, 4935.

Stevens, N. P., Bray, E. E. & Evans, E. D. (1956). Hydrocarbons in sediments of the Gulf of Mexico. *Bull. Am. Ass. Petrol. Geol.* **40**, 975.

Stevenson, F. J. (1956). Isolation and identification of some amino compounds in soils. *Soil Sci. (Soc. Am. Proc.)* **20**, 204.

Stevenson, F. J. (1957). Aminopolysaccharides in soils. I. Colorimetric determination of hexosamines in soil hydrolyzates. *Soil Sci.* **83**, 113.

Stevenson, F. J. (1967). Organic acids in soil. In *Soil Biochemistry*, A. D. McLaren and G. H. Peterson (eds.), pp. 119–46. New York: Marcel Dekker.

Stevenson, I. L. & Katznelson, H. (1958). The oxidation of ethanol and acetate in soils. *Can. J. Microbiol.* **4**, 73–9.

Stopes, M. C. (1919). Studies in the composition of coal. I. The four visible ingredients in banded bituminous coal. *Proc. Roy. Soc. Lond. B* **90**, 470.

References

Stout, W. (1931). Pennsylvanian cycles in Ohio, Ill. *Bull. U.S. geol. Surv.* **60**, 195–216.

Strickland, J. D. H. & Parsons, T. R. (1965). Pigment analyses. In *Methods of Analysis of Sea Water. Bull. Fish. Res. Can.* 117–27.

Struthers, P. H. & Sieling, D. H. (1950). Effect of organic anions on phosphate precipitation by iron and aluminum as influenced by pH. *Soil Sci.* **67**, 3.

Studier, M. H., Hayatsu, R. & Anders, E. (1965). Origin of organic matter in early solar system. I. Hydrocarbons. *Enrico Fermi Inst. Preprint*, pp. 65–115.

Subrahmanyan, V. (1927). Biochemistry of water-logged soils. II. The presence of a deaminase in water-logged soils and its role in the production of ammonia. *J. agric. Sci., Camb.* **17**, 449.

Sugihara, J. M. & McGee, L. R. (1957). Porphyrins in gilsonite. *J. org. Chem.* **22**, 795.

Sugisawa, H. & Edo, H. (1964). The thermal degradation of sugars. I. Thermal polymerization of glucose. *Chem. Ind.* (892), 561–5.

Sverdrup, H. U., Johnson, M. W. & Fleming, R. H. (1942). *The Oceans*, pp. 1–1078. New York: Prentice-Hall.

Swain, F. M. (1944). Stratigraphy of Cotton Valley beds of northern Gulf Coastal Plain. *Bull. Am. Ass. Petrol. Geol.* **28**, 577–614.

Swain, F. M. (1956). Stratigraphy of lake deposits in central and northern Minnesota. *Bull. Am. Ass. Petrol. Geol.* **40**, 600–53.

Swain, F. M. (1961a). Limnology and amino-acid content of some lake deposits in Minnesota, Montana, Nevada and Louisiana. *Bull. geol. Soc. Am.* **72**, 519–46.

Swain, F. M. (1961b). Reporte preliminar de los sedimentos del fonds de los lagos Nicaragua y Managua, Nicaragua. *Boln Serv. geol. nac. Nicaragua* **5**, 11–29.

Swain, F. M. (1963a). Stratigraphic distribution of some residual organic compounds in Upper Jurassic. *Bull. Am. Ass. Petrol. Geol.* **47**, 777–803.

Swain, F. M. (1963b). Geochemistry of humus. In *Organic Geochemistry*, I. A. Breger (ed.), pp. 87–147. Oxford: Pergamon Press.

Swain, F. M. (1964). Early Tertiary fresh-water Ostracoda from Colorado, Nevada, and Utah and their stratigraphic distribution. *J. Paleont.* **38**, 256–80, pls. 41–4.

Swain, F. M. (1965). Geochemistry of some Quaternary lake sediments of North America. In *The Quaternary of the United States*, H. E. Wright and D. G. Frey (eds.), pp. 765–81. Princeton: Princeton University Press.

Swain, F. M. (1966a). Bottom sediments of Lake Nicaragua and Lake Managua, western Nicaragua. *J. sedim. Petrol.* **36**, 522–40.

Swain, F. M. (1966b). Residual organic compounds in Paleozoic rocks of central Pennsylvania. *Coal Science, Am. chem. Soc., Adv. Chem. Ser.* **55**, 1–17.

Swain, F. M. (1967). Stratigraphy and biochemical paleontology of Rossburg Bog (Recent) Aitkin County, Minnesota. In *Essays in Paleontology and Stratigraphy R. C. Moore Commem. Vol.*, C. Teichert and E. Yochelson (eds.), pp. 445–75. Lawrence: University of Kansas Press.

Swain, F. M. (1968a). Humus, geochemistry of. *McGraw-Hill Yearbook of Sci. and Technol*, pp. 212–14. New York: McGraw-Hill Co.

Swain, F. M. (1968b). Properties of Minnesota lake sediments. *Bull. Minn geol. Surv.* in editorial preparation.

Swain, F. M. (1969). Humus, geochemistry of. *McGraw-Hill Encyclopedia of Sci. and Technol.* New York: McGraw-Hill Co. (In Press.)

References

Swain, F. M., Blumentals, A. & Millers, R. (1959). Stratigraphic distribution of amino acids in peats from Cedar Creek Bog, Minnesota and Dismal Swamp, Virginia. *Limnol. Oceanogr.* **4**, 119–27.

Swain, F. M., Bratt, J. M. & Kirkwood, S. (1967). Carbohydrate components of some Paleozoic plant fossils. *J. Paleont.* **41**, 1549–54.

Swain, F. M., Bratt, J. M. & Kirkwood, S. (1968). Possible biochemical evolution of carbohydrates of some Paleozoic plants. *J. Paleont.* **42**, 1078–82.

Swain, F. M., Bratt, J. M. & Kirkwood, S. (1969). Carbohydrate components of Upper Carboniferous plant fossils from Radstock, England. *J. Paleont.* **43**, 550–3.

Swain, F. M., Bratt, J. M., Kirkwood, S. & Tobback, P. (1969). Carbohydrate components of Paleozoic plants. In *Advances in Organic Geochemistry* pp. 167–80. 1968. Braunschweig: Fried. Vieweg. und Sohn.

Swain, F. M. & Gilby, J. M. (1965). Ecology and taxonomy of Ostracoda and an alga from Lake Nicaragua. *Staz. Zool Napoli,* **33** *suppl. Ostracods as Ecological and Palaeoecological Indicators,* 361–86.

Swain, F. M. & Kraemer, S. A. (1969). Amino acid components of some Paleozoic plant fossils. *J. Paleont.* **43**, 546–50.

Swain, F. M. & Meader, R. W. (1958). Bottom sediments of southern part of Pyramid Lake, Nevada. *J. Sedim. Petrol.* **28**, 286–97.

Swain, F. M., Pakalns, G. V. & Bratt, J. M. (1969). Possible taxonomic interpretation of some Palaeozoic and Precambrian carbohydrate residues. In *Advances in Organic Geochemistry,* 1966, G. A. Hobson and G. C. Speers (eds.). Oxford: Pergamon Press (In Press.)

Swain, F. M., Paulsen, G. W. & Ting, F. T. (1964). Chlorinoid and flavinoid pigments from aquatic plants and associated lake and bog sediments. *J. Sedim. Petrol.* **34**, 561–98.

Swain, F. M. & Prokopovich, N. (1954). Stratigraphic distribution of lipoid substances in Cedar Creek Bog, Minnesota. *Bull. geol. Soc. Am.* **65**, 1183–9.

Swain, F. M. & Prokopovich, N. (1957). Stratigraphy of upper part of sediments of Silver Bay, Lake Superior. *Bull. geol. Soc. Am.* **68**, 527–42.

Swain, F. M. and Prokopovich, N. (1969). Biogeochemistry of Delta–Mendota Canal, Central Valley Project, California. *U.S. Bur. Reclamation, Res. Rept* 20, pp. 1–42.

Swain, F. M., Rogers, M. A., Evans, R. D. & Wolfe, R. W. (1967). Distribution of carbohydrate residues in some fossil specimens and associated sedimentary matrix and other geological samples. *J. sedim. Petrol.* **37**, 12–24.

Swain, F. M. & Venteris, G. (1964). Distribution of flavinoids and some other heterocyclic substances in lake sediments. In *Advances in Organic Geochemistry,* U. Colombo and G. D. Hobson (eds.), pp. 199–214. New York: Macmillan Co.

Swain, F. M., Venteris, G. & Ting, F. (1964). Relative abundance and order of stability of amino acids in some aquatic plants and associated fresh water sediments. *J. sedim. Petrol.* **34**, 25–45.

Sweet, J. M. (1964). Developments in Alaska in 1963. *Bull. Am. Ass. Petrol. Geol.* **48**, 1035–49.

Swindale, L. D. & Jackson, M. L. (1956). Cheluviation in soils. *Trans. 6th Int. Congr. Soil Sci. E,* 233.

Takijima, Y. (1960). Studies on behavior of the growth-inhibiting substances in paddy soils with special reference to the occurrence of root damage in the peat paddy fields. *Bull. natn. Inst. agric. Sci., Tokyo Ser.* B, **13**, 117–52.

Talling, J. F. (1966). Comparative problems of phytoplankton production and photosynthetic productivity in a tropical and a temperate lake. *Memorie Ist. ital. Idrobiol.* **18**, suppl., 399–424.

Taylor, M. B. & Novelli, G. D. (1961). Analysis of a bacterial sulfated polysaccharide. *Bact. Proc.* **61**, 190.

Teichmuller, M. (1958). Metamorphism of the coal and prospecting for petroleum. *Coll. Int. Pét., Rev. de l' Ind. Min. Num. Spéc.* **99**.

Thaidens, A. A. & Haites, T. B. (1944). Splits and washouts in the Netherlands coal measures. *Meded. geol. Sticht, Ser. C.* II, **1**, 1–51.

Thiele, H. & Kettner, H. (1953). Huminsäuren. *Kolloidzeitschrift.* **130**, 131–60.

Thiessen, R. (1920). Compilation and composition of bituminous coals. *J. Geol.* **28**, 185–209, pls. 3–11.

Thiessen, R. (1925). Origin of the boghead coals. *U.S. geol. Surv. Prof. Paper* **132**, 121–38, pls. 27–40.

Thom, W. T. (1934). Present status of the carbon-ratio theory. In *Problems of Petroleum Geology*. Tulsa, *Am. Ass. Petrol. Geol.* 69–95.

Thomson, P. W. (1950). *Braunkohle Wärme Energie* (3rd Quart.), 39.

Ting, F. T. (1967). The petrology of the Lower Kittanning Coal in western Pennsylvania. Doctoral Dissertation, The Pennsylvania State University, 1–138.

Titov, N. (1932). Über die Bitumina des Sphagnumtorfes. *Brennst. Chem.* **13**, 266–9.

Tokuoka, M. & Dyo, S. (1937). Organic constituents of humus in Formosa from the viewpoint of social plants. II. Organic constituents accompanying humus in the soils. *J. Soc. trop. Agric. Taiwan.* **9**, 26.

Tollin, G. & Steelink, C. (1966). Biological polymers related to catechol; electron paramagnetic resonance and infrared studies of melanin, tannin, lignin, humic acid, and hydroxyquinones. *Biochim. biophys. Acta* **112**, 377.

Trask, P. D. (1932). *Origin and Environment of Source Sediments of Petroleum*, pp. 1–323. Houston: Gulf Publishing Co.

Trask, P. D. (1939). Organic content of Recent marine sediments. In *Recent Marine Sediments*. Trask (ed.), pp. 428–53. Tulsa, *Am. Ass. Petrol. Geol.*

Trask, P. D. & Wu, C. C. (1930). Does petroleum form in sediments at time of deposition? *Bull. Am. Ass. Petrol. Geol.* **14**, 1451–63.

Treibs, A. (1934a). Organic mineral substances. II. Occurrence of chlorophyll derivatives in an oil shale of the upper Triassic. *Justus Liebigs Annln Chem.* **509**, 103–14.

Treibs, A. (1934b). Organic minerals. III. Chlorophyll and hemin derivatives in bituminous rocks, petroleums, mineral waxes and asphalts. Origin of petroleum. *Justus Liebigs Annln Chem.* **510**, 42–62.

Treibs, A. (1935a). Organic mineral substances. IV. Chlorophyll and hemin derivatives in bituminous rocks, petroleums, coals and phosphorites. *Justus Liebigs Annln Chem.* **517**, 172–96.

Treibs, A. (1935b). Organic mineral substances. V. Porphyrins in coals. *Justus Liebigs Annln Chem.* **520**, 144–50.

Treibs, A. (1936). Chlorophyll and hemin derivatives in organic mineral substances. *Angew. Chem.* **49**, 682–6.

Troels-Smith, J. (1955). 'Karakterisering af løse jordarter' (Characterization of unconsolidated sediments). *Danm. geol. Unders.* IV. Raekke. Bd. 3 Nr. 10, 1955.

References

Tschamler, H. & De Ruiter, E. (1966). A comparative study of exinite, vitrinite, and micrinite. In *Coal Science, Am. chem. Soc., Adv. Chem. Ser.* **55**, 332–43.

Udden, J. A. (1912). Geology and Mineral Resources of the Peoria Quadrangle, Illinois, *Bull. U.S. geol. Surv.* (506), 1–103.

Urey, H. C. (1952). *The planets: their origin and development.* New Haven: Yale University Press.

Vallentyne, J. R. (1955). Sedimentary chlorophyll determination as a paleobotanical method. *Can. J. Bot.* **33**, 304–13.

Vallentyne, J. R. (1956). Epiphasic carotenoids in post-glacial lake sediments. *Limnol. Oceanogr.* **1**, 252–62.

Vallentyne, J. R. (1957). The molecular nature of organic matter in lakes and oceans with lesser reference to sewage and terrestrial soils. *J. Fish. Res. Bd. Can.* **14**, 33–82.

Vallentyne, J. R. (1963). Geochemistry of carbohydrates. In *Organic Geochemistry*, I. A. Breger (ed.), pp. 456–503. Oxford: Pergamon Press.

Vallentyne, J. R. (1964). Biogeochemistry of organic matter. II. Thermal reaction kinetics and transformation of amino compounds. *Geochim. cosmochim. Acta* **28**, 157–88.

Vallentyne, J. R. & Bidwell, R. G. S. (1956). The relation between free sugars and sedimentary chlorophyll in lake muds. *Ecology* **37**, 495–500.

Vallentyne, J. R. & Craston, D. F. (1957). Sedimentary chlorophyll degradation products in surface muds from Connecticut lakes. *Can. J. Bot.* **35**, 35–42.

Vallentyne, J. R. & Swabey, Y. S. (1955). A reinvestigation of the history of Lower Linsley Pond, Connecticut. *Am. J. Sci.* **253**, 313–40.

van der Heide, S. (1950). Compaction as a possible factor in Upper Carboniferous rhythmic sedimentation. *Rep. 18th Int. geol. Congr.* (4), 38–45.

van Houten, F. B. (1966). Grains of black oxides in red beds. *Progr. geol. Soc. Am. 79th Mtg., San Francisco*, 229.

van Krevelen, D. W. (1961). *Coal*, pp. 1–514. Amsterdam: Elsevier.

van Krevelen, D. W. & Schuyer, J. (1957). *Coal Science*, pp. 1–375. Amsterdam: Elsevier; Princeton, N.J.: D. Van Nostrand Co.

Varier, N. S. (1950). Fossil resin from lignite Beds at Warkalay. *Bull. cent. Res. Inst. Univ. Travancore, Ser. AI*, **1**, 117–18.

Velikowsky, I. (1956). *Earths in Upheaval*, London: V. Gollanz, Ltd.

Vernon, L. P. (1960). Spectrophotometric determination of chlorophylls and pheophytins in plant extracts. *Analyt. Chem.* **32**, 1144–50.

Visser, S. A. (1964). The decomposition of *Cyperus papyrus* in swamps of Uganda in natural peat deposits as well as in the presence of various additives. *E. Afr. agric. For. J.* **29** (3), 268–87.

Vlasyuk, P. A., Dobrotvorskaya, K. M. & Gordienko, S. A. (1957). The intensity of enzyme activity in the rhizosphere of certain agricultural plants. *Dokl. vses. Akad. sel'-khoz. Nauk* **22** (3), 14.

Wade, A. (1929). Madagascar and its oil lands. *J. Instn. Petrol. Technol.* **15**, 2–29.

Waksman, S. A. (1938). *Humus: Origin, Chemical Composition and Importance in Nature*, pp. 1–526. Baltimore: Williams and Wilkins Co.

Walters, E. H. (1917). Isolation of p-hydroxybenzoic acid from the soil. *J. Am. chem Soc.* **39**, 1778–84.

Wanless, H. R. (1947). Regional variations in Pennsylvanian lithology. *J. Geol.* **55**, 237–53.

References

Wanless, H. R. (1950). Late Paleozoic cycles of sedimentation in the United States, *Rep. 18th Int. geol. Congr.* (4), 17–28.

Wanless, H. R. & Shepard, F. P. (1936). Sea level and climatic changes related to late Paleozoic cycles. *Bull. geol. Soc. Am.* **47**, 1177–206.

Wanless, H. R. & Weller, J. M. (1932). Correlation and extent of Pennsylvanian cyclothems. *Bull. geol. Soc. Am.* **47**, 1003–16.

Watts, W. A. & Winter, T. C. (1965). Plant microfossils from Kirchner Marsh, Minnesota—a Paleoecological study. *Bull. geo. Soc. Am.* **77**, 1339–60.

Weber, M. M., Hollander, T. C. & Rosso, G. (1965). The appearance and general properties of free radicals in electron transport particles from *Mycobacterium phlei*. *J. biol. Chem.* **240**, 1776.

Webley, D. M., Eastwood, D. J. & Giningham, C. H. (1952). Development of a soil microflora in relation to plant succession on sand dunes, including the rhizosphere flora associated with colonizing species. *J. Ecol.* **40**, 168–78.

Welch, P. S. (1952). *Limnology*, 2nd edn., pp. 1–538. New York: McGraw-Hill Co.

Weller, J. M. (1930). Cyclical sedimentation of the Pennsylvanian Period and its significance. *J. Geol.* **38**, 97–135.

Weller, J. M. (1931). The conception of cyclical sedimentation during the Pennsylvanian Period. *Bull. Ill. geol. Surv.* **60**, 163–77.

Weller, J. M. (1956). Argument for diastrophic control of late Paleozoic cyclothems. *Bull. Am. Ass. Petrol. Geol.* **40**, 17–50.

Weller, J. M. (1960). *Stratigraphic Principles and Practice*, pp. 1–725. New York: Harper and Bros.

Whetten, J. T. (1966). Sediments from the Lower Columbia River and origin of graywacke. *Science, N.Y.* **152**, 1057–8.

White, C. D. (1915). Some relations in origin between coal and petroleum, *Wash. Acad. Sci. J.* **5**, 189–212.

White, C. D. (1933). The role of water conditions in the formation and differentiation of common (banded) coals. *Econ. Geol.* **28**, 556.

White, E., Tse, A. & Towers, G. H. N. (1967). Lignin and certain other chemical constituents of *Phylloglossum. Nature, Lond,* **205**, 285–6.

Whitehead, D. C. (1963). Aspects of the influence of organic matter on soil fertility. *Soils Fertil.* **26** (4), 217–23.

Whitehead, D. C. (1964). Identification of *p*-hydroxybenzoic, vanillic, *p*-coumaric, and ferulic acids in soils. *Nature, Lond.* **202**, 417–18.

Williams, C. H. & Steinbergs, A. (1962). The evaluation of plant-available S in soils. I. The chemical nature of sulfate in some Australian soils. *Pl. Soil* **17**, 279.

Woolley, D. W. & Peterson, W. H. (1937). The chemistry of mold tissue. XIV. Isolation of cycline choline sulfate from *Aspergillus sydowi. J. biol. Chem.* **122**, 213.

Wright, H. E., Livingstone, D. A. & Cushing, E. J. (1965). Coring devices for lake sediments. In *Paleontological Techniques Handbook*, B. Kummel and D. Raup (eds.), pp. 494–520. San Francisco: Freeman.

Wright, H. E., Winter, T. C. & Patten, H. L. (1963). Two pollen diagrams from south-eastern Minnesota; problems in the regional late-glacial and post-glacial vegetational history. *Bull. geol. Soc. Am.* **74**, 1371–96.

Yoshida, R. K. (1940). Studies on organic phosphorus compounds in soil; isolation of inositol. *Soil Sci.* **50**, 81–9.

Young, R. C. (1957). Late Cretaceous cyclic deposits, Book Cliffs, eastern Utah. *Bull. Am. Ass. Petrol. Geol.* **41**, 1760–74.

References

Zemlyanukhin, A. A. (1959). Effect of organic acids on physiological processes and yield. *Vest. sel.-kleoz. Nauki, Mosk.* **9** (1), 26–9.

ZoBell, C. E. (1946). Studies on redox potential of marine sediments. *Bull. Am. Ass. Petrol. Geol.* **30**, 477–513.

ZoBell, C. E. (1963). Organic geochemistry of sulfur. In *Organic Geochemistry*, I. A. Breger (ed.), pp. 543–78. Oxford: Pergamon Press.

Zscheile, F. P. & Comer, C. L. (1941). Influence of preparative procedure on the purity of chlorophyll components as shown by absorption spectra. *Bot. Gaz.* **102**, 463–81.

Zubovic, P. (1966a). Physiochemical properties of certain minor elements as controlling factors in their distribution in coal. In *Coal Science, Am. chem. Soc., Adv. Chem. Ser.* **55**, 221–31.

Zubovic, P. (1966b). Minor element distribution in coal samples of the Interior Coal Province. In *Coal Science, Am. Chem. Soc., Adv. Chem. Ser.* **55**, 232–47.

Zubovic, P., Sheffey, N. B. & Stadnichenko, T. M. (1961). *U.S. geol. Surv. Prof. Paper* **424**D, D345–8.

Züllig, H. (1955). Sedimente als Ausdruck des Zustandes eines Gewässers. *Memorie Ist. ital. Idrobiol.*, suppl. 8, 485–530.

Züllig, H. (1956). Sedimente als Ausdruck des Zustandes eines Gewässers. *Schweiz. Z. Hydrol.* **18**, 5–143.

Zumberge, J. H. (1952). The lakes of Minnesota, their origin and classification. *Bull. Minn. geol. Surv.* (35), 1–99.

Subject Index

Subject Index

amino acids (*cont.*)
in peat deposits, 186; stabilizing effect of mineral particles, 204; thermal stability, 69; zwitterion properties, 207
aminobutyric acid, 174, 193, 201
amino compounds, reactions, 38
amino sugars, 24, 174; in soil, 317, 319
ammonia, from urea, 40; in amino acid synthesis, 39; in Wohler Reaction, 40
ammonium, clay-fixed, 316
amorphous polysaccharides, 211
Amphidinium carteri, 288
Amphora ovalis affinis, Lake Patzcuaro, 95
amygdalin, 27
amylase, 239, 305, 328
amylene, 11
amylopectin, 25
amylose, 25, 28
α-amyrin, in amber, 170
Anabaena, carbohydrates, 211, 217, 218; saturated hydrocarbons, 118
Anacystis marina, fatty acids, 117
Anacystis montana, olefins, 118, 120
Anacystis nidulans, fatty acids, 117
anaerobic bacteria, Rift Valley lake sediments, 92; conditions in sediments, 54, 74
anemone, carbohydrates, 211
anil, 40
aniline, invertase inhibitor, 329
ankerite, Cedar Creek peat, 108
Annularia sphenophylloides carbohydrates, 242
Anodonta cygna, sterols, 116
Anoka Sand Plain, Minnesota amino acids of lakes, 187
anomers, 442
Anomoeonis polygramma, Lake Patzcuaro, 95
Anomoeonis sphaerophora, Lake Patzcuaro, 95
anoxic substrate, in methane production in plants, 113
anthocyanins, 246
anthoxanthins, 246
anthracene, 14; absorption spectrum, 61; in sediment extracts, 126
anthroquinones, 246
anthraxylon, 350-2
antiseptic propeties of peat, 279
ants, undecane in, 114
apatotrophic lakes, 92, 177; amino acids, 210
apo-3-norbixinal methylester, 277
Appalachian geosyncline, organic productivity, 369
aquatic plants, chlorinoid pigments, 258-62

aquatic vegetation, source of organic sediments, 73
araban polysaccharides, 25
arabinose, 22, 24, 211; stability in sediments, 243
L-arabofuranose, 26
Araucariaceae, resin producers, 168
Araucarioxylon arizonicum, resin receptacles, 169
arborescent lycopod, carbohydrates, 240
Archeosphaeroides barbertonesis, 167
arginine, 20, 42, 318
arginine, difficulty in paper chromatography, 206
argulite (bituminous sandstone), 155
aromatic organic acids, from soils, 315; carboxylic acids, 293; ethers, in humic acids, 309; fractions of recent sediment extracts, 129; hydrocarbons, chromatographic analysis, 59; type formula, 11
Arrhenius equation, 38
arthropods, chitin in, 25
Artemesia pollen curve, 108
arylethers, in humic acids, 309
ash falls, 111
asparaginase, 305, 329
asparagine, 35, 305, 319
aspartate, 305
aspartate–alanine transaminase, 305
aspartic acid, 19, 35, 318, hydrolytic degradation, 39; in pyrimidine synthesis, 41
Aspergillus clavatus, ethylene from, 121
Aspergillus nidulans, in cystine formation, 45
Aspergillus niger, 316; organic acids from, 293; polysaccharides, 26
Aspergillus terreus, 316
asphaltic chromatographic fractions, 59
Assam oil seeps, 162
Asterionella formosa, 289
Athabasca Lake, 101
attritus, 350-2
aurochrome, 277
auroxanthin, 277
authigenic non-marine sediments, 74
auxins, 47, 121, 294, 315; possible source of naphthols in peat, 144
Avicenna pollen curve, Everglades, 79, 80
avocado, carbohydrates, 23
azafrin, 255

Bacillus spp., carbohydrates, 26
Bacillus stearothermophilus, 122
bacteria cultures, hydrocarbon substrates, 122
bacteria, sources of pigments, 274, 275

Subject Index

Subject Index

epiphase, in carotene extraction, 72, 266, 267

Epithemia sp., Clear Lake, 213

Equisetum arvense, carbohydrates, 239

equivalent weight, humic acid, 308

ericads, pollen curve, North Carolina, 78

Escherichia coli, in sulfate reduction, 44

esters, 10, 16, 35

Esthwaite Water, England, chlorinoid pigments, 265–7; pigment contents of sediments, 268, 269

ethanal, 6

ethane, 6, 8

ethanol, 6

ethanolamine, 201, 317, 319

ethene, 11

ethers, 16; aromatic, 28

ethonium ion, 9

ethyl alcohol, 6, 14; enzyme-catalyzed oxidation, 35

ethylamine, 201

ethylcarbinal, 14

ethylene, 11, 34; dicarboxylic acid, 36; production in plants, 121

ethyl ether, 16

ethyl group, 16

etioporphyrin, 279, 281, 287; spectrum, 285, 286

eucalyptus oil, lipids, 115

eucaryotic cells, appearance of, 366

eutrophic lakes, amino acids, 209

eutrophic lake sediments, chlorinoid pigments, 258–72

eutrophic lakes, bitumens, 146

eutrophic lakes, examples, 103

even-numbered fatty acids, 297

even-numbered paraffins, 297

Everglades, sediments, 77

exchangeable cations in soils and humics, 306

excretion products, glucuronic acid, 24

exinite, 351, 352; structure of, 355; hydroxyl content, 362

extraterrestrial synthesis of biphenyl, 46

facies of peat, Cedar Creek Bog, 108

Fagus pollen curve, 78

Fannie Lake, Minnesota, 187, 188

Farlowia mollis, chlorophyll content, 248

fat acids, synthesis, 35

fatty acid calcium salts, 121, 124

fatty acids, 293; in Recent sediments, 296; in Russian lake, 114;in sediments, 10; in soil, 314; methyl esters, 64, 65; Mud Lake, 130; source organisms, 117; source of hydrocarbons, 145

fermentation, oxygen level required, 367

ferulic acid, 294, 315, 344

fichtelite, 116

Fig Tree chert, 368; hydrocarbons, 168

first-order reaction, 9, 38

Fischer peptide synthesis, 39

Fischer–Tropsch synthesis, 33

flagstaffite, 115

Flaig humic acid model, 326

flavineadenine dinucleotide, 279, 280

flavinoid pigments, 72, 256

flavins, 29; free radical contents, 324

flavonal, 31

flavone, 29

flavonoid pigments, 246, 279

flavoproteins, 72, 246, 279

Flintshire coalfield, England, 164

flotation of ilmenite, 331

fluorene, 15

fluorescence of bitumens, 58, 127, 141

fluvial sediments, 109, 110

forest peat, 106, 108

formaldehyde, 50, 43

formic acid, 293, 314

Formica rufa, undecane content, 114

förna, 73, 74, 106

fossil resins, geochemistry, 168

fossil wood, lipid contents, 115

Fraxinus pollen curve, 108

Fredericella sultana, carbohydrates, 232

free amino acids, 69, 203, 318

Freeport coal, 352–4

free cystine from rhizosphere soil, 321

free radicals, 8, 36, 168, 308, 322, 345

free sugars, 213, 218, 219, 243, 244, 320

freshwater algae, carbohydrates, 211

freshwater animals, carbohydrates, 211

freshwater diatoms, on Middle Atlantic Ridge, 74

freshwater plants, amino acids, 174; carbon isotopes, 173

freshwater plant fossils, amino acids, 206

freshwater saw grass marsh, 78

friedelan-3-β-ol, 116

Friedel–Crafts Reaction, 46

fructose, 22, 23, 305

fucose, 23, 24, 211

fucoxanthin, 255, 288

Fucus furcatus, chlorophyll contents, 248, 250

fulvic acid, 309

fumaric acid, 36, 293, 314

fungi, in ethylene production, 121; organic acids from, 293; trehalose in, 24

furan, 29, 362

furfural, 6, 16, 42, 232, 233

fusain, 349, 351, 352

Fusarium, 316

Subject Index

Subject Index

Subject Index

reducing conditions in sediments, 32, 54
reducing sugars, 24
relict fauna, alewife, 122
Reno Lake, Minnesota, amino acids, 182, 190
reproductive power of living cells, discussion, 365, 366
resenes, 168
resin(s), 168–70
resistance to insects and fungi, in *Gingko*, 331
resinite, 351, 352; infrared spectra, 354, 355
resonance, 4
respiration, oxygen levels required, 367
retene, 115
retinite, 169
rhamnose, 23–26, 211, 243
rhenish lignites, 334
Rhizophora pollen, Everglades, 79, 80
Rhizopus, organic acids from, 293
rhodopol, 255
rhodo-porphyrin, spectrum, 286
rhodopurpurin, 255
rhodoviolascin, 255; in lake sediments, 272, 273, 278
rhodoxanthin, 255
Rhopolodia, sp., Clear Lake, 213
Rhodopseudomonas spheroides carotenoids from, 278
Rhynia gwynnevaughani, carbohydrates, 239
riboflavin, 24, 31, 279–82; absorption maxima, 72
ribonucleic acid, 24, 319
ribose, 22, 24, 211, 222, 224
D-ribotol, 24
D-ribulose, 22, 24
rice paddy soil, organic acids, 315
river water, pH, 51
rock surfaces, bacterial organic acids, 294
Rodomella larix, chlorophyll content, 248
rosin, 12
Rossburg Bog, 140, 141–5, 193, 226,
rubacene, 308
Ruhr region, Germany, amino acids of, 204, 205
Ruppia maritima, chlorophyll content, 247
Rush Lake, Minnesota, 181, 187, 188, 208
Russell lignin alkaline degradation, 49
Russian lakes, toxic humus, 330
Rutaceae, α-amyrin in, 170
rutinose, 25

saccharide molecule, 21
Sakhalin Island, humic acids, 161
salicylic acid, 285
salt beds, around Lake Onondaga, 122

Sanna oil, pentadecane content, 114
San Nicolas Basin, fatty acids from, 65
saponification of esters, 35
sapropel, 73–5, 90–2, 106
sapropelite, 74
sapropsammite, 107
Sargassum muticum, 288
Sarothra gentianoides nonane content, 114
saturated dicarboxylic acids, 293
saturated fatty acids, 117
saturated hydrocarbons, chromatography fraction, 58, 118, 148, 226
Schiff Reaction, 50
Schizophyllum commune, in synthesis of mercaptans, 45
Sclerotium rolfsii, 316
Schmidlin Ketene Synthesis, 34
Scholl Condensation, 46
Schurmann's Rule, 349, 350
Schweitzer's Reagent, 320
sclerotinite, 352
scyllitol, 24
seasonal variation of chlorophyll in aquatic plants, 261, 262
secondary alcohols, 14
secondary amines, 17
secondary wood, resinous material, 169
second-order reaction, 9
sedge peat, 106, 108
sedimentary chlorophyll degradation product (SCDP), 71, 256
sedimentary history of lakes, 77
sediments, Lake Ohrid, composition, 84, 85; of Blue Lake, Minnesota, 104; of dystrophic lakes, 104; of Everglades Swamp, 79; of Great Slave Lake, 101; of Kandiyohi Lake, 104; of Lake Baikal, 101; of Lake Ballivian, 90; of Lake Minnetonka, 103; of Lake Nicaragua, 85; of Lake Patzcuaro, 92, 94; of Lake Tanganyika, 92; of Lake Victoria, 90; of Leech Lake, 102; of Linsley Pond, 106, 107; of Mexico City Basin lakes, 85, 87; of Mille Lacs Lake, 104; of Pyramid Lake, 92, of Rainy Lake, 102; of Silver Bay, Lake Superior, 95; of Zumbro Lake, 82
D-sedoheptulose, 22, 23
Sedum family, 23
seminal fluid, carbohydrates, 23
semi-fusinite, 352
semi-quinone polymers, 325
senescent-stage bog, 107
separating gel, in disc electrophoresis, 311
Sephadex, analysis of humic acids, 309
sequestration, 316
serine, 19, 318; in formation of cystine, 44;

Author Index

Author Index

Author Index